Encounters across Difference

Anthropology of Tourism: Heritage, Mobility, and Society

Series Editor

Michael A. Di Giovine (West Chester University of Pennsylvania)

Mission Statement

The Anthropology of Tourism: Heritage, Mobility, and Society series provides anthropologists and others in the social sciences and humanities with cutting-edge and engaging research on the culture(s) of tourism. This series embraces anthropology's holistic and comprehensive approach to scholarship, and is sensitive to the complex diversity of human expression. Books in this series particularly examine tourism's relationship with cultural heritage and mobility and its impact on society. Contributions are transdisciplinary in nature, and either look at a particular country, region, or population, or take a more global approach. Including monographs and edited collections, this series is a valuable resource to scholars and students alike who are interested in the various manifestations of tourism and its role as the world's largest and fastest-growing source of socio-cultural and economic activity.

Advisory Board Members

Recent Titles in the Series

Encounters across Difference: Tourism and Overcoming Subalternity in India by Natalia Bloch

Study Abroad and the Quest for an Anti-Tourism Experience edited by John Bodinger de Uriarte and Michael A. Di Giovine

Tourism and Language in Vieques: An Ethnography of the Post-Navy Period by Luis Galanes Valldejuli

The Ethnography of Tourism: Edward Bruner and Beyond edited by Naomi Leite, Quetzil E. Castañeda, and Kathleen M. Adams

Capoeira, Mobility, and Tourism: Preserving an Afro-Brazilian Tradition in a Globalized World by Sergio González Varela

Rethinking the Anthropology of Love and Tourism by Sagar Singh

Encounters across Difference

Tourism and Overcoming Subalternity in India

Natalia Bloch

Translated by Katarzyna Byłów

LEXINGTON BOOKS
Lanham • Boulder • New York • London

Published by Lexington Books
An imprint of The Rowman & Littlefield Publishing Group, Inc.
4501 Forbes Boulevard, Suite 200, Lanham, Maryland 20706
www.rowman.com

6 Tinworth Street, London SE11 5AL, United Kingdom

British Library Cataloguing in Publication Information Available

Library of Congress Cataloging-in-Publication Data Is Available

Names: Bloch, Natalia, author. | Byłów, Katarzyna, translator.
Title: Encounters across difference : tourism and overcoming subalternity in India / Natalia Bloch ; translated by Katarzyna Byłów.
Other titles: Bliscy nieznajomi. English.
Description: Lanham : Lexington Books, [2021] | Series: Anthropology of tourism : heritage, mobility, and society | Translation of: Bliscy nieznajomi : turystyka i przezwyciężanie podporządkowania w postkolonialnych Indiach. | Includes bibliographical references and index.
Identifiers: LCCN 2021009324 (print) | LCCN 2021009325 (ebook) | ISBN 9781793624710 (cloth) | ISBN 9781793624727 (epub)
Subjects: LCSH: Tourism--Social aspects—India. | Marginality, Social—India.
Classification: LCC G155.I4 .B56 2021 (print) | LCC G155.I4 (ebook) | DDC 306.4/8190954—dc23
LC record available at https://lccn.loc.gov/2021009324
LC ebook record available at https://lccn.loc.gov/2021009325

For Leonard and Łukasz

Contents

Preface

My ethnographic journey into the informal tourism sector in India took place a few years before the COVID-19 pandemic hit the world. This black swan event has affected all spheres of our lives—health, economy, politics, climate, the everyday—but tourism, along with related sectors, has been the most seriously afflicted. Closed borders, canceled flights, quarantines, lockdowns, curfews, securitization, and physical distancing have resulted in an almost total freeze of international touristic mobility, leaving only little space for domestic tourism. The informal sector—and majority of people working in tourism in postcolonial countries operate within it, including the protagonists of this book—has been even more stricken, left without any governmental support. The Indian poor have suffered in particular. At the beginning of the pandemic we saw hundreds of thousands of desperate migrant daily wagers leave locked-down cities on foot on the way to their home villages and towns, where they could at least find some shelter and food (but mostly no medical care). The pandemic made the everyday struggles of the workers and small entrepreneurs in the informal tourism sector in Hampi and Dharamshala—the two field sites where research for this book was conducted—even harder. It added another dimension to their subaltern, vulnerable position. For the villagers of Hampi, it resulted from forced displacement from a UNESCO World Heritage site, while for Tibetans in Dharamshala, theirs was a story of exile and statelessness. International tourists were perceived by many of my research partners as allies in their struggle for recognition and self-determination: a struggle against state authority in terms of the right to heritage, housing, and benefiting from tourism. On a more individual level, direct, interpersonal relationships developed with tourists "across difference" often bore the potential to overcome structural constraints through closeness, reciprocity, and responsibility. This emancipatory potential of tourism, which is the theme of this book, has been suspended by the pandemic. But we, anthropologists, know very well people's ability to resilience and I strongly believe in the necessity to practice what Arjun Appadurai calls the "anthropologies of aspiration and possibility."

Acknowledgments

This book resulted from a research financed by the National Science Center, Poland, project no. 2011/03/B/HS2/03488. This generous support enabled me to engage in a long-term fieldwork. I wish to thank those who positively assessed my proposal.

This book would not have been possible without the involvement of many people. First of all, I wish to thank my loved ones, not only for their emotional support, so crucial during the process of writing, but also for the more tangible forms of help. I started to write this book when my son had just turned one; the baby and the book were both seeking my attention and competing for it. Fortunately, Łukasz stepped in (combining the roles of a father and a mother at the expense of his own academic work), and so did my parents and Łukasz's mother, who provided care and watched Leonard grow; without them, I would not have been able to write a book at that stage of my life. I wish to thank you with all my heart. I also thank my friend Magdalena, a tourist guide herself, for countless conversations about life and tourism.

I express my gratitude to colleagues from the Institute of Anthropology and Ethnology and the Center for Migration Studies at the Adam Mickiewicz University in Poznań, Poland, for chummy and intellectually inspiring working environment. I wish to thank the head of the institute, Michał Buchowski, for his unwavering support, including partial funding of this book's translation. I also thank the leadership and staff of the Department of Ethnology and Cultural Anthropology at the University of Warsaw, since I finalized this book during my postdoc fellowship in their institution. I wish to thank fellow anthropologists: Łukasz Kaczmarek, Anna Wieczorkiewicz, Anna Horolets, Tomasz Rakowski, Monika Baer, Emilia Sułek, Marek Pawlak, Magdalena Sztandara, Zbigniew Szmyt, Agata Stanisz, Mariusz Filip, Anna Romanowicz, Danuta Penkala-Gawęcka, and Wojciech Dohnal. I thank

Małgorzata Kowalska and Robert Rydzewski for commenting on parts of this book. I send my gratitude to the scholars in India, with whom I had the pleasure to work and exchange ideas, among others: Ananya Dasgupta, Philip Varghese, Kavitha M., Tsering Tsomo and Tenzin Nyinjey. I owe much of my anthropological knowledge to encounters with Noel B. Salazar, Antonio-Miguel Nogués Pedregal, and the postcolonial critic Dipesh Chakrabarty.

I wholeheartedly thank the management of the Supertramp travel agency from Cracow, for their friendliness over the years and for enabling me to do my mobile ethnography during the tours they organized. I also thank the leaders of these tours for taking me "on board." I realize that my presence, as a senior colleague and a former guide, could be initially stressful, but I hope that I was not much of a burden. I also thank the tourists who accepted me as a researcher and decided to share their experiences.

This book would not have materialized were it not for all the people in India, who agreed to become my research partners. I am grateful to permanent, transient, and already-displaced residents of Hampi for their trust. It would be impossible to mention them all here, but their stories are present on pages of this book. I feel especially thankful toward the inhabitants and tourism workers of the main street in Janata Plot, which succeeded the demolished main bazaar of Hampi, where I lived for several months. I thank and send my best wishes to, starting from the riverside, Manu Singh and his family (also for delicious *dosas* with scrambled eggs), Vicky and his family (he was my first guide into the complex situation in Hampi), Mani, Ganga, Raj Shekar and Yogananda, Nirez (also for my portrait), Chomphel, Ringzin and their little son (for all those long, lazy afternoons and endless talks about everything), Mr. Shankar (for evening walks, weekend trips, and long conversations; I regret we will never do it again), Muni and her daughters, Mr. Ram (for an extraordinary life story), Manu (for exquisite "spaghetti 50-50" and for my evening "office") and his boys, Raja, Ming and their daughters (for making me feel at home and giving me the opportunity to be part of Kunsung's first days of life), the whole Lambani family, Little Jay, Kali Das, Ganga and, of course, Celia (for trusting me) and her daughter, Ram, Om, and Jay. I also thank Parmesh, Sakhiko, Aslam, Babhu, Sohana and her family, Anjini and Swathi (for their patience when confronted with my endless questions), Surendra (who sometimes thought I was too much, especially since we became neighbors), Govind Raj, Sri Mohan Joshi, Naga, Panpan, Nagesh, Rashmi, Nagaraj, Jaman, Gangadhara, Kiran, Viresh and Bheema, Vajju, Helongo, Basava, Hussain, Lakshmi, Basappa, Vijay, Vital, Shiva, and Małgosia. Also, I wish to thank Robert (for beautiful Saturday evenings and the painting that hangs in my room), Raja and his "she-elephant" Lakshmi (for hosting me, rides, and late afternoons at the temple), Kuberappa, Hamit, Lakshmi and Ian, Krishna Devaraya and Shama Pawar (for their hospitality and time they offered me), as well as many officials (even though I often asked uncomfortable questions). I have fond memories and feel grateful for conversations with all the people involved in the matters related to Hampi and its inhabitants: John Fritz, Ajith, Mela, Ingo, Fabian, and Phillippe. Last but not least, I express my thanks to people who have "taken care" of me: my host Bhavani (for her wonderful home and friendship), her brother Naga, her daughter Sandhya, and granddaughter Amul, and, of course, Chendru (for his constant support).

In Dharamshala, I thank my old friends—you make the anthropologist in me experience, firsthand, what the closeness is: Tsering Tsomo, Gaphel, Nyinjey, Tenzin Tsundue, Anto, and Tsering Thakchoe. My gratitude also goes to the members of a hybrid and fluid community of McLeod Ganj: Dolker (I am very happy we met), Sherab, Surej and all other guys from the Namgyal Restaurant, Tenzin from the Om Guest House, Parwesh Nowrojee, Mr. Ajit, Surendra and Sanjey, Choklang, Tsering and Chomphel, Dawa Rinchen, Sonam Dorje (even though I was sometimes tiresome), Tenzin Rabgye, Karma Sonam, Lhamo, Kelsang and Kim, Sangmo (for trust, offering an interesting perspective on many things and for those delicious cappuccinos), Dhakpa, Sangye Tashi, Tsering Dorjee also known as "Snow Lion," Riko, Mina (for warm socks), Phurbu, Jampa, Tseyang, Migmar Lhamo and Tsering Migmar, Showkat, Tenzin from the Tibet Ink, Phurbu from the library (as usual, for his tremendous help), the whole team at the Students for a Free Tibet – India (especially Jyotsna and Rashi), the volunteers (especially Namreta and Karma) and participants of the Little Lhasa Tour (for accepting me in their team), Tashi Lhamo from Tong-Len, Kelsang from Rogpa, Norbu from the Learning and Ideas for Tibet and the team of French volunteers, Kunsang Tenzin, Lhakpa Tsering and Lhamo from the Tibet Hope Centre, the team of Black Tent, and especially Thibu (and his girlfriend Yuki), Yeshi and Gina (for long, sincere conversations and spending time together), Ian, Namreta, Jason and Ann, and Elly.

I think about you a lot.

Introduction

Anthropology and Tourism: Dangerous Liaisons?

The moment when—after several years of conducting research in Tibetan refugee settlements in India and among immigrants in Poland—I decided to do a project on tourism, my colleagues, fellow anthropologists, were not very enthusiastic (apart from some of them being overly enthusiastic about "doing state-sponsored tourism under the cover of research into tourism"). I felt like a traitor, abandoning my role as a researcher of human suffering in favor of turning into a tourist lounging at the hotel pool. To be honest, while doing my PhD project among the Tibetan diaspora, I had been avoiding tourists myself. As a beginner anthropologist, I only wished to deal with the "locals," which, in my case, meant the refugees. I pretended not to notice the phenomenon of Tibetans transforming Dharamshala—a refugee settlement—into a popular tourist destination. I was sarcastic about the "tourists-cum-activists" and "tourists-cum-researchers," including anthropologists, who seemed to flock to Dharamshala because "its climate is friendly, telephone and Internet connections work fine, accommodation is not a problem and, on top of it, you can go out and have a beer and a piece of *lasagne* in the evening after work, choosing from a plethora of restaurants and cafes (not to mention a disco on weekends)" (Bloch 2011, 266). It took me over a decade since my first trip to the Tibet-in-exile to notice the potential of tourism to explain how the refugees positioned themselves vis-à-vis the world from which the tourists originated and how they actually benefited from the fact that they had become "an attraction."

In my defense, I would like to remind the reader that researchers who authored the classics in social studies of tourism—such as Erik Cohen—committed the very same sin: "At the initial stage, tourism appeared during fieldwork as an exogenic and unwelcome disturbance of 'my' community by outsiders" (Cohen 2007, 59). For Cohen, who did his fieldwork on the

Arab-Jewish relationships in Acre, Israel, at the end of the 1960s, the "intruders" were young female tourists from Scandinavia. Only after some time, he realized that they were important players participating in power games through which young Arab men, marginalized in an Israeli town, could build connections enabling them to change their subaltern position (Cohen 1971). This book deals with the same topic: the possibilities to overcome limitations that stem from belonging to a subaltern group in the context of a postcolonial state; the possibilities facilitated through close, grassroot encounters with tourists hailing from the so-called Global North. It is a story of the emancipatory potential of tourism, both for communities that have become tourist attractions and their political aims, and for specific people—workers and small-scale entrepreneurs in the informal tourism sector in India.

Why is tourism, as a field of study, treated so contemptuously by (at least some) anthropologists (see Hall 2004, 143–144)? To some extent, it results from the tendency to essentialize research topics. We approach some issues—such as exile or migration—as *essentially* serious and thus important problems, worthy of scientific inquiry and bringing prestige to those who study them. On the other hand, the phenomena such as tourism appear *essentially* trivial, unworthy of scholarly attention, or easy and pleasant to research (see Hall 2011, 39). However, as Cynthia Enloe noted, "Tourism is not just about escaping work and drizzle; it is about power, increasingly internationalized power. That tourism is not discussed as seriously by conventional political commentators as oil or weaponry may tell us more about the ideological construction of seriousness than about the politics of tourism" (2000, 40). Hence, from my point of view, I have not radically changed my interests. I still deal with diverse forms of people's spatial mobility, especially in the postcolonial context. I am interested in power relations in which those mobility regimes are grounded, as well as the ways in which people overcome them. I scrutinize the contacts enabled by mobilities, between people coming from different social, cultural, and economic backgrounds. I study the mutual representations of different parties in such encounters, the processes through which self-representations are created, how status is being negotiated, and what coping strategies are being developed.

Thus this book is about tourism, but not only about it. Tourism provides an opportunity for a broader reflection on contemporary forms of supranational, intersocietal contact between people. By looking at the ways in which the inhabitants of countries sight-visited[1] by tourists perceive their "guests"—at

[1] By introducing this term I want to suggest that many tourists aim not only to see places (thus to do sightseeing) but also to visit the host communities. Such an understanding better reflects the character of tourists' presence in the places where I did my research, the presence that is not easily reduced to just a gaze.

how they position themselves vis-à-vis these visitors, how they interact with them, and what self-representations they put forward—we can gain insight into their own perception of their economic, social, and political situation in the postcolonial reality marked by complex power relations. This is also possible because tourism has inscribed local places into the global world order, rendering this oft-abused conceptual binary still fuzzier. Antonio-Miguel Nogués Pedregal calls such an approach "tourism as a context" (2008, 144).[2]

WHY DO WE FEEL CONTEMPT FOR TOURISTS?

The anthropologists' animosity toward tourism as an object of study seems to have yet another important source: our fear of being seen as a tourist. We look at tourism as "an illegitimate child, a disgraceful simplification, and an impostor" (de Certeau 1984, 143, after Bruner 1989, 439). We fear that the thin line that separates us from tourists could be challenged. We really try to "distinguish ethnography from tourism, for tourism is an assault on our authority, and our privileged position" (Bruner 1989, 439). But we do have so much in common, just to mention colonial nostalgia (Bissell 2005), characterized by mourning cultural change and extinction of traditional cultures (to which both tourism and anthropological "discoveries" were to contribute) or the desire for authenticity (MacCannell 1976), especially in its "hot" variety (Selwyn 1996a), consisting in fraternizing with the natives, that is, establishing an intimate, direct relationship with the Other. Both of us, anthropologists and them, that is, tourists, like to think that we managed to arrive where others had never been, we are both contemptuous of well-trodden paths. Even if we, (self)-reflexive anthropologists, attuned to the mechanisms of power relations, manage to debunk all the above insinuations, there is still one thing that we and tourists have in common. In the eyes of the people who host us, we tend to be one and the same—the Other: "Although for us tourism is an embarrassment, from the perspective of native peoples who are sometimes confused by the social distinctions that are apparently so important to us, what we label as colonialism, ethnography and tourism are be experienced in basically in a comparable manner" (Bruner 1989, 439).

We thus prefer to avoid the areas so accessible as to endanger our elitist professional identity (Crick 1995, 207–208; Burns 2004, 7; see also Bloch

---

[2] The only part of this book that directly stems from texts published elsewhere is chapter 4, "Recovering the Subaltern Voices: Tourism and Engagement." The chapter sprang from the research paper "Making Tourists Engaged by Vulnerable Communities in India." In *Anthropology of Tourism in Central and Eastern Europe: Bridging Worlds*, edited by Sabina Owsianowska and Magdalena Banaszkiewicz (2018).

2014b). By the same token, we started to perceive tourists in the same way as the people who consider themselves to be travelers—as a distinctive category in an identity project, more epistemological than ontological. As noted by Dennis O'Rourke—an ethnographic filmmaker focusing on tourism—tourists tend to treat tourism with the same contempt: "To be a tourist is in part to dislike tourists. Tourists can always find someone more 'touristy' than themselves to sneer at" (1999, 19; see Welk 2004).

Stian Krog, an anthropologist who did research in Hampi—the second site where I collected material for this book—considered "distancing himself from the tourists" as a prerequisite for "getting inside." This needed to be done through a series of "distinguishing" acts (Krog 2007, 67, 57–61). Some of these looked like "going native," for example, adopting the local style of clothing[3] and hairstyle. Other ways of distancing oneself from tourists built on the academic authority of a scholar from "the West," through, for example, refraining from using addictive substances. Unfortunately, as Krog admitted himself, all these strategies proved ineffective and most locals never perceived him as "anything else" than "an average tourist": "To my disadvantage, I was in many ways bound to activities distinctly reserved for tourists, both because they were highly convenient and also because they were in part expected of me. Drinking bottled water, using laundry services, eating at restaurants and visiting internet centers put me in the 'tourist' category" (2007, 57). The paradox was created by the fact that the "locals" themselves, especially men with whom Krog used to hang out, were quite eager to drink alcohol in the evenings (due to lack of other entertainment) and young men were into the Internet. The first of these activities was an excellent occasion to "get inside" (in my case, it was initiated by my lending men money for a drink behind their wives' backs). It thus seems that Krog tried to be more local than the locals, based on the oppositions he created, that is, local versus tourist and tourist versus anthropologist. It is also possible that our fear of being identified as a tourist stems from our professional, self-reflexive shame. The tourist reminds us of what we used to be, something many of us would gladly forget.[4] The world as it is now presented by tourism marketing is the world that the early anthropologists described when our field of study was in its infancy: "Ideas of old-style colonial anthropology—objectifying, reifying,

[3] It is interesting that the "local" male attire that Krog used to distinguish himself from tourists were "Western-style" carefully ironed trousers and buttoned shirt, universally appropriated in India.

[4] This is comparable to the shame felt by the anthropologist Marek Pawlak in the context of nationality and social class in his ethnography among Polish labor migrants in Norway: "I did not want to be identified with 'Polishness.' I felt ashamed of my nationality and did everything I could to 'hide' it in everyday situations and while dealing with members of the Norwegian society. . . . I crossed the street when I 'saw' Poles carrying plastic shopping bags. I recognized them as 'my people' and feared that I was easily recognizable as 'one of them'" (Pawlak 2018, 101–102).

homogenizing, and naturalizing peoples—are widely used by a variety of tourism shareholders, staking claims of identity and cultural belonging on strong notions of place and locality" (Salazar 2012, 874).

WHY DOES TOURISM MATTER?

According to the estimates of the United Nations' World Tourism Organization (UNWTO), the number of tourists visiting other countries will reach 1.8 billion people by 2030.[5] Tourist trips to foreign countries have become a massive-scale global phenomenon, also because of air travel that became commonplace after World War II, due to the air fleet and a pool of experienced pilots that the war has left behind. This coincided with gradual rise of the middle classes—whose members had extra money to spend on the so-called higher-order needs—and the universal right to a paid holiday, which established the division between the working time and leisure time.[6] Due to mass travel, tourism as a form of people's spatial mobility has become the most important—alongside migration and exile—contemporary field of the direct, interpersonal contact in the intercultural context.

Already in the early 1980s, Jafar Jafari (1982) noted that almost every community and every nation, small or big, already developed or still developing, was affected by tourism in varying degree. Tourism has become an integral part of social and economic life in many societies. Tourism industry, as it is frequently called (Eadington and Redman 1991), is one of the main sources of income in many non-European countries' budgets, amounting to 10.3 percent of the world gross domestic product (GDP) and filling one in ten jobs.[7] In this sense, the fate of many regions that do not possess other market-value resources depends on the presence of tourists. The loss of tourists—caused, for instance, by the fear of a terrorist attack or a natural disaster—may push

---

5 https://www.unwto.org/archive/global/press-release/2011-10-11/international-tourists-hit-18-billion-2030.

6 Scholars have considered tourism to be a product of European modernity that manifested since the mid-nineteenth century. Its emergence was made possible by four great revolutions in transport technology, that is, railways, steamships, cars, and airplanes (Turner and Ash 1975). These inventions created the possibility to move groups of people across long distances at a relatively low cost (the beginnings of an organized mass travel were described by Withey 1998). The same inventions, especially the steamers, contributed to the development of another form of spatial mobility that appeared in the nineteenth century, that is, international migrations (see Hatton and Williamson 1992), and, at the same time, accelerated the process of European colonization of the world (see Headrick 1981). In this book all these processes intertwine.

7 Data for 2019 according to the World Travel and Tourism Council (WTTC): https://wttc.org/Research/Economic-Impact. The data includes people who provide services indirectly and who manufacture goods for the tourism sector. In fact, these numbers are even bigger, if we add to them all those people who work in the informal sector, just like the protagonists of this book.

these places toward an economic collapse. This is especially true for the "tourist flows"—as they are called in the macro perspective literature (e.g., Williams and Zelinsky 1970)—from the so-called Global North to the postcolonial countries of the so-called Global South.[8]

From this point of view, tourism is neither trivial nor innocent. It is inscribed into the asymmetrical power relations and by itself produces them. It is, as Nogués Pedregal said, "one of the names of power" (2008, 141). Scholars are interested in its "hard" form, manifesting as environmental degradation or economic exploitation that deepens global inequalities through imposing the neoliberal regime of a supposedly free market (Azarya 2004; Croall 1995; Freitag 1994; Patullo 1996, 28–51; Higgins-Desbiolles 2007).[9] But they are also interested in the social and cultural issues, in line with Foucault's approach, according to which tourism is a tool for constructing Otherness, a practice of cultural, discursive dominance (Cheong and Miller 2000; Hollinshead 1999; see Foucault 1980, 2002).[10] This power also holds sway over the tourists, led like children by the tourism operators and other cultural intermediaries through the reality they sight-visit, controlling the way they perceive it (Urry 2002). The industry manufactures tourism imaginaries—and thus performs the "tourismification" (Salazar 2009)—under the guise of tourist guidebooks, travel literature, TV shows, and so on. It aims to make even those tourists who see themselves as independent travelers follow the established interpretation schemes (Barthes 1972; Salazar 2010; Salazar and Graburn 2014; Selwyn 1996b; Skinner and Theodossopoulos 2011).[11] Moreover, even though the emergence of tourism is linked to the idea of free

---

[8] I am aware that this division is a simplification, and in this book, I often criticize it (e.g., see the section "(Un)obvious Economic Inequalities" in chapter 2).

[9] In practice, this most often involves dictating rules to the countries that are less economically privileged and that frequently are tourist destinations. One of the main study topics here is the so-called leakage, that is, the outflow of money spent by tourists to corporations of the Global North (which are mostly the tourists' origin countries). Scholars also study the negative impact of tourism on the local employment structures, that is, the drainage of employees from the more prestigious professions (such as teachers or doctors) in favor of better paid jobs in the tourism sector (such as maids, waiters, or drivers). The inflow of tourists has been seen as the reason of rising inflation and speculation on the real estate market (Freitag 1994), and the development of tourist areas as an excuse to dispossess and displace local inhabitants (Mowforth and Munt 2003) to make space for big investment ventures, mostly beneficial for the Global North corporations (Tribe 2005).

[10] Thus it does not refer to the narrowly understood political and state power. See, for example, Fogelson and Adams (1977) and Cheater (1999) on how power has been understood in anthropology and Macleod and Carrier (2010) on anthropological take on power and tourism.

[11] According to the approaches inspired by Foucault, "the ways we think and talk about a subject influence and reflect the ways we act in relation to that subject" (Karlberg 2005, 1). The power of discourse consists in the fact that it creates the object about which it speaks (Foucault 2002; see White 1987). Internalized, for instance, as something that appears "normal," it remains invisible to us.

time, the tourism itself is a fuel for capitalism: the pattern of spending holidays by traveling makes people work harder (Turner and Ash 1975).[12]

Within tourism studies, however, anthropologists look at power mainly in relation to the other side of the tourist contact, that is, the employees of the tourism sector, or more broadly, from the perspective of the visited communities. The literature review shows that even though scholarly conceptualizations of tourism oscillate between "blessing and blight" (Young 1973; also Tuting and Dixit 1986), from the very beginning of academic reflection within the field, that is, the late 1960s and early 1970s, tourism has been critically interpreted through the lens of class differences and the center-periphery binary (see de Kadt 1979; Høivik and Heilberg 1980). The first—according to Georgette Leah Burns (2004, 8)—anthropological take on tourism was concerned with the damaging impact, in the form of acculturation, of the influx of tourists recruited among the urban middle class into the rural areas in Mexico (Nuñez 1963). Anthropologists touched upon the problem of the commodification of culture, understood an inescapable and negative consequence of tourism, endangering the authenticity of local traditions (Greenwood 1977; Rossel 1988). They used the metaphor of "cannibalism of difference" to show that the touristic consumption of another culture involved its destruction (MacCannell 1992, 17–73). Tourists themselves, especially within the context of the fast-growing mass package tourism, were presented as privileged inhabitants of industrialized metropoles—contemporary "golden hordes" invading less developed regions and transforming them into the "pleasure peripheries" in order to satisfy their needs (Turner and Ash 1975; see also Britton 1982).[13] The instrumentality of the relationship between tourist and host, based on mutual stereotypes and distance, was emphasized (Nash 1977). Within such an approach tourism became an imperialistic power, a legacy of colonial relationships (Nash 1977; Lea 1988) and the mirror for racism (MacCannell 1992, 121–146, 158–171), since colonialism, as the practice of exercising cultural dominance, was based on hierarchical construction of

---

[12] Since tourism studies were initially developed by sociologists, they focused on tourists, their typologies (e.g., Cohen 1972, 1979), their modes of experiencing the world, and constructing particular identity (e.g., MacCannell 1976 and his concept of a tourist as a modern pilgrim searching for authenticity but being trapped within a tourist spectacle). This bias, manifesting in the studies of tourists—their motivations, attitudes, ways of "domesticating the Otherness"—was quickly balanced through the work of anthropologists, who turned toward the other side of the tourist contact. The key moment was the publication of *Hosts and Guests: The Anthropology of Tourism* (1977), a book edited by Valene L. Smith, which focused on relations between tourists and their hosts and on the impact of tourism on the local communities and their cultures (the book itself was a result of the first anthropological symposium devoted to tourism studies, organized by Smith as part of the activities of the American Anthropological Association, three years prior to the publication).

[13] Louis Turner and John Ash are sociologists, but their highly influential book, *The Golden Hordes*, set the direction for the first conceptualizations of tourism also in anthropology. Stephen Britton, on the other hand, represents an analogous perspective within social geography.

Otherness. Consequently, contact within tourism was interpreted through the lens of confrontation or even conflict (see Robinson 1999), and the behavior of tourists as disgrace (Larsen 2000).

THEORETICAL FRAMEWORK AND OBJECTIVES: THE NECESSITY OF (RADICAL) RETHINKING

This perspective—which I call counter-hegemonistic, since it condemns the international tourism industry as a form of neocolonialism—was critiqued in the 1990s as oversimplifying, and thus unable to account for complex, subtle, and ambiguous relationships between the host communities and the tourists. However, it still is popular among some scholars (e.g., Reid 2003; Nowicka 2009; Burszta 2001, 2008),[14] and is vividly present in public debates: so-called experts, journalists, filmmakers, and NGO activists sound an alarm bell about the destructive impact of tourism (e.g., Kok, Timmers 2011; Srisang 1992; Routledge 2001)[15] and call for a more moral tourism, achieved through its alternative, more ethical varieties (Poon 2003; see Butcher 2003). I see this approach—even though its motives are noble, and it is, in many respects, reasonable—as carrying the risk of discursive objectification of both parties of the tourist contact. This binary perspective, demonizing tourists as rich ignorant hedonists, and victimizing the host communities as passive victims of acculturation processes and exploitation, prevents us from noticing subtle forms of negotiation, collaboration, resistance, and appropriation that take place between those parties (Cheong and Miller 2000, 373–374; see, for example, Joseph and Kavoori 2001; Kanemasu 2008). Nogués Pedregal pointed out that this emphasis "has kept tourism researchers focusing on dialectics rather than on dialogics; in so doing, both the anthropological

---

14 In the introduction to his book, Donald G. Reid wrote about tourism as "a dynamic force homogenizing societies and commodifying cultures across the globe" and compared it to "a major force in the 'cancer stage of capitalism'" (2003, 1, 6). Ewa Nowicka, on the other hand, listed negative consequences of the development of trekking tourism in Nepal on its local economy, social relationships, architecture, and environment, interpreting the relationship between tourist and porter/guide according to the "servant–master" scheme (2009, 40). Wojciech J. Burszta lamented the commodification of culture—"the boutique multiculturalism" in international tourism (2008, 21) and the artificial nature of the tourist spectacle "well-designed but disconnected from any kind of local tradition" in the form of agritourism (2001, 177–178).

15 See, for example, documentaries: *Cannibal Tours* (dir. Dennis O'Rourke 1987) is about ethnic tourism in Papua New Guinea, where affluent white Europeans and Americans go down the Sepik River on a luxury yacht called Melanesian Explorer, looking for villages inhabited by cannibals; *Thailand Backpackers: Full Moon Party* (dir. Richard Parry 1998) is about European backpackers and their drug and sex-fueled all-night parties on the southern beaches of Thailand; *Framing the Other* (dir. Ilja Kok and Willem Timmers 2011) is about ethnic tourism in Ethiopia, where Dutch tourists watch Mursi women as if they were exhibits in a museum.

understanding of sociocultural processes in tourism contexts and the production of knowledge have been limited" (2008, 148).

This is why, just like Tomasz Rakowski (2016) did before me, I perform an "anthropological perspective shift" and treat both parties of the tourist encounter as active makers of meaning and relationships, and not as passive recipients of the frameworks of meaning that have been imposed upon them.[16] In his study of Polish regions stricken with post-socialist poverty, Rakowski showed unemployed miners, factory labors, and impoverished inhabitants of rural areas not as "passive subjects" lacking initiative, succumbing to the "transformation trauma," but as agents—"practitioners of powerlessness," who, within the imposed systemic constraints, created their own interpretations and acted accordingly. Similarly, Jacek J. Pawlik, in his analysis of coping strategies of the inhabitants of Togo capital, instead of repeating the mantra of the unbearable poverty of the African metropolis, presented everyday life experience of Lomé residents through the lens of creativity, innovation, and entrepreneurship (in the form of an economy of sharing), depicting them as individuals whose "restraint, akin to ascetic zeal, was accompanied by inventiveness, initiative and searching for opportunities" (2016, 188). Maciej Kurcz, based on his research in Southern Sudanese Juba, demonstrated that, despite harsh life conditions in an African city, the migrants and refugees manifested "initiative and the ability to create . . . their own social and political niche" using "a variety of ways, both formal and informal" (2012, 14–15).

I apply the same approach in my work, because perceiving members of the visited communities as passive objects of tourists' projections and practices does not allow us to notice that due to long-term experience, they have learned to recognize expectations and needs of tourists and to use them in achieving their own goals (see, for example, Boissevain ed. 1996; Wilson 1997a; Adams 1999; Little 2004). In a later edition of *Hosts and Guests*, published over a decade after it first appeared (1989), when Valene L. Smith asked the authors to review their papers, she sparked reflection that tourism may not be the main factor causing the negative cultural change and not always it is the source of unilateral exploitation (see Smith's reflection in Nash and Smith 1991, 15; for example, Deitch 1989; McKean 1989; or even Greenwood). Scholars started to notice that tourists' interest in local culture could serve to revive it by providing material means of its survival (Cohen 1988), or by valorizing the elements that before, due to colonial policies, were considered inferior to the universalized European patterns (see, for example, Picard 1990, 1996). Moreover, cultural production for the tourism market was shown to have the potential to politically and economically empower in

[16] Cf. Smith (2006, 32) for a similar take on the subject in cultural heritage studies.

particular marginalized groups (Healy 1994; Azarya 2004). Eventually, there appeared studies demonstrating that the members of the visited communities could perceive their relationships with tourists in terms of reciprocity (Adams 1992) or friendship (Tucker 2003b; Simoni 2017).

The tendency to interpret (especially international) tourism as a neoliberal force responsible for a series of undesirable phenomena is part and parcel of a wider trend, which Sherry B. Ortner, in her famous article (2016), called "dark anthropology." According to Ortner, since the 1980s we have tended to perceive the world through the lens of the ubiquitous "power, exploitation, and chronic pervasive inequality" (2016, 50), focusing on lives lived in the conditions of marginalization, exclusion, and uncertainty.[17] On the one hand, this is a response to neoliberalism and a recognition that it prolongs colonialism, but it is also a legacy of the powerful impact of early Foucault's theory and neo-Marxist approaches. An "anthropology of the good" was a reaction, or rather a means of resistance, toward dominating anthropological narratives of injustice and suffering. It would focus on solidarity, commonalities, and resilience (Robbins 2013). The intention of the proponents of this turn was moving beyond the "suffering subject" living in pain, poverty, and violence, and directing the attention of anthropologists toward the categories of "morality, well-being, imagination, empathy, care, the gift, hope . . ." (Robbins 2013, 448), happiness, and good life (see also Fischer 2014, 17 and his "positive anthropology"). Such proposal was also formulated in the field of tourism studies as "an academy of hope," situated within the critical turn (Ateljevic et al. 2007, 3). This project, encompassing a series of interpretative, critical, and emancipatory approaches, postulated such studies of tourism that would also affect change toward social justice, equality, and resisting exploitation, or to put it differently, a "hopeful tourism" (Pritchard et al. 2011). Alternative, responsible, sustainable, and just forms of tourism that are promoted within this framework are compared against mainstream capitalist tourism.

Ortner (2016, 60), however, suggests that instead of seeing these tendencies as opposite, we should attempt to integrate them, that is, not only not to ignore wider contexts of power and inequalities (and thus get enmeshed in microhistories) but also not to let criticism overwhelm us to the point as to render us powerless. The author of "Dark Anthropology and Its Others" finds a way to achieve this through an engaged study of resistance and activism in an anthropology of social movements, inspired by Gramsci's famous phrase: "Pessimism of the intellect, optimism of the will" (Ortner 2016, 66).[18] My

[17] See James Laidlaw's commentary to Ortner's paper: http://www.journals.uchicago.edu/doi/10.14318/hau6.2.005.

[18] Ania Loomba does exactly the same in postcolonial studies (2015, 230). Ortner quotes the article by Veena Das (2010) about the relationship between a Hindu and a Muslim in a poor neighborhood

own idea of how to break through this dialectics—my redefined "academy of hope"—puts accents in a different way. Being aware of power relations and diverse forms of subalternity, I see not only (or not always) resistance—since it is, in my view, another dialectic trap (also, resistance is frequently conceptualized from an *etic* point of view, since, as engaged anthropologists, we are very keen to see it)—but rather possibilities: for empowerment, political struggle, or just human understanding "across difference" (Tsing 2005, 3–5). Arjun Appadurai, while referring to a participatory project conducted with activists and dwellers of Indian slums, writes about "the ethics of possibility," grounded in "those ways of thinking, feeling, and acting that increase the horizons of hope" (2013, 295). Following the "ethics of probability," that is, focusing solely on constraints, carries the risk of killing off grassroots optimism in relation to life and the world. People have the "capacity to aspire," which frequently sparks their agency, and studying the possibilities to develop this capacity may contribute to overcoming the existing limitations in the future (within an anthropology of the future, developed, among others, by Appadurai).[19] The ethics of possibility may, therefore, "offer a more inclusive platform for improving the planetary quality of life and can accommodate a plurality of visions of the good life" (Appadurai 2013, 299–300; see, for example, Mahony 2018). In a commentary to Ortner's article, Appadurai writes about the necessity to practice "anthropologies of aspiration and possibility" (2016, 2). To put it differently, the scholars are to situate themselves somewhere between totalizing power and an uncritical apotheosis of agency.

My conceptual area in this book has to a great extent been set by the postcolonial critique (Gandhi 1998; Loomba 2015). Such a "theoretical confession," as Paul Willis calls it (2000, 113), does not mean "tightly structured positions looking merely exemplification" (2000, 114) but mean a point of departure "which is then overturned, or perhaps strengthened, or positively diverted, or fulfilled in unexpectedly elegant ways" (Willis 2000, 113). I do not revere theory—rather, I see it as a fertile resource that needs to be creatively modified when faced with what is available empirically. Such dynamic, theoretical bricolages "pushes our thinking in unknown directions, creating trajectories that we could not predict" (Buchowski 2005). They allow us to create unbound interpretations, shedding new light on the phenomena that we used to take for granted. I also agree with Willis, when

---

in Delhi as an example of such a successful integration. Das shows how love and mutual care develop in the shadow of a long history of Hindu-Muslim animosity, through an analysis of everyday contacts between the fiancées and their families within the wider context of politics and violence.

[19] Cf. remarks by Pawlik in his study set in Lomé: "Inspired by ideals, pondering their own possibilities, following their knowledge, skills and intuition, the inhabitants of Lomé activate different levels of entrepreneurship and creativity in order to develop and achieve wealth" (2016, 176).

he says that in the best ethnographies "theory is mainly beneath the surface, informing but not interrupting it" (Willis 2000, 117). Research, according to such an understanding, requires ethnographic imagination, which Paul Willis and Mats Trondman define as follows:

> Most basically, we are interested in recording and presenting the "nitty-gritty" of everyday life. . . . These are moments where new understandings and possibilities are opened up in the space between experience and discourse, at the same time deconstructing and reshaping the taken for granted in a particular response to the shape of the social order, a response that transcends dichotomies such as public/private, social/individual. Aha effects fuse old experiences with new ones, thus opening up readers' minds toward new horizons. (2000, 399)

If we combine such methodological stance with the postcolonial perspective, inspired, among others, by hermeneutic orientations and poststructuralism, and thus deconstructing the image of "science as a human activity seemingly outside of various historical and cultural conditioning" (Kotowa 2008),[20] the resulting project would enable serendipity, that is, openness to the unexpected,[21] in stark opposition to the closed-circuit theories that restrain development and put us in an intellectual cage. Even though negotiating "fertile tensions" between different theoretical approaches is far from easy and sometimes feels like stunt horsemanship (Mallon 1994, 1515), theoretical flexibility and creativity seem to be the way forward.

The postcolonial perspective is an intriguing source of inspiration because it questions its own theoretical status—it learned from post-structuralists to distrust meta-narratives[22] (Lyotard 1984). At the same time, it is a combination

[20] Theoretical reflection, as Kirsten Hastrup highlighted, is always done "in a context of which the practitioners are irrevocably part" (1995, 182).

[21] The term "serendipity" is used in situations where accidentally we stumble upon a fortunate discovery, especially when we have been looking for something completely different. It comes from the old Arab name for Sri Lanka, that is, Serendip, and was first used in 1754 by Horace Walpole in a letter to his friend Horace Mann. After reading an old Persian tale, "The Three Princes of Serendip," Walpole decided that its main motive—the accidental circumstances leading to the happiness of the unexpected—should be captured in the concept he named "serendipity." Haim Hazan and Esther Hertzog translated the concept for anthropology in their edited volume *Serendipity in Anthropological Research* (2012). In the introduction they stated that anthropological research is a lifelong nomadic quest for discovery, during which the world brings an infinite number of topics to study and numerous ways to study them, while each stage breeds its own questions and requires a separate methodology. Within the context of the anthropology of tourism, Donald V. L. Macleod and James G. Carrier wrote that long-term research increased the odds that the researcher would accidentally come across something that they had not predicted (i.e., when preparing for the project), which would turn out to be crucial for the understanding of both people and phenomena under study (2010, xii). See also Rivoal and Salazar (2013).

[22] Here, I refer to scientism and theories understood as based on apparently universal category of rationality stemming from formal logic and, hence, proclaiming themselves to be the only justified

of very diverse, and often mutually exclusive theoretical approaches, including the "catachrestic combination of [among others] Gramsci and Foucault" (Prakash 1994, 1490).[23] This, in my view, resonates with the postulate to overcome "dark" tendencies in anthropology and open up toward new possibilities.[24] Secondly, the postcolonial critique is not only about relationships between the former colonies and the former metropoles. This perspective has been applied to the analysis of all sorts of situations where domination and subordination resulting from global power relations are observed (see Mishra and Hodge 1991, 407). That is why one of the key categories developed within postcolonial studies, that is, the subaltern[25] (because of race/ethnic origin, gender and class, and more specifically their intersectionality), has been extended to include all those groups, whose voices have been silenced by dominant discourses, preventing them from creating their own self-representations within public debates—the groups for which somebody else speaks and decides. Lacking political representation—ignored, infantilized, or demonized—they become marginalized socially and economically. Currently, such studies also encompass research into the areas of post-socialist poverty, local communities pushed into the margins of nation-state projects, people with disabilities, non-heteronormative persons, and so on. Within such an approach, tourism from countries with higher GDP to countries that are less developed "situates workers [of the tourism sector] as the subordinate party not merely because of their occupational role but also because of their status as a formally colonized, (largely) working-class people (and often women) of color" (Kanemasu 2008, 125).

Such understanding of tourism to postcolonial countries—as the anthropologist Benoît de L'Estoile (2008, 267) pointed out—is part of the colonial legacy, just like global capitalism (Ross and Trachte 1990) or the new international division of labor (Fröbel et al. 1980; Dirlik 1994). The concept of

forms of knowing. In other words, these are close-circuit statements that aim to provide a universal explanation of the way things work in the world.

23 Within the postcolonial critique, the Foucauldian concept of an overwhelming and all-encompassing network of power relations from which there is no escape, and of which people are usually not aware, was used to conceptualize diverse challenges posed to status quo (Loomba 2015, 59). Ortner also points out that in some of his later work, Foucault began to move away from the concept of relentless power, noticing different ways in which discourse could be widened (2016, 51; see, for example, Martin et al. 1988; Laidlaw 2014, Ch. 3).

24 Gramsci's maxim about the "pessimism of the intellect, optimism of the will" as in Ortner, above, is one of the most quoted phrases in postcolonial studies.

25 Subaltern is a category developed by members of the Subaltern Studies Group at the beginning of 1980s, to name social groups that have been excluded from political representation in the set structures of reality. In the colonial context—especially in India—these were not all colonized groups, but mainly peasants, whose voices had been appropriated by the elites (see Guha 1982) and women, on whose behalf spoke men (both the colonizers and the colonized, see Spivak 1993). The concept of the subaltern was borrowed from Antonio Gramsci (1996), who used it in the letters he wrote in prison, to refer to working class ("proletariat").

colonial legacy has pejorative connotations as, at least since the 1970s,[26] it has been most often used in public debates in the context of negative phenomena occurring in contemporary world, phenomena that are thought to stem from colonialism: underdevelopment of former colonies in comparison to metropoles, ethnic conflicts leading to ethnic cleansing, "stabilizing" military interventions, corruption, and authoritarian rule in postcolonial countries. It has also been used by the postcolonial leaders to avoid responsibility for political, economic, and social failures and has given rise to demands in a game involving moral compensation (de L'Estoile 2008, 267–268), for example, in the context of aid programs or migration policies. However, just as L'Estoile emphasized, we should not treat this category as deterministic in our analysis. Colonial legacy is not something that can be passed down in an unchanged form but something that is negotiated: questioned, appropriated, remembered, forgotten, or selectively applied. It is something that is both shared and that divides the "heirs," something that reflects the ambiguity of the term "share" in the English language (de L'Estoile 2008, 270). In this sense, the anthropology of colonial legacy promoted by L'Estoile is not yet another subdiscipline but a necessary element of every anthropology of contemporary world: material culture studies (museum collections, archives, monuments, architecture, food, music), flows of people, goods and representations (in migrations, tourism, and economics), international relations, art, science, and so on. Colonial past and the discourses that it shaped impact mutual perceptions and relationships in contemporary world: Europeans vis-à-vis migrants, migrants vis-à-vis host societies, the inhabitants of the so-called Global North vis-à-vis development programs and the beneficiaries of aid programs, or mutual representations and expectations of tourists and members of the host communities (de L'Estoile 2008, 273, 267).

In summary, which I wish to highlight, the postcolonial perspective—contrary to the anti-colonial view and "dark anthropology"—does not picture the world as simply black and white. Just as the history of colonialism was not only the confrontation of colonizers/oppressors and the oppressed/colonized victims, tourism to postcolonial countries is not just a story of evil tourists/colonizers' heirs and the good locals/victims. Both groups are heterogeneous, internally divided, and often create their own forms of subordination. Meanwhile, dichotomous narratives obliterate this complexity. Therefore, contrary to what some may think, especially when they mistake

[26] In the period directly following decolonization, the former colonies experienced optimism based on newly regained independence and hope related to modernization and development programs supported by former metropoles. However, the crisis of the 1970s ended the era of postwar prosperity and welfare state in Europe, which ushered in the stage of increased rise of neoliberal capitalism, resulting in limitation of workers' rights and social care, which were also imposed on the postcolonial countries in exchange for loans (Ortner 2016, 52–53).

postcolonialism for anti-colonialism, the postcolonial critics are not telling the story of domination over and over again and are not creating a binary world image, according to the scheme "West and the Rest" (Moore-Gilbert 1997, 67). The point is to complicate this image by indicating that the division lines may run in pretty unexpected places and that people may take unexpected stance vis-à-vis subordination (De Boeck 1996). Postcolonialism reads the contact resulting from asymmetrical relations of power in terms of interaction and mutual influence, as "a process involving complex negotiation and exchange," which not only destroys but also creates (Trivedi 1993, 15, after Gandhi 1998, 125; de L'Estoile 2008, 268). According to such understanding, tourism is one of the "contact zones"—a space of encounters and interaction, even if it is marked by difference and grounded in unequal power relations (Pratt 1992).

However, scholars of tourism only infrequently use interpretative tools provided by postcolonial studies. Only the categories of Orientalism (Said 1979) and imperial gaze (Pratt 1992)—as sources of colonial nostalgia—seem to enjoy an unwavering popularity as the foundations, upon which tourists build their representations of (not only) postcolonial countries (e.g., Echtner and Prasad 2003; Hutt 1996; see Horolets 2013 and Bachórz 2013 as very interesting examples of application of postcolonial perspective in the analysis of contemporary Polish tourists' perception of Russia and countries under the former USRR). Local responses to these representations are also interpreted as manifestations of strategic self-Orientalization (e.g., Gregory 2001; Minca and Borghi 2009). Slowly, there appear studies, mainly authored by scholars from outside of the Euro-Atlantic area, written in the spirit of decentralization of the dominant perspective of "Western" academia (e.g., Teo and Leong 2006). At the same time, postcolonial critics—those who deal with contemporary relations of power and subordination, and not with a new reading of colonial history—tend not to pay attention to tourism. Instead, they focus on more vivid forms of neocolonialism, such as domination of global capital or political hegemony of "the center" which is increasingly enforced through military interventions (see, for example, Dirlik 1994; Lazarus 2006).

Scholars of tourism noticed this gap and in 2004 they built the first "bridge," in the form of a book titled *Tourism and Postcolonialism* (Hall and Tucker 2004).[27] The editors noted that grafting analytical categories developed in one area into a completely different field may prove to be a very fertile and mutually beneficial practice (Hall and Tucker 2004, xiv). It enabled posing new research questions, just like the one that forms the basis of this book: If

[27] Even though the majority of texts in the volume edited by Colin Michael Hall and Hazel Tucker relate to the question of representation-making in tourism, these studies do not diverge from earlier analyses based on the observations by Edward Said (1979).

and in what situations can tourism be a space where subalterns regain their voice, following the profound question asked by Gayatri Chakravorty Spivak in her seminal essay "Can the Subaltern Speak?" (1993, originally published in 1988; cf. Aitchison 2001)? In other words, can local communities, despite being enmeshed in global power relations, create their own representations and achieve their (political etc.) goals within, through and thanks to tourism? Does this apply to individuals, to specific persons, who make up these communities? Can tourism be an area where people exercise their agency and (re)build their subjectivities, despite the apparent continuation of colonial relationships in an encounter between a white tourist from "the West" and a service-providing inhabitant of a postcolonial country? Can tourism be a source of empowerment? As Anna Horolets already asked (2012, 127), "Can tourist contact transform into interpersonal contact?" How does it influence mutual representations? Is it possible to move beyond Orientalism's and Occidentalism's binary views? How do people build bonds? Who benefits and what benefits stem from such contact?

In other words, the aim of this book is to search for an answer to the following question: Does tourism have emancipatory potential? I study this by looking at everyday encounters between tourists and employees in the informal tourism sector[28] in two small-scale, popular Indian locations: Hampi in the Southern state of Karnataka and Dharamshala in the Northern state of Himachal Pradesh. They share not only their status as tourist attraction but also their vulnerability.[29] Hampi is a village whose residents have been gradually evicted in the name of protecting the material cultural heritage, while Dharamshala is de facto a Tibetan refugee settlement.

TOURIST INDIA: SPACES OF ENCOUNTER

India is the most populous postcolonial country in the world,[30] and invariably a popular destination of tourist "pilgrimages," as the manifestation of Said's orientalizing representations, offering a multitude of attractions at relatively

---

[28] The informality of the tourism sector follows the distinction made by Eric Cohen (1982), who wanted to distinguish big, formalized tourist enterprises (such as all-inclusive resorts) from small, grassroots businesses providing services to individual, low-budget tourists, mostly backpackers. Obviously, the distinction between these two sectors is blurred and subject to different interpretations (e.g., according to business size, whether customers can pay with credit cards or whether employees are paid salaries, see Oppermann 1993). I will further discuss local businesses of this kind in my field in chapter 1, and the specificity of backpacking tourism in chapter 2.

[29] This category will be discussed in chapter 1, in the context of the communities I study.

[30] It was not by accident that postcolonial studies were institutionalized, at the beginning of the 1980s, among Indian postcolonial intellectuals, even though they usually worked outside the country (see Prakash 1994).

low prices.[31] According to the World Tourism Organization, approximately 2.5 million tourists visited the Republic of India in 2000, and just one decade later their numbers reached almost 6 million (in 2013, when I started research for this book, this number grew by another million visitors[32]). The country is ranked twentieth on the list of countries with the biggest income from foreign tourism (14.5 billion U.S. dollars in 2010—a fourfold increase to 2000—in 2013 this income increased by further 4 billion USD) (UNWTO 2015; 2017). The important role of tourism in Indian politics and economy is reflected in the Indian government's having a separate Ministry of Tourism.[33] According to the Ministry's data, income from tourism in 2012 to 2013 accounted for 6.88 percent of the country's GDP, and the share of tourism sector and its related industries in the labor market reached 12.36 percent (Ministry of Tourism 2017, 88).[34] According to estimates, tourists spend more time in India than in any other country in the world—27 days on average—which is not only due to the multitude and variety of attractions on offer but also due to the all-year tourist season[35] (Singh 2001, 141). Also the very beginnings of tourism phenomenon are related to India—the son of Thomas Cook[36] transformed his father's project of democratized travel into a flourishing business, establishing branches in India, organizing Queen Victoria's visits to the "Jewel in the Crown," as well as pilgrimage tours for Indian Muslims to Mecca (Podemski 2005, 16). According to the *Encyclopedia of Tourism* (Jafari 2000, 302), it was the British colonizers who laid the foundations of tourism in India, by constructing an extensive network of railway and roads and leisure infrastructure in the Himalayan regions (including the hill station in Dharamshala), which also reinforced the ancient tradition of pilgrimage to holy sites of different religions of the Indian subcontinent (such as Hampi,

---

[31] India promotes especially those sites which offer diverse tourist attractions (in terms of nature, history, religion, culture, etc.; see Singh 2001, 144). Within the country's territory there are thirty-eight sites listed as UNESCO World Heritage, including thirty cultural sites (Hampi is one of them), seven natural sites, and one mixed one (http://whc.unesco.org/en/statesparties/in). Geographic and climate diversity of the country creates a lot of opportunities, starting from trekking in the Himalayan regions, through desert trips in Rajasthan, to relaxing on the beaches of the Arabian Sea and the Bay of Bengal. India also attracts tourists looking for spiritual advancement in multiple religious centers—Hindu, Buddhist, Jain, Sikh, and Muslim.

[32] By the end of my research, the number of tourists exceeded eight million in 2015 (UNWTO 2016), and the Indian Ministry of Tourism noted a 10 percent increase the following year (Ministry of Tourism 2017, 4).

[33] http://tourism.gov.in/. Moreover, each state of the Indian federal system has its own department responsible for tourism promotion.

[34] In 2016 the income from this sector already amounted to 9.6 percent of GDP (WTTC 2017).

[35] When the tourist season ends in Southern states, it begins in the North of the country. We can assume that such a long average stay in India is due to the country's popularity among the backpackers, who often travel across the country for several months.

[36] Cook was an English pastor regarded as the originator of organized group travel.

not only listed as the UNESCO World Heritage but also listed as an important Hindu site).[37]

However, even though India had been one of the travel destinations for Europeans even before the colonial period,[38] and although it attracted hosts of hippies in the 1960s and 1970s,[39] mass foreign tourism only developed there in the 1990s, when the Indian government announced the "decade of tourism" (Routledge 2001, 225).[40] Today, contrary to what happened in many postcolonial countries that developed a "monoculture" of enclavic tourism (see Britton 1982; Freitag 1994; Poirier 1995), which reinforced dependency of the peripheries in relation to the centers, India is characterized by a large diversity of travel options, ranging from stays in luxury resorts, through sight-seeing tours, to the very popular backpacking, that is, individual, long-term,

---

[37] See, for example, Bhardwaj (1973), Jha (1985), Kaur (1985). Pilgrimages and trips to the hill stations (resorts) are among the most popular forms of domestic tourism in India (see Richter 1989, 104–106; Singh 2001, 139), which I discuss in chapters 1 and 2. Other popular forms are business tourism (i.e., business-related trips combined with visiting sites and recreation) and the so-called VFR tourism (Visiting Friends and Relatives). According to Linda K. Richter, such a strong growth of domestic tourism makes India less vulnerable to changing supply of international tourists (1989, 106). At the same time, an important group of tourists coming to India from abroad—most often as VFR tourists—are the so-called Non-Resident Indians, that is, Indians permanently residing outside their country of origin.

[38] See, for example, Podemski (2005, 208–316), who studied diaries written by Poles traveling through the Indian subcontinent (the oldest one is dated 1596). Sixteenth-century routes to India were set by traders, and since the seventeenth century they were set by missionaries. These two groups paved the way for British colonization, which was in full swing from the mid-eighteenth century (except in the Portuguese colony in the Southern part of the west coast, existing since the sixteenth century and the French colony established in the seventeenth century on the Southern part of the east coast). In the nineteenth century and the beginning of the twentieth century, subordination of India by the European governments resulted in an influx of specialists (such as engineers or educators), researchers, and travelers coming mainly from the upper classes of European societies (see, for example, a diary by a Polish countess, Ewa Dzieduszycka, of her travel to India in 1909, published in 1912, or a best-selling novel *A Passage to India* by Edward Morgan Forster, published in 1924).

[39] The 1960s were the time of hippie movement in Northern America and Western Europe, which encompassed the fascination with a broadly understood, romanticized, and exoticized "Eastern culture" (Cohen 1973). Trips to India—often road trips via Istanbul, Iran, Afghanistan, and Pakistan—were its integral part (MacLean 2006), as demonstrated by the history of tourism development in Hampi (which I describe in chapter 1). Certainly, India's popularity with hippies was linked to the fact that pop-culture icons were staying in Indian meditation centers (in 1968, The Beatles spent several weeks in Rishikesh, currently situated in the same state as Dharamshala; see Norman 2011 on spiritual tourism to Rishikesh).

[40] "Visit India Year" campaign was announced in 1991. It was, to a large extent, a reaction to pressures exerted by the International Monetary Fund and the World Bank to internationalize the market within the New Economic Policy (Hutnyk 1996, 10). The beginning of the 1990s marks the beginning of a period of economic liberalization and globalization in India. Opening the country up for international tourism was part and parcel of these processes. The growing popularity of fundamentalist movements in India in recent years—mainly Hindu nationalism—may be interpreted as a reaction to the consumption-based society and increasing secularization that resulted from these changes (Pawlik 2011, 29). I discuss this further in chapter 2, alongside Occidentalism.

and low-budget stays.[41] Such diversification, combined with multisectoral domestic economy makes India much less susceptible to the so-called leakage, amounting to mere 10 percent (WTTC 2013, 13).[42] In comparison, in Botswana the leakage reached 70 percent (Mbaiwa 2005, 164), and in the mid-1990s in Bahamas it amounted to as much as 90 percent (Meyer 2006, 7).[43] Also, when compared with authoritarian states—such as Cuba or Myanmar—tourism in India is far less controlled by the state. This has given rise to a flourishing informal tourism sector, that is, small, local enterprises, usually in a form of family-run simple guesthouses, restaurants, shops, street stalls, and other grassroots initiatives. It is their owners and employees that are the main protagonists of this book. The informal sector is characterized by a relatively democratic access to the tourist—which includes beggars, street tea sellers, or self-appointed guides—which creates a multitude of possibilities.[44] By the same token, many sites that tourists visit in India, including those depicted in this book, constitute "heterogenous spaces" (Edensor 1998), in which tourism developed in a haphazard, unpremeditated way, and as a result, where the worlds of tourists and locals tend to overlap.

[41] This partly results from the federal character of the country: particular states have the possibility to implement their own tourism policies, adjusted to their specific circumstances and needs. In line with official guidelines, central government is responsible for foreign tourism, and state authorities deal with domestic tourists (Singh 2001, 139–141). However, in practice, these two forms often coincide, which often leads to a lack of clarity as to who should do what, leaving many areas outside of the state supervision.

[42] Due to a comprehensive economic growth, India does not have to import goods and services to develop tourism. It also matters that much foreign investment into infrastructure is based on the principle of public-private partnership (Jafari 2000, 302). Of course, supranational hotel chains such as Hyatt, Sheraton, or Holiday Inn are present in India, but their facilities are mainly located in big cities (often nearby international airports) and are not significant in the scale of the whole country.

[43] This means that for every 100 dollars spent by a tourist, 90 leaks out to the corporations of the so-called Global North. Such leakage is made possible through travel agencies, booking package holidays for tourists, airlines, hotel chains, insurance brokers, bank charges, imports of goods that are in demand among tourists, salaries of foreign management staff, or foreign investment in local tourist infrastructure. The phenomenon of leakage is one of the main reasons why international tourism, seen according to the dependency theory, has been blamed for deepening global inequalities (Oppermann and Chon 1997; see a similar approach in Plange 1989; Harrison 2001). This phenomenon is especially damaging for small countries whose economy is largely dependent on tourism. Leakage may be limited through the so-called linkages, that is, links created between the tourism sector and local economy.

[44] Even though many Indian researchers working on tourism sector harshly criticize central government for inconsistency, negligence, and indolence (e.g., Singh 2001), Richter, who studied the beginnings of mass tourism in India, noted that the hectic style of implementing tourism policy in India, partly a planned economy, and partly a private venture, was surprisingly effective for a democratic rule (Richter 1989, 103). However, from the beginning of the twenty-first century, the Indian government has been trying to shape the tourism sector, in order to adjust to the imagined expectations of more affluent customers (see Bloch 2016c for Hampi or Lobo 1991, Routledge 2001, and Saldanha 2007 for Goa). Many critics point out the role that wealthy Indian businessmen and corrupted politicians play in this process (Saldanha 2007, 202). I discuss this further in the conclusions.

Even though I have worked in India since 2002, I consider myself to be a post-exotic anthropologist (Clifford 1997, 351, footnote 2), in the sense that I am interested in human beings grounded in different configurations of power relations and agency, and not the view focusing on ethnic/cultural difference. Looking at how anthropology situates itself in public debate, I become less and less of "an expert in other cultures" and, generally, move away from the Boasian take on cultures as "bounded entities comprising 'groups' that share basic values and customs" (Eriksen 2001, 132). Such a "culturalist rhetoric" reinforces differences between people, which may then be used as a fodder for conflicts (Turton 1997). The misleading ontology of culture made ethnology into "the art of viewing natives in their natural environment in order to identify, classify and reduce them to so many laboratory specimens" (Eriksen 2001, 138). For more than two decades, anthropology has been breeding calls to abandon the "exotic" view of culture (Hastrup 1995, 6), to exercise seeing what is "ours" from a distance, and seeing the familiar in what, heretofore, appeared "exotic."[45] As part of the project of decolonizing anthropology, we critically deconstruct the process of "othering," in which we, as a discipline, have been so much enmeshed (Harrison 1991). Writing "ethnographies of the particular," that is, recounting stories of particular individuals and their experiences, have potential to reveal aspects of human solidarity above or across cultural difference (Abu-Lughod 1990, 157), since "difference always comes along similarity, and it is this dialectics that [. . .] provides an interesting context for anthropological inquiry" (Pawlak 2018, 13). This book is about people whose paths crossed through tourism: about people grounded in very different contexts, who in other circumstances would have never met. It is a story of these direct, grassroot encounters in spaces created through global processes: the story of trajectories of such encounters and of what they bring into the lives of particular people.

As a prelude to this story I offer excerpts from my field notes made during the participant observation in one of Tibetan cafés in Dharamshala. They demonstrate the potential of the encounters' diversity, enabled by the democratic access to tourist spaces; global encounters "across difference," as Anna L. Tsing calls them, often "awkward, unequal, unstable, [but] creative" encounters, that can lead to unexpected "new arrangements of culture and power" (2005, 4–5).

> Day 1: There is a young Indian couple sitting in the terrace with a little girl, they talk in English, and after a while they invite the café owner's young daughter to sit at their table; at the next table, there is an elder woman from France, a

[45] Other "life-forms," as Nigel J. Rapport suggests, can be found in the world of our children, our neighbors, our political opponents, people of different social status, and so on (2000, 72).

sponsor of a Tibetan boy aged 12–13, she is telling her protégé that she used to be poor in math's in school and that he should consider who he wants to be in the future, so that he would not have to worry whether he is going to be able to provide for his family (the boy remains silent and keeps playing with a box in which his gift was wrapped; eventually, the boy shyly states that he would like to be a teacher); further back, there is a young couple—she is blonde, he is Tibetan, wearing a fashionable shirt; after a while two English-speaking women come onto the terrace, both try to speak Tibetan with the waitress, the elder one is much worse at it, she is carrying some papers, on which she is drawing and writing something, evidently learning Tibetan; the younger woman is drinking hot water for breakfast, according to the instructions of Tibetan doctors. . . .

Day 2, morning: A 30+ Tibetan man is sitting in the terrace, he looks like a newcomer, wearing grey T-shirt and jeans, with a *mala*[46] around his neck. He is talking to a Western female tourist, in her thirties, telling her about the significance of Buddhist teachings for a personal growth (she has finished her breakfast and is now drinking coffee, he is drinking tea with lemon and ginger); he offers to teach her about Buddhism, she writes down an address and tells him that she is paying the bill, he is protesting a little bit, but not really; he feels she is slipping away, and becomes sad, so she suggests they should meet tomorrow; but generally, she does not seem interested, does not encourage the discussion, just mumbling "mhm" and nodding, and after he leaves, she changes tables, joining a very handsome man, a tourist, and then starts to speak fervently ("Haven't we met before, somewhere?")—this time he is not interested and keeps using his smartphone and, after a while, he leaves. Then, the girl who drunk hot water the day before enters the café with a young man, probably American, who brings a small drum as a gift for the café owner's little daughter. . . .

Day 2, afternoon: Situation on site is as follows—an elderly couple of tourists, possibly English, an elderly Indian man is animatedly explaining something in fluent English to his European companion; a young Tibetan woman, accompanied by two male backpackers (she is very much at ease, seems to enjoy it); an elder Tibetan man, probably with his son and a granddaughter (the girl is completely focused on her laptop); a young cool-type Indian man managed to strike a conversation with a blond girl sitting at the next table; alone young woman—a tourist—is sitting at the last table; there are three monks by the window; after a while the sponsor I saw yesterday comes in with another protégé, a bit older, they look at something on a tablet, and later a friend joins them, she is also around 60+, accompanied by the boy who was here yesterday (it seems that both women are sponsors); soon after that, another elderly woman comes in, with a Tibetan girl of about 13; there is also a young Tibetan monk

[46] Sanskr. *mālā*, a Buddhist rosary.

meeting a tourist from Spain (they are of similar age), the Spanish shows the monk something on his laptop (they seem to be filling a questionnaire, maybe a visa application form), he also explains how to charge a mobile phone and encourages the monk to learn English (the monk barely speaks the language), they agree to meet by the temple the next day; two young Tibetan women come in, I saw one of them yesterday (they are exceptionally fashionably dressed: wearing high heels, leather jackets, tight jeans)—for the next hour and a half they drink one bottle of mineral water, the aim seems to be to show oneself, and be around tourists (they know the waitress who keeps joining them every now and then). . . .

Day 3: Two elderly French women again, with their sponsored children; a young Indian and a Tibetan man discuss a project while working on a laptop; a middle-class Indian man seems to be on intimate terms with a 30+ blond woman; an elderly monk; a young monk accompanied by a non-monastic cool-type friend; a tourist woman in a baseball cap and a tattoo on her back is chatting to the waitress; elderly Germans, who are just about to leave the town, are offering small gifts, a kind of lace accessories, to the staff and the owners' children. Etc.

READER'S GUIDE TO THE CHAPTERS

This book consists of five chapters. Chapter 1, "In a Circle of Mobility: Field Sites, Research Partners, Methods," offers a description of the sites where I conducted research for this book—Hampi and Dharamshala. I describe their specific qualities and commonalities that they share despite apparent differences. I focus on mobile histories of those sites, which made them into popular destinations that tourists desire to visit, spaces where people coming from different cultural, social, economic, and political backgrounds can meet. I paint the images of these sites with mobile stories of people, my research partners. In chapter 1 I also focus on questions related to methodology of anthropological inquiry in such fields and with such highly mobile topic as tourism. I present methods and research techniques I chose to apply in collecting the ethnographic material for this book. I also discuss challenges I had to face, and roles that were ascribed to me. Employing self-reflection, I try to reveal the dynamics of the field and in this particular kind of research, so as to give the reader the possibility to imagine the process that led to the production of knowledge, which determined the character of this book.

The following chapters constitute the main body of the book, presenting an analysis of the material collected during the research. In chapter 2, "The World Map as Seen from the Peripheries: The Tourist as an Object of Gaze," I depict tourists as being one of the parties of the tourist contact, and—more specifically—how their hosts imagine them to be and what kind of expectations they face. I show that, contrary to popular belief about an unanimous perception of tourists in the non-European countries through the lens of Occidentalism—demonizing "the West" as materialistic, hedonistic, and ego-driven—employees of the tourism sector are very much aware of the heterogeneous nature of the category of "the tourist," hold different opinions and apply different approaches to specific groups that constitute it. I also demonstrate how global economic inequalities are far from obvious in the relationships between the inhabitants of the so-called Global South and tourists from the so-called Global North (and how blurred this separation line may sometimes be). In chapter 2, I also write about Indian tourists. I believe that foreign tourism has been too often analyzed separately from the domestic tourism, which made us forget that the Other is also mobile, and also a consumer. This gives me the opportunity to reflect on complex relationships that result from interpersonal contact, when the Self becomes the Other.

In chapter 3, "Not-So-Empty Meeting Grounds: Self-Representations and Relationships," I reflect on tourism's capacity to be a platform for creating one's own representations that are not mere reproductions of tourism imaginaries, and for establishing interpersonal relationship that goes beyond the service provider/customer pattern. I critically address the categories of commodification and authenticity, the classics of social studies of tourism, wondering to what extent they spark tourist desire and regulate actions of the hosts. Can overcoming this conceptual binary opposition be a point of departure for a completely new view on the matter? I review the practices of ascribing authenticity to goods and services and look at the competition for rights to cultural heritage, including the right to profit from it. I present various ways in which workers and small entrepreneurs in the tourism sector position themselves within their relationships with tourists. I point out the role of such categories as closeness, partnership, reciprocity, cosmopolitanism, responsibility, and tranquility in shaping the interpersonal contact grounded in tourism, through which both parties attempt to move beyond the framework of transaction-oriented capitalism.

In chapter 4, "Recovering the Subaltern Voices: Tourism and Engagement," I ask the key question: To what extent can tourism be the space in which vulnerable communities can regain their voice and agency? I look into two specific types of tourism developed by the communities in Hampi and Dharamshala, to achieve their collective goals through the skillful use of tourist attraction status: the tourism that I call political, and the volunteer tourism. To put it differently, I write about the tourists' involvement in political matters, such as the international struggle for recognition and self-determination, struggle against state authority in terms of right to heritage, or activities aiming to abolish economic inequalities. Even though I agree with many critical remarks on volunteer tourism, I show what other, less obvious benefits it can bring to its individual beneficiaries.

Finally, in chapter 5, "Tourism as a Source of Individual Empowerment: Stories of Encounter," I look at the impact of encounters enabled by tourism on individual people's lives and their social and economic status. Here, I recount the stories of my research partners, which demonstrate that tourist contact may lead to empowerment—under the guise of social mobility or improvement of material conditions—especially for people who belong to the most excluded groups due to their gender, social, or political status. I thus focus not on the often criticized (see Romanowicz 2017) empowerment programs implemented by NGOs or government agencies but on grassroot, informal forms of empowerment. I discuss different kinds of support that employees and small business owners receive from tourists, that is, sponsoring, supporting local entrepreneurship, or the opportunity to migrate. It is a tale of friendships and love relationships in the time of tourism, a tale of what these close relationships bring to both parties.

The names of most people who are this book's protagonists have been changed to protect their privacy, the exception being cases where I refer to public activities of these people or where they explicitly stated that they wished to be known by their real name.

ON NAMES AND TERMS

In the main body of this book I use popular English transcript of non-English words, however, in order to enable specialists to identify specific terms, each time such a term appears in the text, I provide a footnote elucidating its origin and, in case of Tibetan terms, its transliteration according to the system developed by Turrell V. Wylie (1959). For Indian languages, I applied the international Latin transliteration standard ISO 15919 for languages using the Devanagari script or its derivatives.

In terms of geographical names, I use their decolonized variants. The Indian authorities have been changing colonial names of many sites (Bombay is now Mumbai, Madras became Chennai, Calcutta became Kolkata, Bangalore became Bengaluru, etc.). It is worth remembering that not all of these are local names (sometimes the change only affected the pronunciation of the English name) and that many inhabitants still use their previous versions.

I use "Indian" while referring to all citizens of India, regardless of their ethnic origin or religion, and "Hindu" to refer to the followers of Hinduism (so "Indian" for all that is related to India and "Hindu" for all that is related to Hinduism).

Chapter 1

In a Circle of Mobility

Field Sites, Research Partners, Methods

INTRODUCTION TO THE FIELD

Hampi and Dharamshala are 2,500 kilometers away from each other, and they offer very different kinds of tourist attraction, but they do have much in common. Firstly, it is their cosmopolitan character *despite* being local. When compared to other sites of the same size, for example, villages and small towns that had not yet got enmeshed in the web of tourism, both display markedly urban heterogeneity or even cosmopolitanism,[1] and high cross-cultural skills of their permanent and transient inhabitants (cf. Cohen 2006). This is a result not only of an influx of tourists but also of seasonal workers and small entrepreneurs, which challenges the classic dichotomy between mobile tourists and the place-bound locals (see Glick Schiller and Salazar 2013). Secondly, both locations give tourists an opportunity to mingle, as they cater for the needs of the so-called low budget, mostly individual tourists within the informal tourist sector, featuring no institutionalized enclavic spaces.[2] Thirdly, in both places, tourism developed in basically the same

[1] It is a common view among both tourists (one of the volunteers said that McLeod Ganj was "very Western" when compared to the "rest of India") and Indian citizens. Tibetans, especially those who did not live in Dharamshala permanently, often called it "the Tibetan New York." See Shekhawat 1995; Samphel 2006, 29; for Hampi see Krog 2007, 46.

[2] Tim Edensor (1998, 41–68, 149–180), based on his research in Agra—the city being the home of the probably most famous tourist attraction in India, that is, the Taj Mahal tomb—distinguished two types of tourist spaces: enclavic and heterogeneous. The former, highly standardized, is created for and almost exclusively used by tourists, with locals present only as service providers. These are mostly all-inclusive resorts, owned by national tycoons or global corporations. The tourist there is both protected from the outside world through Boorstin's "bubble" (1962) and subjected to control, patterned according to the Goffman's concept of "total institutions" (1961). A heterogeneous space, on the other hand, is shared by locals and tourists alike, and so they can interact outside of the customer versus service-provider roles, set by the tourism sector. In such spaces, tourism is just one

way, even though the resources were different. Last but not least, both Hampi and Dharamshala are examples of politically subaltern communities—as a consequence of exile and statelessness for Dharamshala, and displacement and demolition in the case of Hampi—which made both of them vulnerable (figure 1.1).[3]

Hampi: Pilgrims, Hippies, and UNESCO

Hampi is a small village in the Southern Indian state of Karnataka, which underwent transformation from a community of agricultural laborers into a hub of small entrepreneurs almost entirely reliant on tourism. The scenario of this metamorphosis was underpinned by three presences. The first one relates to Hindu pilgrims. The center of Hampi is marked by the "eternally alive" Virupaksha temple devoted to the local emanation of Lord Shiva.[4] Moreover, the whole area surrounding the village is identified with the episodes of the great Hindu epic of *Ramayana*, which makes it a sacred land and a destination of numerous pilgrimages from all of the south of India (see Das 1996). In the 1970s, the Indian Hindu followers visiting Hampi were joined by the North American and European hippies who looked for an alternative to the party-driven and crowded Goa, located just 350 kilometers away (see Saldanha 2007, 11–20, 37–38; D'Andrea 2007, 176, 184–185; Odzer 1995). For them, Hampi became a chill-out zone. They lived in small groups of several people, staying in caves and abandoned temples, smoking marijuana and hashish, practicing free love and contemplation. Some of the villagers provided them with food, others with a range of drug-related necessities, such as the clay *chilam*[5] pipes. By doing so, they became the first cultural intermediaries (see Edensor 2001) (figure 1.2).

The hippies plotted the routes for tourism, which in Hampi developed the most rapidly—just like in the whole of India—in the 1990s. A major impulse for this development was the designation of remnants of the Vijayanagara kingdom—a vast complex of temples, palaces, and fortifications with its heart in today's Hampi—as a UNESCO World Heritage site in 1986. Vijayanagara was the last great Hindu empire which ruled over the southern part of the

of the activities, and tourist infrastructure (in the informal sector that typically means small, cheap, and often family-run guesthouses, restaurants, shops, and service points) stands side by side with residents' homes and their local businesses. Both types of spaces can coexist within the same site, but there have so far been no enclaves in Hampi or in Dharamshala.

3 The term "vulnerable," that is, sensitive, prone to being hurt, unable to defend itself (from Latin: *vulnerare*), was first used in relation to the victims of natural disasters (Cutter et al. 2003). Nowadays, social studies of vulnerability focus on people suffering from poverty, people with disabilities, the elderly, children, and diverse minorities subjected to social exclusion.

4 Sanskrit *Śiva*—one of the major Hindu gods, manifestation of both the destructive and constructive force.

5 Persian *cilam*.

Figure 1.1 Map Showing the Location of Dharamshala and Hampi. *Source*: © kosmozoo/DigitalVision Vectors/Getty Images. Edited by Łukasz Kaczmarek.

Indian Peninsula. Established in the fourteenth century, it was abandoned two centuries later after being raided by local sultanates (see Fritz and Michell 1991, 2009, 2014; Kotraiah 2008; Verghese 2002). Most of the tourists visiting Hampi in the time of my fieldwork were backpackers and small organized groups with relatively low budgets and moderate expectations as to the range

Figure 1.2 Hampi, Virupaksha Temple. *Source*: Photo by Natalia Bloch (2017).

of services they are offered. They usually stayed in the village for several days, even though some backpackers prolonged their visit for weeks or even months, becoming long-term tourists. More wealthy visitors used to see the ruins as part of one- or two-day trips departing from the beaches of Goa,[6] staying overnight in larger and more upmarket hotels in Hospet,[7] twelve kilometers away, or along the route between Hampi and Hospet. Along the influx of tourists, there was always a steady stream of pilgrims, part of whom combined religious motivations with sightseeing and leisure. There was also a weekend tourism, which became popular a few years earlier and which was fueled by young people from dynamically growing Indian middle class (see Varma 1999), hailing mainly from the metropolises of Karnataka and Maharashtra states.

The characteristic feature of the informal tourism sector in Hampi was that many services were provided not by the villagers but by the seasonal migrants (see Williams and Hall 2002, 10). The restaurants were pretty much "monopolized" by the Nepalese who served French onion soup, Mexican enchiladas, and "Israeli" hummus to tourists. Tibetans, whose ancestors had

---

[6] Goa—a former Portuguese colony on the west coast of the British India—is now the smallest of India's states. However, its coastline of more than 100 kilometers spotted by sandy beaches shaded by coconut trees has been a popular destination for foreign visitors ever since the arrival of the hippies—I elaborate on that in chapter 2. The international beach tourism in Goa is thought to have developed in the space created in an aftermath of the late decolonization of the region, in 1961. Because of this, Goa was "labeled 'India for Beginners'; . . . considered less dirty and impoverished than other parts of India . . . with pork and beef more readily available as was alcohol" (Gupta 2014, 98; see Newman 2001).

[7] The official name of the place is Hosapete.

settled in the north of India before the Tibetan exodus, sold Buddhist statues, "yak wool" shawls, and wall hangings with aphorisms from the Dalai Lama. Domestic migrants from deserted northwestern state of Rajasthan, located 1,500 kilometers away, traded in artisanal jewelry, textiles, and leather goods. People who in Karnataka are most often called Lambani, a nomadic tribe originating from the northwest of the Indian subcontinent, specialized in selling garments. Tourism created a space, in which they could continue their nomadic trade traditions, even though they invested the money they made in purchasing agricultural land.[8] Finally, there were also, although not numerous, tourists-turned-migrants from the so-called Global North (figure 1.3).

Figure 1.3 Market Street in Hampi Village. *Source*: Photo by Natalia Bloch (2014).

[8] Sanskr. Lambāni. The name has different local phonetic versions (e.g., Lamani, Lambadi), and derives from the Sanskrit word *lavan* for salt, although in other states of India the term *Banjara* (Sanskr. Banjārā from *ban*—jungle and *jārā*—the one who wanders; another etymology of the group's name refers to the word *banaj* used for trader; Naik 2010, 1) was more prevalent. *Banjaras* probably come from the northwestern areas of the Indian subcontinent, mainly from Rajasthan. In the past, they raised cattle and were itinerant traders in salt, grains, and other goods. The group claims that the European Roma are their descendants (this topic is discussed in Mróz 1989). Railway construction and salt tax rises imposed by the British colonizers contributed to the fall of the traditional *Banjara* occupation and forced them to settle, which they usually did in the forests, far from other villages, where they sold wood for fuel (while some of them become robbers, because of the general impoverishment of the group). The governmental aid programs, run after India regained independence, became an additional impulse for settling down. Today, *Banjaras* are mostly farmers, cattle herders, and traders, but also wage laborers. They live in separate villages or clusters at the edge of other villages (Naik 2010;

The villagers of Hampi rented out premises for the seasonal businesses—shops, restaurants, and workshops—as well as living quarters. They themselves operated small family-run guesthouses,[9] travel agencies, bike and scooter rental shops, transport services (auto-rickshaws), Internet cafés (until Wi-Fi became widely available), and so-called general stores, offering not only basic local goods but also vital goods for foreign tourists (i.e., bottled drinking water, toilet paper, cigarettes, or shower gel). But they also had settled in Hampi no longer than eighty years ago. In the mid-1940s, the region was struck by a major drought and many poor, mostly landless farmers were forced to leave their villages. Some of them came to Hampi, which they knew to be a pilgrimage site located by a sacred and life-giving river. The dam that was constructed there in the late 1950s and early 1960s significantly increased the possibility to irrigate land, offering work on the fields owned by the Virupaksha temple authorities. Moreover, the steady influx of pilgrims provided a complementary source of income. Another impulse that attracted new inhabitants to Hampi from the less potential regions of Karnataka and neighboring Andhra Pradesh state was the increasing influx of tourists that started in the 1990s (figure 1.4).

Figure 1.4 Hampi Village with the View Over Virupaksha Temple. *Source*: Photo by Natalia Bloch (2014).

Burman 2010). Lambanis who entered the tourist market first used to sell their ethnicity—colorful women's clothes (embroidered with many patterns, including those made of glass beads and little mirrors), and jewelry. Currently, they mostly sell factory-made clothes for everyday usage by the backpackers, that is, "hippie-style" trousers, skirts, dresses, or shirts.

[9] Guesthouses consisting of several to less than twenty standard-quality rooms rented at relatively inexpensive prices. In many cases, especially in Hampi, the owner's family lived in the same building, so they often were in fact homestays (even though they were not formally registered as such).

Figure 1.5 Demolished Houses and Small Businesses in Hampi. *Source*: Photo by Natalia Bloch (2013).

Hampi is not an isolated case. The Nepalese, Tibetans, Rajasthanis, and Lambanis are omnipresent in the tourist spaces across India. The only ones missing in Hampi were the Kashmiris. After the terrorist attack in Bengaluru, the capital of Karnataka, which took place in 2008, the police supposedly found a link that led them to a Kashmiri man from Hampi. The man had been, unfortunately for other Kashmiri traders, elected by them as the president of the Association of Kashmiri Merchants in Hampi. He was arrested, and the other Kashmiris were ordered to leave the village. In this way, Hampi became one of the very few tourist destinations in India with not a single Kashmiri living in there.

In 2011, Hampi underwent what Michael Herzfeld (2006) called "spatial cleansing," carried out in the name of protecting the tangible cultural heritage and boiling down to forced displacement and demolition of houses, guesthouses, restaurants, and shops, which, according to the local authorities and state archeologists, posed a threat to the monuments' "integrity" (see Bloch 2016c, 2017a, 2017b). The inhabitants and merchants from the main bazaar leading up to the Virupaksha temple were the first to be removed. This "spatial cleansing," continued over the next years, led to a substantial shrinking of the village[10] and brought about a kind of liminality, since the official plans for the rest of the settlement were disclosed to the inhabitants. However, during my fieldwork in Hampi, conducted between 2013 and 2014, life went on as usual—tourists used

[10] Before the first displacement took place, Hampi had approximately 2,500 inhabitants. Today, it has just over 800. However, due to the influx of mobile workers and small entrepreneurs, this number grew substantially in the tourist season.

to come at the beginning of October, and along with them came the mobile workers and small entrepreneurs of the informal tourism sector (figure 1.5).[11]

At the end of March, when it becomes unbearably hot and the season ends, the village emptied. Most of the seasonal migrants, however, did not go back home, but traveled further north, to the foothills of the Himalayas, for the summer season there. They usually spent the next few months in the northern state of Himachal Pradesh. The Nepalese left Hampi for Manali and Kasol, where they ran restaurants. Tibetans went to Spiti, further up the mountains, and Rajasthanis moved to Dharamshala, where I continued my research with them.

Dharamshala: A Refugee Camp, Dalai Lama Superstar, and a Cricket Stadium

Dharamshala—or more precisely its upper part, that is, McLeod Ganj—became home to Tibetan refugees in 1960, when the Indian government allocated it as the residence of the 14th Dalai Lama and the Tibetans who accompanied him, fleeing Tibet after it had been invaded by the People's Republic of China. Earlier, in the mid-nineteenth century, the British colonizers built a hill station here for their soldiers and their families,[12] bringing the Zoroastrian Parsi family as supply experts (see Kamerkar and Dhunjisha 2002).[13] Their shop, although not operating anymore, is still the landmark of the main square of McLeod Ganj. It was them who suggested that Tibetan refugees should settle in the hill station that had been abandoned due to decolonization and as the aftermath of an earlier earthquake. At the same time, the area was also used as grazing lands by the pastoral Gaddi ethnic group inhabiting nearby villages (see Wagner 2013).

In the 1970s, the first hippies and Western students of Tibetan Buddhism showed up in McLeod Ganj. Due to Tibetan spiritual teachers, who started traveling aboard, Tibetan Buddhism became known in North America and Europe (Tseten 1993; Lopez 1998).[14] The fact that in 1989 the 14th Dalai

---

11 Due to the continued warnings issued to the villagers by the authorities to stop their commercial activities, many mobile small entrepreneurs decided to give up their businesses in Hampi. Also the number of tourists significantly decreased, as a result of rumors that Hampi village "is being demolished." The COVID-19 pandemic affected the local tourism market even further.

12 Hence the name of this part of the settlement: David McLeod was a colonial governor of Punjab, which at the time for administrative purposes also included Dharamshala. In this book I use the names Dharamshala and McLeod Ganj interchangeably, but it should be mentioned that the Lower Dharamshala is a town located in the valley, approximately seven kilometers downhill from McLeod Ganj.

13 Parsi are Zoroastrians who in the eighth century were forced to leave Persia invaded by the Arabs and came to India. Today they live mostly in Mumbai and its surroundings and specialize in industry and trade (one of the most affluent Indian families, the Tata, are Parsi).

14 In this way, Dharamshala became a new destination on the Indian map of Buddhist-related tourism, next to the more traditional sites linked to important episodes in the life of Buddha Sakyamuni

Lama was awarded the Nobel Peace Prize was for Dharamshala an equivalent of the inclusion of Hampi in the list of the UNESCO World Heritage. As a result, the spiritual and political leader of Tibetan refugees became a globally recognizable "spiritual celebrity" (Iyer 2008)[15] and the main tourist attraction of Dharamshala. From the 1990s, the settlement has become a "Mecca" for international backpackers attracted by their imaginations of Buddhism and Tibetan freedom struggle.[16] Many of them spent longer periods in the hill town, enrolling themselves in different courses from a wide range of spiritual development options or getting involved in the activities of numerous NGOs. The construction of a national cricket stadium in Lower Dharamshala in 2001 made the place famous also among domestic tourists.[17] Narrow streets of McLeod Ganj got crowded with families from Indian metropoles and young Sikhs from the nearby wealthy Punjab state, who got a liking to pop into this hill town for weekends, driving their Land Cruisers. Even though the labor market in Dharamshala is much more diversified than the one in Hampi, tourism is an important source of income for numerous local social actors (figure 1.6).

However, in the context of exile and statelessness, Tibetans are denied rights enjoyed exclusively by Indian citizens, which is perceived as a natural superiority of belonging to a nation-state versus lack of such ascription. As a result, Tibetans can officially purchase neither land nor real estate, which is a heavy blow to their economic rights; they do not have passports and are not entitled to consular provisions based on agreements concluded by the Indian state, which limits their right to move freely. This also concerns generations born in exile, who stand by their stateless status.[18] Moreover, since India is not the signatory of the 1951 UN Convention Relating to the Status of Refugees, Tibetans do not enjoy international protection (figure 1.7).

Today, it is not just Tibetan refugees (and the next generations born in India) who compete for attention and resources of tourists visiting Dharamshala. They

---

(today located mainly in Uttar Pradesh and Bihar states), to the later period of Buddhism advances in India, mainly during the reign of the emperor Asoka (e.g., in Madhya Pradesh state) and Buddhist Himalayan regions that in the past were under cultural influence of Tibet (such as Ladakh, Spiti, or Sikkim). All these places are visited by both pilgrims (i.e., the followers of Buddhism) and spiritual tourists (those who are still looking for their new spiritual center, see Norman 2011 for this distinction).

15 Pico Iyer uses this expression in relation to people who, while being "icons," also represent something more valuable and durable (2008, 72).

16 See the subchapter *Nelson Mandela as Tourist Attraction* in Marschall (2004, 106–107) for a similar phenomenon.

17 I also heard the opinions that Dharamshala gained fame in India due to a growing popularity of the TV cooking shows, which presented McLeod Ganj as culinary heterogeneous space. The restaurants serving Tibetan, Italian, Korean, or Japanese food attract tourists interested in tasting different cuisines.

18 This complex issue is discussed in Choedon (2018).

Figure 1.6 McLeod Ganj, the Upper Part of Dharamshala. *Source*: Photo by Natalia Bloch (2013).

Figure 1.7 The Main Market of McLeod Ganj. *Source*: Photo by Natalia Bloch (2003).

have been joined by Nepalese citizens (who, apart from running restaurants, mainly trade in trekking gear, "ethnic" clothes, and handicraft), the local pastoral Gaddi (who run tourist agencies and all-purpose shops), the Rajasthanis (who mostly sell jewelry), Kashmiris (who run handicraft, precious stones, and textile shops and, recently, also have started to offer accommodation), and

Figure 1.8 McLeod Ganj. *Source*: Photo by Natalia Bloch (2013).

Punjabis (who run pubs, restaurants, and hotels). There are also the "invisible" workers of the tourism sector (Lozanski 2010, 754–755),[19] that is, internal immigrants from one of the most economically challenged states in India, Bihar, who work for Tibetan service providers as unskilled workers. Around Dharamshala there are also makeshift campsites of internally displaced persons from the poorest, drought-stricken regions of Rajasthan, who have found their niche on the Dharamshala's tourism market—begging or scavenging.[20] Finally, a number of people who originate from the so-called Global North countries also live, more or less permanently, in McLeod Ganj (figure 1.8).[21]

The season in Dharamshala is quite long, but during two-three winter months, the settlement becomes a ghost town—even the numerous NGOs take a break. At those times, some tourist-related businesses and their workers move to the south of India.

MOBILE LIVES

There are some patterns that can be discerned in the mobile lives of the informal tourism sector workers in Hampi and Dharamshala, even though they are

[19] I write more about the category of "(in)visibility" in the first section of chapter 5.

[20] The biggest of these informal settlements was demolished in 2016 (see Bloch 2019).

[21] Dharamshala is inhabited, according to data collected by the Tibetan administration, by approximately 5000 Tibetans. However, there are no collated data including all inhabitants of McLeod Ganj. For sure, their number grows substantially during the tourist season.

changeable according to the situation. The Rajasthani and Kashmiri rarely migrate alone. As a rule, they do business as male teams of two, so that they can take turns and visit their families during the trading season. They take one or several helpers along—young, unskilled boys, usually distant relatives or sons of neighbors or acquaintances. The helpers watch over the goods and do auxiliary jobs, such as cleaning, delivering endless cups of chai,[22] or preparing meals. They are paid peanuts but have the opportunity to learn how to do trade, deal with customers, and speak English. There are exceptions to this rule, especially among the Rajasthani, who tend to have less financial capital and sell goods that are less expensive or easier to keep an eye on, such as textiles or clothes—they can be sole traders, even though they travel in groups of several people each.

The Nepalese and Tibetans do mobile business in a very different way, as they often migrate with their families. Marriage is meant to support each other in business, which entails a different distribution of social roles. Rajasthanis also sometimes travel to do business with their wives and young children, but they differ from Tibetans and Nepalese in that when the children of the latter reach school age, they are usually left with the relatives or sent to residential schools, while married couples continue to go on business together. Rajasthani couples tend to separate, and wives stay at husband's natal home with children, in line with the patrilocal system. Even if they are together, however, it is the woman who takes care of children and their mobile home, without getting involved in business.[23] I have never come across families from Kashmir migrating for business.[24]

Sometimes, and not very rarely, the Nepalese or Tibetan mobile businesses are run by women alone, even though they are married. Husbands from the refugee Tibetan families—when forced to do it by the situation—take over running homes or raising children. Also Lambani women often do business by themselves, which translates into a relatively greater freedom of tribal women when compared to the Indian families which operate in the caste system.[25] It seems that the fact that Nepalese and Tibetans are external migrants

---

22 Hindi: *cāy*—sweet tea made with milk instead of water.

23 The style of doing business by Nepalese and Tibetans was often criticized by the migrating Rajasthani men ("A woman should stay with children"; "A family cannot live in such conditions long-term"). However, I also heard opinions, such as Udai's, a jewelry trader in Hampi and Dharamshala, who said, "It is good that many people in the family earn money, it helps."

24 Many Kashmiri Muslim women can only participate in the public tourist spaces to a limited extent, due to pardah rules of separation between the world of women and men. In Kashmir, such rules apply to a different degree in different social groups (Rukhsana and Amin 2012), but working in the tourism sector, which involves traveling and interacting with male customers, was—according to my Kashmiri interlocutors—out of the question.

25 The majority of Lambani are Hindu, but they tend to practice their faith in an undogmatic syncretic form, just like part of my Nepalese research partners (the rest of Nepalese, just like Tibetans, are Buddhist).

in India and the nomadic traditions of the Lambani made these groups more egalitarian in terms of gender relations. The necessity to earn a living as a migrant, a refugee, or while traveling resulted in a lack of segregation in public space and lack of clear-cut division of work.[26] Nepalese and Tibetans in particular seem to implement the model of permanent mobility, without any point of reference that one could call "home." Phurbu, a Nepalese restaurant owner, when asked why they were traveling with families for business, while the Rajasthani were not, said, "They have homes where they can leave their wives and children. We don't, so we move together."

For the majority of mobile small-scale entrepreneurs who became my research partners, doing business in the north of India was the major part of their activities, because of a longer tourist season[27] and the fact that this part of the country is more popular among foreign tourists. I often heard that going down south was not profitable, that it was really done as "a time pass." Why do it then? Why not spend this time having some rest?[28] When asked, the entrepreneurs gave three reasons. First, expressed mainly by Nepalese and Tibetans, concerned climate: "to escape the chill" of winter months in the north. Second, was economic: "not to eat the money away" (winter business in Hampi was not particularly profitable, but it covered the cost of living). The third reason—"not to sit idly"—sheds some light on the way work is perceived by the informal tourism sector workers. It suggests that profit is not the main driving force but instead it is the activity itself that brings the sense of value and agency.[29]

---

[26] For sure, geographical and economic conditions in the origin countries also played a role here, along with the social and cultural practices that they had shaped. Harsh life conditions in the Himalayas in Tibet and Nepal (most of the Nepalese mobile workers come from the central-northern part of the country) forced all members of families to actively earn a living—it was simply impossible to push women to the domestic sphere only. Charles Bell, a British diplomat who traveled through Tibet in 1920, noted that many women in Lhasa, the capital of Tibet, run shops and stalls on their own (1928, 157). I elaborate on the relationship between tourism and social status of women in chapter 5.

[27] At the end of March and the beginning of April, the temperatures in Hampi already exceed forty degree Celsius. Monsoon comes around June and lasts well until September. In Hampi it is not too troublesome, but it is worth remembering that tourists are attracted to the south of India mostly by beaches, and heavy rainfall in the western coast excludes beach tourism. In the hill towns of the north, such as Dharamshala, the tourists stop coming only for two winter months, while in the summer the Himalayan foothills provide respite from the bothersome heat in the plains.

[28] There are, of course, exceptions to this rule. In my field they related to young, unmarried beginner entrepreneurs from Rajasthan doing business alone. They were still independent, as far as the financial obligations went, and they could afford two to three months of holiday in their family homes. They were aware that this would change as soon as they started their own families.

[29] See, for example, an excerpt from Phurbu, who was commuting with his family between Hampi and Kasol in Himachal Pradesh: "Otherwise, we just sit there and eat the money away. In winter there is nothing to do in the north, just sleeping and eating. . . . When I do nothing, I do not feel well. Besides, here we can escape from the cold. We earn a little and it's all right."

With all this mobility, it is striking that people get so attached to their place of work, that is, the space such as a service point, a shop, or a stall. It happened sometimes that the employees of the Nepalese restaurants in Hampi did not leave the rooftops where they worked for days or weeks in a row.[30] They slept in the back rooms adjoining kitchens or on mattresses that tourists sat on during the day, just adding mosquito nets hung over the chairs. Helpers working in shops, after closing the business spread blankets on the floor, and those working in guesthouses slept at the reception desk. Only entrepreneurs traveling with families rented an extra room. Some mobile workers and small entrepreneurs had sedentary plans for the future. The Lambanis invest the money they earn in land, just like the Rajasthanis. Some consider moneylending outside of the banking system, so popular in India, as a way to earn a living. Shopkeepers often dream about having a wholesale warehouse, ideally one that exports goods to "the West." Such examples demonstrate the importance of the category of place in research into mobility conducted in mobile areas, that is, avoiding the fetishization or romanticization of mobility as boundless, liberating liquidity—something that new mobility paradigm adherents warn us about (Sheller and Urry 2006, 210–211). Spatial mobility cannot be read as individual freedom or chance. People working in the informal tourism sector, the protagonists of this book, move along specific routes, plotted by social networks, since starting a mobile business requires grounding and familiarity. In most cases, tenants and landlords have remained bound by tenancy agreements for years, even though these relationships may often be complex and challenging. In case of a breakup, finding new premises depends, to a large extent, on the "good name" the tenant managed to earn in a given locality and the system of referrals.

Mobile tourism workers do not only travel for work. They often become tourists and consumers themselves, which makes the classic relationship between the host and the guest dynamic and equivocal. This is true of mainly small-scale entrepreneurs, working as sole traders. While traveling as tourists, many of them take the advantage of contacts with other tourism sector workers which they made through mobility. Such connections were used by, for instance, the owners of a family-run guesthouse in Hampi, who combined the Himalayan Amarnath Yatra—a pilgrimage to one of the most important Shivaite cult sites—with a trip through Kashmir made on the invitation of a Kashmiri trader who used to rent a shop in Hampi. Ever since a daughter of one of the Hampi residents married a Nepalese restaurant owner, who rented the premises from her family, different sets of relatives traveled to Nepal for holidays; despite very limited funds they could afford it since they could count

[30] These are popular open-air restaurants built seasonally on the flat roofs of buildings.

on free accommodation and lower prices than those in India. These are cases of the VFR tourism mentioned in the introduction, which itself is a product of migrations generated by the tourism sector (Boyne et al. 2002). Moreover, dispersed families usually meet up and take few-day trips at the beginning or the end of the season, just to spend some time together, visit temples, go shopping, and do sightseeing. Often, they visit locations where they work, which for some time turn into recreational spaces for them. One form of spatial mobility can thus generate other mobilities and coexist with them—the migrant/refugee becomes a tourist, the tourist becomes a migrant—and the conceptual boundaries separating these categories turn out to be less obvious than it is generally assumed (Hall and Tucker 2004, 14–15; see also Hall and Williams 2002; Duval 2003; Coles and Timothy 2004; Lenz 2010).[31]

Below I present several biographies which illustrate mobile lives. I shall return to these stories and their protagonists—my research partners—on many occasions in this book.

Norbu and Dolma with Their Son Tsering

They belong to the group that Tibetan refugees call *gyakar Khampa*,[32] that is, the Indian Khampa, as their ancestors left Tibet before the Chinese People's Republic took over the country. The Himalayan border between India and Tibet was then still unsealed, and people often crossed it (without even realizing it) for trade, to reach pastures or go on pilgrimage (see Aziz 1978).[33] As a result, some Tibetans settled in today's border states of Himachal Pradesh and Uttarakhand, as well as Sikkim, situated between Nepal and Bhutan. Differently from Tibetan refugees, who intentionally maintain their stateless status, even in generations born in exile (Bloch 2011), *gyakar Khampa* hold Indian citizenship, although, as "non-natives," they have limited rights to

---

[31] In the relevant literature scholars use, for instance, a category of "migrant tourist-worker," which combines motivations and practices linked to free time, sightseeing, and paid work (Williams and Hall 2002, 5, 13; Bianchi 2000). This relates, for instance, to backpackers from Australia and New Zealand spending months traveling in Europe, who supplement their tourist budget in this way (Mason 2002), but also to backpackers from Japan, South Korea and Taiwan, who are temporarily employed by tourism sector entrepreneurs originating from those countries but born or settled in Australia and New Zealand, targeting tourists from their countries of origin (Cooper 2002). Malcolm J. Cooper calls this practice "working holidays." On the other hand, many backpackers travel for more than twelve months, which, theoretically, is the limit distinguishing tourism from migration (Hall and Williams 2002, 13).

[32] Wylie: *rgya gar* means "Indian" and is used here as an adjective, while *Kham pa* refers to an inhabitant of one of the three original Tibetan provinces—Kham (Wylie: *Khams*). It is difficult to say why these people are called Khampa. My interviews show that their ancestors had not necessarily come from this part of Tibet, even though it needs to be noted that the intergenerational transmission regarding the place of origin is weak within this community.

[33] This was confirmed by another research I conducted, in Dolpo region in the Nepalese-Tibetan border area in 2008.

purchase land. Dolma attended Tibetan primary school in Manali, Himachal Pradesh. She can read Tibetan[34] and considers herself to be Buddhist. Norbu was educated within the Indian schooling system. His Tibetan is quite poor, and he feels closer to Shiva than to Buddha. They came to Hampi when the Kashmiri had left the site. They used to run a shop with Tibetan handicraft and jewelry for five to six months and then set on a five-day journey to Tabo in the Himalayan region of Spiti (first making the 24-hour train journey from Hospet, connecting at Hyderabad to Delhi; then a 12-hour night bus ride from Delhi to Manali; and in the last two days covering the remaining 250 kilometers of high-altitude road). They built a beautiful, big house—with hot water in the bathroom—near Manali, where both of them were born, but had no time to live there. Norbu liked to say that they were "like Gypsies." Earlier, they had migrated with their little son, Tsering. But when Tsering turned six, they put him in a monastery school in Tabo. Tsering usually joined his parents in Hampi for part of the season—when the school was closed due to relentless cold—and there he attended a local private school. He spoke Hindi with his parents; in his school in Tabo, which had a Tibetan headmaster; he studied in the local Tibetan dialect; and in the school he attended near Hampi the language of instruction was Kannada[35] and English, the latter for science. When I asked him about his best friend, he asked me back, "Where?"

A few years earlier, a second shop in Hampi selling the same kind of goods had been run by Dolma's mother (until her retirement). Meanwhile, Dolma's younger sister and her husband were trading in one of the Tibetan bazaars in Goa. Their two sons attended school in Dharamshala, staying with the older sister of Dolma, Metok, or—more precisely—with her husband originating from Ladakh,[36] since Metok spent the whole days keeping an eye on three stalls in McLeod Ganj (for a year, the stall next to Metok's was run by a Himachali man, who then moved to Hampi to sell jewelry, one street down from Norbu and Dolma's shop; he worked for a businessman from Hampi, whom he met when the latter visited Dharamshala as a tourist).

Udai and Mohan

They used to say that they were "karmic brothers."[37] They belonged to the caste of blacksmiths and met in a jewelry factory in Rajasthan, where they came from. Then, they worked as helpers for other traders in Goa. Finally,

[34] Dolma's parents had to pay for her education, contrary to Tibetan refugees who get free schooling offered to them by the government-in-exile with the help of foreign funds.

[35] The official language in Karnataka.

[36] A Buddhist region within the administrative borders of Jammu and Kashmir.

[37] That is, they were connected through previous lives (from Sanskrit *karman*).

they decided to try their luck and start their own business. It worked. When we met, they had been running a small jewelry shop for seven years—in Hampi during summer and in Dharamkot, a village located three kilometers above McLeod Ganj, in winter months. Part of their merchandise was handmade by them in the shop, which also served as a workshop. They had helpers: one or sometimes two boys from Rajasthani villages. They were husbands and fathers, and so twice per season visited their families, even though a one-way journey from Hampi took two days (forty-five hours by train that run just once a week—from Hubbali, through Goa, and Gujarat). They spent a month and a half every year—during the school holidays their children had between May and mid-June—with their families, at home or in Himachal Pradesh ("so that my wife could have some rest," as Mohan explained).

Mohan was well educated, he held a bachelor's degree in social sciences (history and political science). He dreamt about a better kind of job: "I imagined myself sitting at a desk, with a pipe, reading books, I love reading!" However, after his father's business had gone bankrupt, he was forced to quickly start earning money and that is how he ended up in a factory.

Phurbu and Padma with Daughters Aiya and Lhamo

They are both Nepalese—from Tamang and Magar ethnic groups—even though they were born in India. Their fathers migrated for work and earned a living as construction workers in Himachal Pradesh. When Phurbu finished sixth grade, they went back to Nepal for a few years but could not make ends meet and once again left for India. In Nepal, there have not been many jobs: farming cannot sustain a family, there is practically no industry, just tourism. But tourism—contrary to India, where "there is enough room for everyone," as Phurbu says—is concentrated in just several specific locations, which generates a lot of competition and high costs, while unstable politics makes the business a high-risk enterprise.[38] Moreover, at the time, Phurbu and his younger brother were of "risky age." Maoist guerrillas, who had been losing public support, started to forcefully recruit boys, especially from remote villages, away from the centers of power, such as Phurbu's. Padma visited the village where her family comes from just once. Today, Phurbu's father is supervising construction work around Manali and Padma's dad runs a small "Chinese-style" eatery in Kasol.

They lived their lives between the restaurant they run in Hampi in winter for the past ten years and its summer version in Kasol, Himachal Pradesh. Before they got there, for a few years they had been running an eatery in

[38] In its recent history, Nepal experienced a Maoist insurgency, civil war, and numerous general strikes (see Bloch 2010–2011).

Dharamkot, near Dharamshala, just above the seasonal shop run by Udai and Mohan. Phurbu started his career in tourism by cleaning rooms in a cheap hotel in New Delhi. But he was determined to advance and got enrolled in a cooking course. Before he started his own business, he used to work in Goa—first as a helper in a restaurant, and finally as a chef. It was there that he learned secrets of "European cuisine," because at the course he attended he had only learned to cook Indian dishes.

Phurbu and Padma were accompanied by their Nepalese employees: the cook and two helpers (the latter were cousins, whom Phurbu took with him to India at the request of an elder brother of one of them, who had worked for him for several years; their fathers also used to come to Himachal Pradesh every year and do seasonal jobs in the fields, forests, quarries, or carpentry workshops). Since 2014, Phurbu's brother and his wife, Padma's sister, were also working with them: he ran the so-called German Bakery and she was responsible for juice and fruit cocktail bar. Earlier, for several seasons, Phurbu's brother had worked in Kerala, the second—after Goa—most popular coastal state in the south of India.

Both daughters of Phurbu and Padma were born in Hampi—Aiya (a Japanese name, which Phurbu heard a lot from tourists and liked it), who was four years old when we met, and the little Lhamo, whose birth I witnessed during my fieldwork (I was in the fourth month of my pregnancy then). Aiya just started attending the same private school as Tsering. In the future, Phurbu and Padma wished to send her to a residential school in Bengaluru, the capital of Karnataka.

Nelli and Anita

Nelli is Swedish. She came to India at the beginning of the 1990s, when backpackers in huge numbers were discovering this part of the world. She fell in love with the fable-like Rajasthan more than with the charming auto-rikshaw driver, but when she became pregnant after a decade of living in India, they married. She gave birth to a girl, Anita. However, the marriage was not going well. Nelli left her husband for the first time when Anita was eight. They ended up in Hampi, through a British man who was Nelli's acquaintance and was then, together with a local tourist guide, setting up an NGO for children from poor Hampi families. Nelli got very much involved in this work. After a while, however—benefiting from his Rajasthani social connections—her husband discovered her there. They decided to give the marriage one more try and set up a seasonal textile shop. But they failed. Four years later, Nelli broke off all connections and never went back to her life in Rajasthan. She rented a small but cozy flat in a town near Hampi and tried to make ends meet by running a tiny shop. Every two months, on average, she embarked on a long journey to Rajasthan to purchase goods, trying not to cross paths with her husband and his friends. Four in five nights that she needed to travel there were in a train.

Even though she lived in India for almost a quarter of a century, she still needed to extend her visa every few years. The Indian law does not allow double citizenship and Nelli did not want to give up her Swedish passport. This is common among many other Europeans who have been living in India for dozens of years in a specifically liminal legal situation. That is also the reason why Anita also only had Swedish passport, which meant that when she was coming back from holidays she sometimes spent in Sweden, she needed to apply for Indian visa. Anita could have had the PIO status (i.e., Person of Indian Origin), but her father was unable to provide the required documents. Paradoxically, Anita was no different than her Indian girlfriends in perfectly ironed local school uniforms. She spoke perfect Kannada, which was her peers' language, she also knew Hindi, which she spoke with her father. Every day, after school, she took a rackety bus and came to take her mother's place in their shop so that Nelli could go home to do daily shopping and prepare dinner.

Dolkar

Dolkar was born in Tibet. She fled to India in 1996, at eighteen, as the only member of her family (we browsed photos sent over MMS messages to her smartphone from Tibet: her brother's arm with a dragon tattoo that he no longer liked and thought about getting rid of it, Dolkar's sister and her daughter, an uncle who turned ninety). She made her way to Dharamshala, where she lived for the last ten years, renting a room with a kitchen and a bathroom. She earned a living by running a stall, like many other *sarjor*[39]—"new refugees," who were coming to India since the beginning of the 1990s, usually as already adults, not falling within the schooling system (see Bloch 2014a). They did not have sufficient financial capital, cultural competences, or social networks to rent shops, and—on the other hand—they did not want or were unable to get hired. Dolkar used to sell inexpensive jewelry, partly made by herself—necklaces, strings of beads and bracelets made of beads and stones bought from itinerant Indian wholesalers. As most of the stall sellers, she moved south for winter—usually to Goa, with its dynamic, tourist-driven Tibetan bazaars.[40]

[39] Wylie: *gsar byor*, usually translated as "newcomers" in English.

[40] Not like in the business commonly known as "sweater-selling"—that is, trading in winter clothes bought from Indian wholesalers—with Indians as customers. Contrary to common belief, the

She had two sons—the four-year-old one was raised by his father, who also took care of their rented flat, while the older one attended Tibetan boarding school in Lower Dharamshala. They traveled together to Goa, where the tourist season coincided with the school holiday period in Northern India. Sometimes, on the occasion of teachings given by the Dalai Lama, Dolkar and her husband set up a make-shift food stall, serving Indian snacks, Tibetan soup, and tea to Tibetan pilgrims and foreign tourists.

Prakash

Prakash comes from Bihar, an underinvested, corruption-ridden state in the north of India. Biharis often work for the same person for many years. They are neither promoted nor start their own businesses. Usually they send all their earnings to the families left behind in Bihar. When we met, Prakash had been working in Dharamshala for twenty years, ever since he turned thirteen. Supposedly, when the owner of one of the oldest Tibetan restaurants in McLeod Ganj was on pilgrimage in Bodh Gaya—a site in Bihar where Buddha Shakyamuni was enlightened—a family from a nearby village talked him into taking a boy to work for him. Together with the boy, they threw in his cousin—Prakash. He said he had really wanted to go. His parents were not able to feed their children, so he would not go to school anyway. Before he came to Dharamshala, he had washed dishes in a roadside eatery in a district town in Bihar for two years, working for peanuts. He was to earn a bit more in his new job, and—as he said—Tibetans were known for "treating well" their employees—firstly, they were to be obliged to do so because of the 14th Dalai Lama's widespread teachings, and secondly, people said that they fed their workers well, with high-protein diet. Prakash joked that half of his village worked in Dharamshala. Many of them already understood Tibetan (I sometimes heard Prakash swear under his breath in Tibetan when customers annoyed him).

After his employer passed away, Prakash joined another Tibetan restaurant owner, on recommendation by another boy from his village. During the season, he worked as much as sixteen hours per day, eating and sleeping in the restaurant. He usually visited his natal village once a year—for two to three months in winter. He provided for his elderly mother, wife, and an eight-year-old son from his first marriage (his first wife died in premature childbirth).

majority of Tibetan refugees do not earn their living by trading in "traditional" handicraft—they do so by pursuing this less visible type of business.

Kelsang and Tamiko

Kelsang, just as Dolkar, was *sarjor*. However, not like Dolkar, he was sent to a school for adult newcomers run by the Tibetan administration. He managed to graduate, but difficulties started afterward. The newcomers—not speaking Hindi, having only basic education and unversed in local realities—tend to miss out on the local labor market, where they have to compete with the second generation of Tibetans already born in India (see Bloch 2014a).[41] For such people, tourism creates a space where they can mark their presence, both socially and economically. Kelsang, along with two other colleagues he had met at school, and with the help of an European wife of one of them, started a Tibetan nomad tent-style coffeehouse (all three of them came from nomadic families). They served vegetarian burgers and held jam sessions every Wednesday, during which Tibetans and Indian and foreign tourists had a cultural exchange, singing, playing and dancing together.

Tamiko was from Japan. Like many Japanese who chose to settle in India, she escaped "fast pace of life and workaholism." Every two to three years she used to resign from her corporate job and set on traveling, to "charge batteries." Among others, she visited the United States and Australia, combining tourism with work (see Cooper 2002). She came to Dharamshala to participate in a yoga teachers' course taught in Japanese, organized by another Japanese woman who lived there. During one of the breaks between classes, she went to Kelsang's restaurant and that was how they met and fell in love. In my second year in the field, they opened "the first Japanese café in McLeod Ganj"—a tiny place with a bilingual menu (in English and Japanese), serving Japanese, Tibetan, and Indian teas and several Japanese snacks (see Snepenger et al. 1995). Tamiko—just like other Japanese and Israelis who chose Dharamshala or Hampi as their temporary residence—decided to cater for tourists from her own country, offering them "homely" stuff. She did not plan to return to Japan, even though Kelsang was thinking about going back to Tibet. For the time being, they decided to try their luck in Dharamshala.

Tenzin

Tenzin is a Tibetan born in exile, in one of the settlements in the north of India. Since his parents used to travel south to Pune in Maharashtra to

---

[41] Newcomers are often not that "new" as the name would suggest. They have often lived in India for several or even more than ten years. The term serves to distinguish this group of Tibetans from, first of all, the first wave of refugees, who came to India right after the 14th Dalai Lama, at the turn of 1950s and 1960s, and, mostly, the born-in-exile Tibetans (Bloch 2011). Not only do the latter speak Hindi and English but they also use a standardized version of diasporic Tibetan (while newcomers speak mainly local dialects). They are well educated and, hence, more competitive on the labor market.

"sell sweaters," he joined a university there. Like many other born-in-exile Tibetans he is well educated (see Bloch 2011, 181–218)—he holds a Masters' degree in Economics. However, his employment opportunities—as for a noncitizen—were somehow limited. All public sector jobs, which in India are thought of as relatively well-paid and socially most secure, are out of reach for Tibetans. Thus, for four years after his graduation he had worked as an accountant for the government-in-exile, in his native settlement. But his remuneration was more a pocket money than a proper salary, so he asked to be dismissed from his duties ("with this money, you can survive if you are single, but you cannot provide for a family").

Tibetan refugees in India rarely look for hired work, and the most common way to overcome limitations imposed on them on the labor market is self-employment. This was also Tenzin's choice. Together with his brother, he ran a small tourist agency in McLeod Ganj. They used to sell long-haul bus, railway and air tickets; make bookings; and offer information. Their customers were not only foreign tourists but also Tibetan refugees, who seemed to always be on the move. The brothers also tried to start Buddhism-related tourism in India, targeting tourists from Taiwan and Singapore with their offer of tours to important places connected with the life of Buddha Shakyamuni. Additionally, they also sold DVDs of Tibet-related movies, mainly about human rights' violations.

ETHNOGRAPHY IN MOBILE FIELDS

Fieldwork conducted in areas such as those described above make us aware that as anthropologists we are not the only ones who are mobile: "Today, the people whom they [ethnographers] are trying to study are increasingly likely to be as mobile if not more so than the ethnographers trying to keep up with them" (Amit 2000, 12). Moreover, as Akhil Gupta and James Ferguson noted already at the beginning of the 1990s, "People have undoubtedly always been more mobile and identities less fixed than the static and typologizing approaches of classical anthropology would suggest. But today, the rapidly expanding and quickening mobility of people combines with the refusal of cultural products and practices to 'stay put'" (1992, 9). In an attempt to deal with these challenges while conducting research for this book in 2013 to 2014, I combined ethnography in the mobile fields of Hampi and Dharamshala with mobile ethnography (Haldrup and Larsen 2010, 154; Schein 2002, 231),[42] employing several research techniques.

[42] Mimi Sheller and John Urry call it an "itinerant" ethnography (2006, 217–218), and Anthony D'Andrea a "nomadic" ethnography (2007, 33).

My mobile ethnography consisted in accompanying tourists in their travels, listening to their expectations and comments on the visited worlds, observing interactions that they had with people who inhabited those worlds and attitudes they assumed in such interactions. I worked in this way with Polish tourists,[43] taking part in the tramping tours to the so-called South India (which itinerary included Hampi) and North India (including Dharamshala).[44] Tramping tours are low-budget trips of small groups, often less than ten people, traveling for several weeks mainly in the non-European countries. They combine elements of ethnic, cultural, and nature-related tourism with adventure (trekking, kayaking, camping, etc.). Tramping, presented by its organizers and participants as an alternative to mass tourism, is a more individualized form of organized traveling is de facto—to use an oxymoron—an organized form of backpacking.[45] Tramping tourists do not use all-inclusive packages or chain hotel. They do not travel by rented buses, but by public transport, sleep in inexpensive guesthouses, and eat in local eateries. In the context of India, they are customers of the informal tourism sector, just like backpackers. However, they do travel according to a set—even though to an extent open-ended—schedule, chaperoned by a tour guide.[46] Mobile ethnography

[43] The workers of the informal tourism sector in India classified Polish tourists as belonging to the collective category of "Europeans." I write about their typology in chapter 2.

[44] Each of those tours lasted for twenty-five days.

[45] The term "tramping" comes from nineteenth-century traveling for education and leisure organized for young people from working class as a way to democratize the popular eighteenth-century *Grand Tour* traveling by the aristocracy. It is these trips, in their romanticized version, that were to be then adapted for tourism by the middle-class youth (Adler 1985). However, in Polish language, the term "tramping" seems to rest on the notion of a vagabond (cf. Vogt 1976).

[46] The participants in tramping tours in which I took part knew who I was and what my project was about, as I had sent out this information in advance. However, I had an impression that the fact that I used to be a tour guide myself in the past made me, in the perception of participants, into a second tour guide on those trips. This previous career was my way to travel to the field before I was lucky enough to secure a PhD research grant. The tramping tour operator was then urgently looking for a guide to lead a group to Northern India and Nepal and that is how we started this perfect symbiosis. My boss used to buy me an open ticket, I worked with the tourists, and after the season ended, I stayed on to do my research. This gave me an exceptional opportunity to look at the mechanisms and impact of tourism from the point of view of a participant (cf. a self-reflexive anthropological essay in Bloch [2016b]; on the role of anthropologists' biography in constructing the field see Amit 2000, 6–8). This role turned out to be quite useful, as it made tourists come to me to discuss their problems and questions, which gave me insight into their perceptions of the encountered reality. I should mention that working in the tourism sector—as a tour guide, guide, or interpreter—had already been used as a research technique by anthropologists. Edward M. Bruner used to work as an interpreter for tourists traveling in Indonesia (often provoking them to place themselves on the other side of the tourism encounter or deconstructing their desire for authenticity, which finally got him fired; see Bruner 1995, 2005). Valene L. Smith used to run a travel agency, herself encouraging the involvement of anthropologists in the tourism sector, so as to shape it from within (Smith 2005; see, for example, Blundell 2005, the author of a program of itinerant educational seminars in Taiwan, aiming to facilitate cross-cultural communication between guests and hosts). In Poland, this technique was applied by Magdalena Banaszkiewicz (2012), studying Polish-Russian encounters during excursions to St. Petersburg. According to Smith, tourism sector shapes and sells key anthropological specialties, including culture, ethnicity, and heritage (2005, 252).

provided an excellent opportunity to participate in the confrontations between tourists' representations of the Other with the people categorized as such, which would not be possible if I were to reconstruct these imaginations only after the encounters. However, at the end of each trip in India, I invited the participants for semi-structured interviews, since I was keen to learn how they reflected on their experiences. I refer to this part of my research mostly in chapter 3, where I write about the need for authenticity in tourism.[47]

As I demonstrated at the beginning of this chapter and in the introduction to mobile fields, it is not only tourists who are on the move. I was, therefore, interested in reversing the perspective, while, at the same time, challenging the binary opposition between a mobile cosmopolitan tourist and a rooted local. I aimed to see how the encounters triggered by tourism look like from the other side. Therefore, I tried to follow a postulate formulated in social studies of tourism, to depart from the "touristcentricity" (Franklin 2003, 132)—that is, researchers' penchant for focusing on tourists and their experiences—and put more weight on the local providers of tourism services, as well as the spatiotemporal context in which they operate (Wilson and Richards 2008, 12, 25). That is why, as seasons changed, I commuted between Hampi and Dharamshala, following the footsteps of mobile workers and small-scale entrepreneurs of the informal tourism sector. Returns to the field offered me insight of the temporal perspective, which rendered visible the dynamics of the phenomena. (Did the budding love between a woman tourist from France and a shopkeeper from Hampi survive until the next season? Did an Israeli tourist and a Rajasthani small entrepreneur strike a business deal based on their common idea?) I spent almost eight months doing such fieldwork.

[47] Since tourism imaginaries are shaped long before the travel begins, this mobile ethnography was preceded by other studies: I did an analysis of visual and textual representations of India and tramping tourism along with its promise to fulfill the quest for authenticity voiced in the travel agencies' brochures; I also used questionnaires, which were distributed among the clients of the tour operator for whom I used to work. As an anthropologist, I am skeptical about using questionnaires, but I decided that when used as a qualitative tool, it could bring preliminary insight into tourism imaginaries and the expectations that were based on them (I admit having been overcome by a temptation provided by access to such a big number of tourists offered me by the tour operator). I extend my gratitude to the sociologist Agnieszka Mróz, with whom I collaborated in creating the questionnaire and in data analysis. Especially that I was a difficult partner, trying to smuggle in, wherever possible, open-ended questions on the perceptions of India or tramping, which may have undermined the whole study (Agnieszka warned me that participants may not be willing to fill in so many open-ended fields). However, our worries turned out to be exaggerated. Tramping tourists, who strongly identified with this way of traveling, were willingly sharing their thoughts (some shared with me directly—when taping in their answers to an online questionnaire, they addressed them to me as an ex-tour guide, which, funnily enough, personalized a tool that was supposed to be anonymous). A total of 179 tramping tourists filled in the questionnaire, 101 of them being people who had participated in a trip to India and who answered questions on the way they experienced the country. The results of this study, as well as a discussion of tramping, have been published in Bloch (2016a).

As part of research I employed one of my favorite fieldwork techniques, called by Renato Rosaldo a "deep hanging-out" (cf. Clifford 1997, 351, footnote 2;[48] see Bloch 2016b). It consists in immersion in the field through spending time with people who had become our research partners, while at the same time following newly emerged social actors, events, and plots. The data, collected in such an informal, casual, and often confidential manner, were of special value to me (see Gillen 2011 for a similar fieldwork in Vietnam). The fact that, given all their mobility, tourism workers are strongly connected to their workplaces worked to my advantage, as I already mentioned in this chapter. My technique, then, consisted in circulating between stalls, shops, eateries, cafés and their back offices, reception desks of guesthouses, tourist agencies, and NGOs, spending time in there (see Stoller 1997 and Crick 1992 and the "street-corner anthropology" of the latter[49]). During countless hours spent in this way, I had the chance to participate in daily activities of my research partners, observe their relationships with tourists, and take part in conversations on all kinds of topics, sparked by different situations.[50] This allowed me to capture both what people did and what they said about what they did (e.g., declared attitudes toward tourism and actual behaviors vis-à-vis tourists). I then attempted to link such daily practices and narratives with global power relations within which tourism to postcolonial countries is set, through moving repeatedly between what is local and what is global, in line with recommendations for a multi-sited ethnography (Marcus 1995).[51]

Stian Krog, the anthropologist mentioned in the introduction, who studied how the residents of Hampi constructed its space, complained that most

---

[48] The term was used by Rosaldo during the conference "Anthropology and 'the Field,'" organized by Stanford University on April 18, 1994. This technique was used, among others, in fieldwork conducted among forced migrants (see Rodgers 2004).

[49] Paul Stoller conducted his research by "hanging-out" with street sellers from West Africa working in Harlem, while Malcolm Crick studied informal tourism sector in Kandy, in Sri Lanka, most of the time sitting with his key research partner in a street corner, thus getting to know other sellers, guides, touts, and prostitutes.

[50] These spaces, especially toward the end of the season, where customers are few and far between so that service providers get bored and have more spare time, gave me an opportunity to collect biographical narratives and trace the histories of specific tourism businesses.

[51] George E. Marcus's idea, based on Immanuel Wallerstein's (2004) world-systems, was an attempt to study ethnographically local impacts of global processes, one of which is, without doubt, tourism. Marcus (1998) also warned against perceiving globalities and localities as separate, binary dimensions, since they are both relational (see also Robertson 1995). I thus followed local and national press, focusing on tourism-related news, read travel books and blogs, and observed social media where tourists described their experiences and took action for the communities they visited. I searched for existing statistical data on tourism in both locations, but these were often patchy, so that in order to get a sense as to where the Hampi tourists came from, I needed to go through police records myself. Finally, I collected data through interviews with officials, which allowed me to sketch a wider, institutional context of tourism in Hampi and Dharamshala, following Gupta and Ferguson (1997, 38) in their call for adding more "vertical" research techniques to the "horizontal" ethnography.

of public space in Hampi was dominated by tourists, animals, sellers, and children, with whom he had nothing to talk about (Krog 2007, 61). I had no such problem, since it was exactly what my field was about. My research partners were owners and workers of family-run guesthouses, restaurants, cafés, bakeries, and clubs (i.e., receptionists, cleaners, waiters, cooks, bakers, kitchen helpers, etc.); shopkeepers; stallkeepers and itinerant sellers; itinerant wholesalers; artisans and other manufacturers (including makers of political souvenirs); owners of tourist agencies; guides; employees; volunteers and beneficiaries of NGOs; auto-rikshaw drivers; tattoo artists; massage therapists; yoga teachers; cooking class teachers; and all other service providers, including their family members and friends.

In the heterogeneous tourist spaces of Hampi and Dharamshala, I came across many tourists from all over the world, including India. Some spent just a few days in the locations under study, others devoted weeks or even months, getting involved in long-term volunteer work; learning yoga, reiki, butoh, or belly dance; meditating; writing a book; making a movie; or coaching a Tibetan women soccer team. Some decided to settle more or less temporarily, or to return every season, forming relationships with the workers of the informal tourism sector or having a go at running a tourist business themselves (such as Korean restaurants, Japanese cafés, guesthouses, or shops),[52] or in the NGO sector. While conducting research among those tourists, I used various research techniques, reacting to the available options. In the case of a few friends from France working as volunteers in one of the Tibetan NGOs, the most suitable technique proved to be an observation of their activities during my frequent visits in the NGO headquarters combined with a focus group interview. To have an insight into political tourism in Dharamshala, I became a participant in a week-long tour organized by Free Tibet activists for the benefit of young Indians. One of the advantages of ethnographic fieldwork, as I already mentioned when discussing the category of serendipity, is its openness toward the unexpected and flexibility in reacting to it, in other words, the possibility to creatively improvise in terms of research techniques, which Crick calls "playing by ear" (1989b, 25). This fieldwork yielded over 750 pages of field notes.

Such a field may seem easy, since I did fieldwork in spaces that are generally accessible, that is—even though they constituted private properties—invited others in. However, shedding the role of a potential customer or competitor,

[52] Once again, this is an example of overlapping migration and tourism, called "tourism-led entrepreneurial migration" combined with the so-called lifestyle migration (Williams and Hall 2002, 31), also defined as "life-style entrepreneurship" (Ateljevic and Doorne 2000; Welk 2010). Such business enterprises often target tourists from a specific national or ethnic group, from which the tourist-turned-migrant originates (as was the case of Tamiko).

and becoming an acquaintance in this context, proved difficult and time-consuming. On the one hand, I was a white woman in her thirties moving through tourist spaces, which to workers and small entrepreneur of the tourism sector looked like a European backpacker traveling alone, and indeed my behavior was not too different from what many other backpackers did (attempts to fraternize with the locals through sitting in their premises, chatting, or hanging out). On the other hand, my inquiries as to the origin of goods, ways of their transportation, costs, income, competition, and so on rose suspicion that I might have wanted to launch my own business in the local market and that I was collecting intelligence. Moreover, doing fieldwork in Hampi proved to be challenging since it was an entire terra incognita to me. During my first visit to the field, I worked hard to establish trustful rapport, which was difficult given the political tensions related to evictions and demolition. People did not trust each other and strangers making every effort to become friends and asking too many questions on sensitive topics were out of bounds. Inquiring about land ownership or the legality of businesses was like walking on a field studded with booby traps. It was only when I returned to the field, people finally became convinced as to the sincerity of my intentions, or at least, my determination, since by that time I was already visibly pregnant (although this also made some people question my sanity). However, multiple attempts to get me involved—like other tourists (I write about this in chapter 4)—into sectarian struggle against displacement, required a lot of skill on my part in navigating the thick forests of competing and often mutually exclusive visions and interests.

In Dharamshala, a seemingly innocent topic of tourism often turned out to be a predicament. I had known Dharamshala ever since I did fieldwork there for my PhD dissertation, and thus I did not expect any trouble. However, even though everyone was perfectly willing to talk to me about political self-determination or preserving Tibetan culture in exile, tourism as an object of study in the context of exile was much less palatable. Tourism was partly seen as banal (in the same way that my colleagues-anthropologists saw it), but mainly, I had the impression that "getting rich on tourists" was regarded as not commensurate with the image of a humble refugee, impervious to materialistic desires. The topic was thus discursively overlooked. Like in Hampi, the informal tourism sector here also involved many activities on the verge of legality, related to restricted right to land and real property purchasing as well as gray zone labor market. My questions related to the Bihari (their age, wages, working times) or land ownership often aroused suspicion and resulted in questions: "Who sent you? How did you come to me?" The fact that Tibetans tend to be very well aware of the politics of positive self-representations to the outside world, and the ensuing self-censorship, made it extremely difficult for me to ask about issues that interested me most.

Moreover, Tibetan refugees, who had been "research subjects" for quite a long time, were used to scholars being interested in their society and culture, but they usually reduced the latter to material and spiritual artifacts of religion and traditional performing arts, and not to contemporary social and economic relationships. My research topic caused a certain cognitive dissonance in their understanding of anthropological inquiry.

However, I have an impression that my presence was also a form of entertainment for those working in the informal tourism sector, often marked with boredom, while waiting for customers. Moreover, my presence attracted tourists, who saw me as another customer confirming the quality of goods or services (I did not even need to explicitly praise the goods, which, however, on occasion I did, especially if I knew that business had been slow[53]). With time, I started being asked to do specific jobs for my research partners and their businesses (I was consulted on foreign tourists' tastes in interior design, goods on offer, or a cultural program prepared by an NGO; my skills were put to use in proofreading leaflets, notices, restaurant menus, and business cards; people tried to use my social networks to gain customers, volunteers, and funding sources). In a sense, I cocreated the reality that I came to study, but this was actually interesting from the research point of view, since all other tourists who expressed the desire to "fraternize with the locals" were being involved in the same way.[54] Also, to a degree, in this way I could apply the principle of reciprocity and show my gratitude for time, trust, and stories bestowed upon me. I often acted as a live book of complaints (that the government did not grant permission to build a restaurant, that tourists are being arrogant, that competitors are cheating, etc.). In other instances, I was put into the role of an influencer, expecting me to apply leverage with local authorities, clearly overestimating my position within local power relations. But first and foremost, I was treated as a source of information. Paradoxically, mobile sellers and service providers, since they were bound to their workplace, had no way to get insight into competitors' businesses, quite unlike me, a mobile researcher moving across different social contexts. Here is an excerpt from my field notes, illustrating such practices: "Raj informed me that 'Manju is

[53] The topic of business (income, customers, sales) cropped up frequently in our discussions—in fact, most of the daily visits started with me asking, "How is business today?" It often did not go very well, especially in Hampi, which—being erased from the landscape—was losing tourists.

[54] On numerous occasions in this book I refer to different roles I was ascribed and how they influenced the research process. For discussion of self-reflexivity, transparency, and positionality in the qualitative research see, for example, Rose 1997; Davies 1999; Alvesson and Sköldberg 2000, and in relation to tourism studies see Ateljevic et al. 2005; Westwood et al. 2006. Reflexivity does not only involve "indulging" oneself through writing oneself into research but also recognizing and being aware of wider conditioning of knowledge production, into which we are "entangled" (Ateljevic et al. 2005, 9).

said to have bought land in Gokarna,'[55] and then he looked at me expectantly, and added 'But I have not asked him about it.' I understood that my role was now to ask Manju about the transaction and pass on that information." The same went for NGOs in Dharamshala, which were competing for volunteers and sponsors. In such cases I employed a tactic of asserting that all the information I collect was confidential (otherwise, I would not be able to win the trust of my research partners).

That I did manage to build relationships based on trust manifested in the fact that I was given access to intimate spaces (e.g., witnessing the birth of Padma and Phurbu's daughter) and to local events, invited to hang out with my research partners even though their free time was limited, and to assist in all kinds of paperwork and dealmaking. My changing status was visible in the changing attitude toward me—starting with a potential customer that one needs to attract, (often) through a potential girlfriend,[56] to a friend one can make fun of, to whom one may confide gossip and secrets, from whom one can borrow money to buy whiskey so that the wife would not know, who is invited to special dinners, to whom one makes a call simply to ask "What's up?" Gradually, I even started to be regarded as a local expert, as when my research partners introduced me to other tourists as follows: "She's a girl from Hampi, she knows more about Hampi than we do." The late Mr. Shankar, a former postman and owner of a guesthouse in Hampi, told me during one of our evening walks, "I am not exaggerating, many people around here speak well about you, saying that you are so kind to everyone." In the field made out of such a tangled net of mutually exclusive interests, it was a huge compliment.

[55] A seaside town in Karnataka, also being an important Hindu pilgrimage site.

[56] Many workers and small-scale entrepreneurs in the informal tourism sector are young men (as I mentioned at the beginning of this chapter), for whom initiating an intimate relationship with female tourists is a practice that allows them to overcome limitations stemming from unequal global power relations (I discuss it in chapter 5). This gender-based role was no longer a threat in the second year of my fieldwork, since I was then visibly pregnant (see Golde 1986).

Chapter 2

The World Map as Seen from the Peripheries

The Tourist as an Object of Gaze

HOW DO THEY SEE US?

Much ink has been spilled on the tourist gaze—a classic category in social studies of tourism since the publication of John Urry's book in 1990 under the very same title (2002). Researchers focused on how Otherness was constructed to fit tourists' needs and how touristic imagination and expectations shaped the reality that constituted their object (Selwyn 1996b; Morgan and Pritchard 1998; Crouch et al. 2005; Salazar 2010; Skinner and Theodossopoulos 2011; Salazar and Graburn 2014). Scholars analyzed specific components of the "image production industry" (Harvey 1989, 290–293), mainly using critical discourse analysis of the visual and textual representations of the Other: promotional materials of tour operators, that is, catalogs, brochures, leaflets (Dann 1996; Echtner and Prasad 2003), guidebooks (Bhattacharyya 1997), postcards (Edwards 1996), travel writing (Calzati 2015), travel magazines or feature movies (Mercille 2005), and so on. It was noted that the industry was supplying us with interpretative schemas, due to which we can navigate the world—especially the world of the Other, in which we look for confirmation of what has already been etched in our imagination (Salazar 2009, 2012; see Ricoeur 1994; Gaonkar 2002). By the same token, the way we look at the world—or, more exactly, what we see and how we interpret it—is being surveilled. Urry stated clearly that his concept stemmed from the Foucauldian (2003) category of medical gaze saying, "And this gaze is as socially organized and systematized as is the gaze of the medic" (Urry 2002, 1).

This leads us directly to a powerful source of touristic imagination, especially in the context of tourism to non-European countries, that is, Orientalism as described by Edward Said (1979; see, for example, Hutt 1996; Echtner and

Prasad 2003). Said based his critique of representations of the inhabitants of "the non-West" on the Foucauldian analysis of the order of discourse (Foucault 1981; see also 1980, 2002). It later became the cornerstone of postcolonial critique, since it demonstrated that "knowledge" about "the Orient" created by colonial Europe legitimized power exercised over it (Loomba 2015, 61). The European representations of the colonized Other present not only in academic writing but also in literature and art—were based on reproducing binary opposition between Europe and its Others, constructed in line with evolutionary thinking, and therefore carrying value-judgment. Such Europe could appear, to oneself and to others, as technologically, politically, and culturally more advanced (Said 1979). If "we," the Europeans, considered ourselves to be rational and balanced, "they" had to be emotional and unpredictable, succumbing to instincts; "they" had to be backward, static, and "out of time" (Thomas 1996)—wild and primitive, or at least decadently ancient[1]—so that "we" could see ourselves as civilized, progressive, and dynamic; "they" lived in villages, in the world of tradition identified as backwardness and "we" inhabited modern metropoles; "they" were still embedded in nature, had myths, beliefs, customs, and shamans, while "we" lived in the world of culture—with science, religion, norms, and medicine. All this amounted to a key dichotomy both for constructing Europeanness and for maintaining and expanding European domination over other territories and their inhabitants. It consisted in producing and reproducing difference, a rigid binary opposition between "us" and "them" (Said 1979). In many cases, the representations of non-European countries created by tourism industry are the continuation of the colonial discourse of Orientalism (Laxson 1991; Hollinshead 1992; Palmer 1994; Wilson 1994; Sturma 1999). It was one of the reasons, already mentioned in the introduction to this book, why international tourism, especially ethnic one, has been referred to as "the mirror image of racism" (MacCannell 1992, 170). Tourist gaze is most often white, heterosexual, and male (Pritchard and Morgan 2000).

Elsewhere, I analyzed how offers posted on the websites of tramping tour operators shaped representations of the *sight-visited* worlds and what were tourist expectations toward them (Bloch 2016a). Within the literature of the subject, there are other, numerous examples of studies focusing on the reception of the Other's image as an object manufactured within tourism, that is, documenting attempts at accommodating, negotiating, modifying, appropriating, or resisting and challenging it (Edensor 2001; e.g. Tucker

[1] It was considered that "the Orient"—by which Said himself referred mainly to today's Middle East, his place of origin—had reached a certain level of civilization, but then faced decline from which it could not emerge by itself. Africa or South America, on the other hand, were realms at the brink of evolutionary development, in the state referred to as wilderness or infancy.

2011; Kanemasu 2008).[2] Here, however, I wish to reverse the perspective and look for an answer to a question that has been asked much less frequently, that is, referring to an article by Deirdre Evans-Pritchard:[3] How "they" see "us"? Evans-Pritchard described how the image of a white man had been constructed by the Navaho as that of a "collector-consumer"—an anthropologist roaming the Native American world wielding a recorder, a tourist—armed with a camera (1989, 91–92). Of course, those on the other side of the tourist encounter have no such powerful tools for representation-making, because it is not them who hold symbolic capital (Bourdieu 1986)[4] or control communication channels. However, they do participate in daily processes of seeing and being seen within the mutually cocreated field of "visual reciprocity" (Schneider and Wright 2006, 18, after Gupta 2014, 36),[5] where it is us who are the object of the gaze. It is thus worthwhile—even though it has to be mediated by an anthropologist—to have a look at oneself in such a mirror (see Pi-Sunyer 1977).

Stian Krog, who—as we already know—found it difficult to mark his separate status vis-à-vis tourists, related the way in which Hampi inhabitants perceived tourists in the following way: "Tourists were expected to: go sightseeing, purchase food, lodging and souvenirs, make international phone calls and use the Internet. Locals attribute them to be: physically and psychologically weak, socially clumsy, rich and yet often stingy, potentially meat/drugs/alcohol-consuming, religiously ambiguous, and slightly unpredictable in general" (Krog 2007, 58). Such an image of a foreign tourist (or an anthropologist) reminds us of an Occidentalist view. Occidentalism, as Ian Buruma and Avishai Margalit suggest (2005), is the way in which "the West" is seen by "the East," that is, a set of essentialized characteristics attributed to all of its inhabitants. In other words, it is a reversed discourse of Orientalism, an anticolonial rhetoric, which can be interpreted as part of the colonial inheritance in postcolonial countries (de L'Estoile 2008). Just like Orientalism refused to see the inhabitants of colonies as humans and likened them to either animals or children (as not-yet-human, see, for example, Arabs as "the childish primitive" in Said 1979, 247), from the Occidentalist point of view the residents

---

2 Kanemasu, for example, using the projection-based method, that is, showing Fijian hotel employees their visual representations created by the tourism industry, captured how selectively they approached them, internalizing specific aspects and firmly rejecting others. I discuss it in more detail in chapter 3.

3 The younger daughter of Edward Evan Evans-Pritchard, a famous British anthropologist, whose work is considered to have paved the way for the interpretative paradigm in anthropology.

4 I discuss capitals as defined by Pierre Bourdieu in chapter 4. Here I just wish to state that it is a cumulative effect of all other capitals, giving the possibility to exercise symbolic power or the power to shape reality through discourse.

5 Darya Maoz (2006), who studied the Israeli backpackers and whose work I shall refer to in this chapter several times, used the term "mutual gaze."

of "the West" are more demons than people.[6] Occidentalism uses the same binary oppositions that Orientalism was shown to use—materialism versus spirituality, modernity versus tradition, individual versus community. They are just valued in an opposite way: a rootless cosmopolitan urbanite and a mechanical society versus an organic community of blood and land; sterile intellectualism and soulless reason versus inner life of the spirit and authentic religiosity. Representatives of "the West" are thus construed as superficial materialists bowing to money as if it were a God, amoral nihilists only looking for more pleasure, vain, alienated egocentrics, and hypocrites.[7] Many critics of international tourism who see it as a neocolonial practice base their arguments on such an image of a "Western" tourist.[8] Within the rhetoric of the religious nationalism in India—for example, under the guise of a discourse formulated by Hindu leaders talking about Hampi—foreign tourism is seen as a Trojan horse of Western influences, and as such appears as a moral contamination (Bloch 2017a).[9]

We need to keep in mind, however, that the image reported by Krog may have been exaggerated through his own dislike of tourists—for reasons I described in the introduction—and the fact that his research partners in Hampi were predominantly people who did not work directly with foreign visitors. My research showed that this image is much more complex and that the tourist as represented in it cannot be reduced to an Occidentalist imagery of a morally suspect "walking wallet" (Krog 2007, 58). Workers and small

---

[6] The authors of this concept firmly stress that Occidentalism—that is, demonizing of "the West"—cannot be seen as purely "Eastern" invention in general and Middle Eastern one in particular. The book was written after 9/11 terrorist attacks in the United States and it has often been understood in such a simplistic way. However, as Buruma and Margalit (2005) show, it is indeed the refusal of modernity identified with "the West," present both in the right wing and the left wing of the political scene in the countries of the so-called Global North (e.g., anti-globalist movements) as well as the Global South. The scholars even suggest that Occidentalism was born in the "Western world" under the guise of struggle against industrialization, counterreformation, romanticism, or national socialism.

[7] Frantz Fanon—born in Martinique and educated in France, a psychiatrist and Algerian freedom fighter, a critic of colonialism—in his famous book *The Wretched of the Earth* (1963) described the way in which the white city-dwellers were perceived by the colonized: "The settlers' town is a strongly built town, all made of stone and steel. It is a brightly lit town; the streets are covered with asphalt, and the garbage cans swallow all the leavings, unseen, unknown and hardly thought about. . . . His feet are protected by strong shoes although the streets of his town are clean and even, with no holes or stones. The settler's town is a well-fed town, an easygoing town" (1963, 38), "for them my brother is my purse, my friend is part of my scheme for getting on" (1963, 46).

[8] See, for example, all movies listed in footnote 15 in the introduction to this book, which not only assume that tourism destroys local cultures but also present tourists as morally degenerate, ignorant hedonists.

[9] Cf. the vision of foreign tourism according to one of the NGOs opposing tourism development in Goa: "Modern Mass Tourism is an invasion by . . . a materialist culture. It is a culture of economic subjugation; a culture of exploitation; a culture of vandalising the earth and environment; a culture of racist arrogance; a culture of leisure, pleasure, hedonism and escape" (Jagrut Goenkaranchi Fouz, that is, Watchful Goan Army, in Routledge 2001, 231).

entrepreneurs in the informal tourism sector were skillfully diversifying tourists and their needs, noticing the heterogeneity of this group. In everyday conversations—both in Hampi and in Dharamshala—tourism workers distinguished three main categories of tourists: "the Europeans," "the Israelis," and "the Russians" (among others who were less visible). Indians, as domestic tourists, constituted a separate group. These categories were far from simply geopolitical denominations and reflected different forms of tourism, associated with specific types of visitors. As I demonstrate in this chapter, "Europeans" were synonymous with noninstitutionalized backpackers, as opposed to "Russians" who were identified with a package charter tourism, organized on a mass scale (in Hampi, "Russians" meant something very different than in Dharamshala, as I shall discuss later on). "Israelis," in turn, were perceived as staying apart "freaks." Below, I present how "they"—that is, the workers and small entrepreneurs in the informal tourism sector at two Indian sites—see "us," that is, tourists, and how the world map looks like from the perspective of the "pleasure peripheries" (Turner and Ash 1975).

Europeans: The West in the Eyes of Its Enemies?

Opposite to—in this context seemingly obvious—the Occidentalist interpretative framework, preferences of people working both in Hampi and in Dharamshala were ranked just like demonstrated above, that is, with "the Europeans" being definitely the most preferred guests. Resorting to the category of "Europeans" does not mean that workers and small business owners in the tourism sector did not recognize particular nationalities. When asked to specify whom they meant by using the term, they listed the most numerous visitors, in line with the local statistical data: German, French, Spaniards, British, Dutch, Swiss, Scandinavians, or Poles. My research partners, however, were noticing some similarities in attitudes and behaviors of all these groups, thus classifying them in a single category. At the same time, when inquired, they insisted that they did not want to make any unjust generalizations. An owner of a family-run guesthouse in Hampi saw it in the following way: "People are different,[10] just like everywhere, but Europeans are generally OK. They are mostly backpackers, who have been traveling in India for a few months already, from the north to the south of the country or the other

[10] I admit that I was surprised by my research partners' reticence in giving opinions on particular groups of tourists. When, for example, hearing of "the European" category, I tried to induce statements about it, they often started with the objections that "people are different, there are good people and bad people everywhere." Of course, I was classified this way myself, so they may have wished not to offend me. Nevertheless, similar resistance to generalizations referred to other groups of tourists as well.

way round, so they have already learned how to behave properly, the Indian way."

"The Europeans" were thus structurally long-term tourists who preferred economy-class traveling style in terms of accommodation, transport, and food, who organized their itinerary by themselves, without any comprehensive support of a tour operator, even though they often followed the same trails as other backpackers (Pearce 1990; Loker 1993; Richards and Wilson 2004; Cohen 2004; Hannam and Ateljevic 2008; see also Loker-Murphy and Pearce 1995; Uriely et al. 2002; Sørensen 2003). These were usually young people doing their gap year between college and university or between graduation and entering the job market (O'Reilly 2006). More and more often these were also people in their thirties who took a break in their professional career, or who worked as freelancers—classified by scholars as the new middle class of late modernity, who live by their post-material values through traveling in a way alternative to mass tourism (Munt 1994). There were also older people, often from the hippie generation, who have been described in literature as "Peter Pan travelers" (Hartmann 1991, after Loker-Murphy and Pearce 1995, 829). Despite their age, they kept traveling in the way they used to. Part of them were the so-called lifestyle travelers, that is, people for whom traveling did not constitute a temporary suspension of everyday life or a passage to the next stage in life but their chosen lifestyle (Cohen 2010).[11] The trails for such traveling were blazed by the hippies in 1960s and 1970s, which triggered the emergence of an informal sector, specializing in services tailored for the needs of this group of tourists (Hampton 2010, 9), of which tourism development in Hampi and Dharamshala are an example.

Cultural sensitivity coming to the fore in the quote above, which manifests in efforts to understand and to adopt part of local customs or in the wish to learn about them, cropped up frequently in my conversations with people working in the tourism sector as the main reason why they held a positive opinion of the "European" tourists (cf. Binder 2004), even though scholars have been highlighting the growing discrepancy between the ideology and practice of backpacking, that is, between the declared wish to get to know the Other and staying among people like oneself (Cohen 2004, 48–49). Such intercultural competence along with the ability to speak English, which made communication possible, was interpreted by my research partners as markers of good education of "European" tourists. In Dharamshala this was all the more important since Tibetans managed to develop a thriving political and volunteer tourism (I discuss this in more detail in chapter 4), which requires the visitors to have basic knowledge of world affairs (related to

[11] Cf. lifestyle migration and lifestyle entrepreneurship in chapter 1.

refugees, human rights, the right of self-determination, etc.). Other, positively evaluated characteristics of this category of tourists were their respectful attitude toward the locals ("as human beings") and their honesty. A former owner of a restaurant in Hampi recollected that she never had to issue bills and that people had been paying her the right amounts of money without her ever having to count it. "And after the meal they cleaned after themselves, bringing the plates into the kitchen." Another frequently mentioned characteristic was their "taste"—"Europeans" were regarded as those who "know their stuff." It was especially the Rajasthani textile traders who liked to stress that they had a "European assortment," that other tourists did not know how to appreciate fully.

In other words, "the Europeans" in the eyes of the workers and small entrepreneurs in the informal tourism sector were the perfect tourists, situated around the middle of the spectrum ranging between familiarity and novelty: the "explorers" as defined by Cohen, curious about the other world, wishing to hang out with people they were visiting, to get off the beaten track, but straying not too far away, so as to enjoy a secure and comfortable accommodation at a guesthouse: "The explorer dares to leave his 'environmental bubble' . . . , but he is still careful to be able to step back into it when the going becomes too rough" (Cohen 1972, 168). In another, later typology, Cohen called them "experiential travelers"—young people raised as part of middle class of postmodern societies, who feel they are missing something in their own axiological center and look for it in the life of others which they perceive as more authentic (Cohen 1979, 186–188). They are to be driven by what Dean MacCannell (1976) called the quest for authenticity. They sometimes "experimented," as Cohen put it (1979), by trying different alternative ways of living (like meditation, or life of an ascetic), but they rarely decided to convert for good, to become "existential" tourists who have found their new "spiritual center" in India.

Nowadays, scholars of backpacking—including Cohen—think that such travels should be considered as instances of "suspension" of everyday life (professional career, routine daily chores), which facilitates personal growth that has become so valued in the individualist capitalist societies, rather than, as was thought before, a rite of passage marking a new stage in life (Richards and Wilson 2004, 255–257; Cohen 2004, 55; Wilson and Richards 2008). However, I have an impression that even this is not a completely correct observation these days. In the past, a backpacker, as the term indicates, traveled with just a backpack, a worn-out copy of the *Lonely Planet* guide, and the company of fellow backpackers. Now, they are literally armed with electronic devices—laptops, tablets, iPods, iPhones, smartphones, and so on—which enable them to stay in touch with the whole world, through daily and real-time use of new technologies. In response to these changes, every tourist site in India, however small—including Hampi and Dharamshala—offers tourists free Wi-Fi, both in restaurants and in guesthouse rooms. Internet cafés and telephone booths have literally disappeared from the tourist landscape of India. In the middle of the first decade of the twenty-first century, the new term "flashpacking" was coined to describe technologically equipped backpackers, who also happened to have more money than their backpacking predecessors (Hannam and Diekmann 2010, 1–2; see, for example, Jarvis and Peel 2010). It is a result of changes that occurred in accessibility of new communication technologies, with the electronic devices becoming cheaper and more compact, and thus easy to be carried along (and so more mobile). Their bigger budgets of today's backpackers stem from the fact that people now tend to travel in such a way when they are older. This is lined with the changes within the societies of the so-called Global North, where the decisions to start families are postponed and a more flexible labor market allows more people to take breaks in their professional careers. As a result, such tourists can now participate in the lives of their relatives and friends back home while traveling. Many of them work and manage their finances remotely, at the same time enjoying the possibility to create their image as a traveler through online posting and uploading photographs and videos on social media (Newlands 2004; Paris 2010). They are both here and there, which also helps them in keeping touch with people they met during travels once they are back home, because those from the other side of the tourist encounter have also become active users of more accessible new technologies, the fact that many scholars seem to overlook.[12]

The literature of the subject also started to feature critical observations on the fact that the majority of backpackers seemed to stick to their own groups, creating "environmental bubbles," and not really looking to create direct relationships with the locals (Murphy 2001). It was argued that the phenomenon of "institutionalization" of backpacking led to practices of spatial separation, ensuring that people feel safe and predictable in the face of Otherness,

[12] For example, Cody Paris, who conducted research on the impact of new communication technologies on backpacking (including their use after the trip), in his survey on backpackers did not take into account the possibility they would keep in touch with the inhabitants of the visited countries or employees of the tourism sector whom they met during the trip but instead included options: "stay[ing] in contact with friends and fellow travelers met during trip" or "stay[ing] connected to backpacker culture online" (2010, 58–60). My own experience is quite different, perhaps because I come from a post-socialist country in Central Europe and I started going to India at the very beginning of the twenty-first century. It was in India, in the Tibetan refugee camp, that I first saw an USB drive and—a few years later—smartphones. It was thanks to my friends and acquaintances in India that I discovered the advantages of online communicators (first Yahoo Messenger, then Skype) or social networking sites (such as hi5 before Facebook became more popular), and finally WhatsApp, enabling free calls, text communication, sending photographs, videos, and sound files (especially WhatsApp proved to be very inclusive, as it can also be used by people who are not entirely literate—instead of text messages, they simply send audio messages).

and so limiting their individual learning and contact with the external world (Hottola 2005; O'Regan 2010). Other scholars pointed to enclave character being a unique feature of backpacking in general (Howard 2007; Wilson and Richards 2008). My research partners, however, did not perceive "European" backpackers in this way, since they were usually those tourists who regularly returned to Hampi or Dharamshala, and who built closer ties with people working in tourism sector[13] (for discussion see chapter 5). They were also actively involved in promoting the well-being of local communities (see chapter 4).

It is interesting that in expressing their good opinion of this category of tourists, my research partners were also aware that they were relatively low-income customers. The workers and small entrepreneurs in the informal tourism sector fully realized both what "type" of customer they were dealing with (i.e., a tourist traveling on a shoe string) and that the economic crisis that has ravaged Europe since 2008 had a negative impact on the financial resources of Europeans visiting India.[14] Even though I heard from them that "Europeans" are no longer such good customers as they had used to be, usually there was also an understanding of their situation, as expressed by Norbu: "They are mainly students who usually work in a pub to earn money to come here. They have a set, limited daily budget, like 500 rupees per day [a bit more than 10 dollars], so they tend not to spend much, especially since they are travelling for a long time, even a few months."[15] One of the tour guides in Hampi described the difference in the tourists' profile as follows: "We provide for tourists who do sightseeing by bikes, not air-conditioned cars." Mohan elucidated to me that there were the following strata of customers: the rich, who "buy in shops that are so expensive that my clients have never even heard of them"; then the upper middle class, captured by the Kashmiri shopkeepers; and then, at the very bottom, the lower middle class, that is, his backpacking customers. The head of one of the Tibetan NGOs that runs a nursery for babies raised by single mothers in McLeod Ganj was

---

13 In the relevant literature there one can find a term "veteran backpackers," that is, those who return, sometimes regularly, to the places and people they know from their previous travels, often with their families (Leon 2000, after Noy and Cohen 2005, 37, footnote 21). See Macleod (2004, 85–86).

14 Norbu noted that since the crisis, European customers have been buying mostly cheap, artificial jewelry, no longer looking for the "authentic" one, that is, silver with real stones, which until now it has been a specialty of so-called Tibetan shops.

15 At the same time, research shows that backpackers make a significant economic contribution to many marginalized communities in less developed parts of the world, and their way of traveling causes much less leakage than mass or elite tourism (Hampton 1998; Scheyvens 2002a; Speed and Harrison 2004). However, many officials from the tourism development agencies believe otherwise, which is evident, for instance, in Hampi, where the regional Department of Tourism is trying to profile the offer for the more affluent tourists (cf. similar reflections by Arun Saldanha on Goa [2007, 35–37] and Cohen in the context of Thailand [2004, 43]).

even convinced that "often, the [foreign] sponsors are poorer than Tibetans. It happens that you don't have money, but you still want to provide support. You quit smoking or drinking Coca-Cola, or some fast-food to save money and support a child." The owner of a small bicycle and motorbike rental shop in Hampi, while referring to a conspiracy theory, according to which owners of the big, luxury hotels in nearby Hospet wanted to take over tourists from Hampi, provided his expert view:

> They know nothing about tourism. They have no clue that tourists here still look for rooms that cost 200–300 rupees [four to six dollars], and sometimes that is even too expensive, and they ask Mahendra [Hampi inhabitant, whose work consists in guiding tourists to the viewpoint, where they can watch sunrise] if he could keep their luggage in his house while they sleep outside. So, he keeps their stuff for 20–30 rupees. There are tourists who rent bikes because they cannot afford motorbikes. There are those who eat *idli*[16] with chutney from the local *dhaba*. How could they pay 100 dollars per room?

One of the textile traders from Rajasthan was trying to convince me that Europe is not really that expensive, that "everybody can go there." A friend from Italy, Gabriel, had told him that "there, you can get pizza for less than in Shanti [a local restaurant]: here it is 400 rupees [8 to 9 dollars], and there it is 2–3 euro." This understanding of the new expectations and opportunities of the not-so-wealthy-anymore tourists was also expressed through a plan made by one of my research partners in Hampi, who intended to establish a dormitory for single tourists, offering clean bathrooms, a safe-storage for documents and money, where people could rent a bed in one big dormitory room for 300–400 rupees (6–8 dollars). Such perception of "European" tourists creates some distortions in the overall picture of the—seemingly obvious—global inequalities (I shall discuss that in more detail at the end of this chapter). It also has a positive impact on the self-perception of small entrepreneurs in the tourism sector, who position themselves as partners in this postcolonial relationship, and sometimes even as somewhat economically superior partners.

Here I wish to stress that in terms of economic situation, the "European" customer in Dharamshala is a bit different than in Hampi. Here, such customers are able to spend more, as they arrive for shorter periods and with the purpose to visit Tibetan diaspora (because of numerous—Buddhist, political, personal—links that will be discussed in chapters 4 and 5). This is reflected

[16] *Iḍlī* (Tamil.)—steam-cooked, round, flattened dumplings with no filling, made from rice flour and fermented lentils. They are often eaten with thick chutney sauce. Cheap and popular in the south of India, eaten especially for breakfast.

in the style of restaurants and cafés catering for "European tastes." During my four-year-long absence in the field, Dharamshala gained numerous fashionable cafés—spacious and with lots of light coming through big windows, with comfortable sofas and "real coffee" made in espresso machines. Their menus often included expressions such as "healthy," "vegetarian," and "vegan"[17] (for instance, smoothies made with soya milk, vegetarian sushi, or quiche with tofu). These cafés were furnished in a way that made reference to natural environment—with wooden chairs, seating pillows in natural linen, rattan screens and lamps, or green curtains. Such places were meant to be coffeehouses and hanging out in them was a popular pastime in Dharamshala, although the prices were quite high as for India.

I need to mention that in conversations about tourists "from Europe," several of my research partners used Occidentalist rhetoric, especially when speaking about several specific topics. The first concerned mixed romantic relationships between women and men. The same Rajasthani shopkeeper who spoke of Europe being "cheap," when I asked him jokingly whether he had ever considered having a tourist wife (he was single, and his friends dated tourist women), responded with great seriousness: "Western women are egocentric, they marry those Indian men who are also egocentric. They do it so that they can have a good time here,[18] and the men think they will go to the West and be rich." He did not wish to live there, because "life there is mechanical." His friendship with Gabriel, mentioned earlier, was significant in this context. It was Gabriel, who had been repeatedly returning to India and who married a woman from Hampi,[19] who was himself disaffected with the materialism and alienation of life in Europe, and he reinforced his Rajasthani friend's Occidentalist views (which confirms, in a very interesting way, the hypothesis by Buruma and Margalit, that the dislike of "the West" was of its own making, exported elsewhere). The same optics was reinforced in workers and small entrepreneurs of the tourism sector by many backpackers, who, in the world they sight-visited, looked for a feeling of belonging to a place and a community.

The second topic that my interlocutors treated with Occidentalist criticism, also related to the sexuality, was homosexualism—shown, once again by the same Rajasthani shopkeeper, as an anomaly normalized in Europe. I also came across worries that "the West" was actually moving toward blurred gender boundaries. I was surprised that such worries were voiced by a Tibetan artist

[17] For example, "We offer only selected, healthy, consciously prepared vegetarian and vegan dishes." Vegetarianism is one of the newer Tibetan traditions born in the diaspora (see Bloch 2011, 466–482). During the research I did for my PhD, nobody mentioned veganism. Still, most restaurants run by Tibetans serve meat.

[18] I write about legalization of stay in India on the basis of marriages to Indian citizens in chapter 5.

[19] I tell this story in chapter 5.

and performer, whose profession involved boundary crossing. On the other hand, he was a Tibetan newcomer, who was used to more normative gender roles. Once, he asked me, "Why do many European women (here he gave me a telling look) look and behave as if they were half women, half men?" What he meant was not only professional emancipation but also unaccompanied travel by female backpackers he had been meeting frequently—autonomous, taking initiative, interacting with others. Another issue often discussed in my conversations about "Europeans" was the attitude of people from "the West" toward the elderly, who were almost all thought to be left to live in care institutions, where they spent their last lonely years. This was due to our fast-paced lifestyle, constant stress, and egocentrism. Finally, the Europeans who settled in Hampi were thought to be well-off due to global inequalities, lazy and heavy drinkers, which combined into the following logical order: "They don't have to do anything here, because there [in their country of origin] they have properties which they rent, or they live off their inheritance . . . , so they start drinking and go downhill."[20] Such examples show that negative characteristics ascribed to tourists' "European culture" were, in fact, expressions of anxiety felt by more conservative individuals about changes associated with modernity which their own societies were undergoing.[21]

Israelis: Grassroots Enclavic Tourism

Young Jews coming from Israel—whom my research partners called "Israelis"—were a fixed element of the tourism landscape in India. Darya Maoz, who did fieldwork among this group, distinguished two categories within it based on the nicknames that young Israelis were giving themselves. The "conquerors" (Hebr. *ha-kovshim*) were aged twenty and over, came to India directly after their military service, while the "senior" visitors, in their thirties, were called "settlers" (Hebr. *ha-mitnahalim*). Both terms come from

---

[20] It is worth noting that exactly the same opinion, but about the inhabitants of Hampi, was expressed by their neighbors from the surrounding villages who considered working in tourism—especially renting premises to migrant entrepreneurs—as easy and lucrative, which was supposed to lead to excessive alcohol consumption. My observations showed that the availability and low prices of alcohol—especially local varieties of whisky—led to its common abuse in the Indian rural areas, at least in Karnataka. The Europeans mentioned above often had a drink with the village inhabitants and their houses were a safe haven for this activity, especially in Hampi, where consumption of alcohol is officially prohibited because of the sanctity of this space.

[21] Kristin, one of Dharamshala's long-term volunteers from Germany, told me about her reflections from language classes conducted in a Tibetan NGO. Her students were mostly newcomers from Tibet, not working in the tourism sector and having previously little direct contact with tourists. Their views on "Europeans" were a classic example of Occidentalism, which is worth mentioning here for comparison: "They told me: 'I heard, that in Europe . . . is that true?'—that we have no moral values, that we have no culture, that we have no religion, . . . that we do not care about our parents or friends, that we cheat and lie . . . and that we're very open, you know, to everything." All this pushed Kristin to prepare "Western culture course," about which I write more in chapter 4.

the Israeli army and political lingo and relate to the issue of physical invasion and Israeli settlements set up in the West Bank and Gaza to confirm the Israeli State's jurisdiction over these contested territories (Maoz 2004, 110). Even though Chaim Noy and Erik Cohen (2005, 17–18, 27–28) emphasized that these expressions were self-ironic and were a form of play on words, according to the cultural intimacy concept put forward by Michael Herzfeld (2005), the dominant attitude expressed through them is meaningful. Maoz claimed that for "conquerors," the experience of traveling to India was just a continuation of their military service experience, hence the tendency to stay together (collectivism), hostile attitude toward the world, and focus on physical and mental exercise, mainly in the form of difficult treks, climbs, and other challenges (especially in the Himalayan regions of Kashmir and Ladakh). Such activities—involving powerful experiences and risk-taking—"reflect a nostalgia for the romantic and 'authentic' past of the Israeli pioneers" (Noy and Cohen 2005, 7). The "conquerors" were supposed to move around more and "tick off" more sites than the "settlers" who, according to Maoz and their own characterization, were less closed-off, more open to meet the locals and to get to know Indian culture, especially its spiritual aspects (2004, 112–114). The latter used the term "conquerors" to refer to the former group in an ironic way, making fun of the fact that after all the inconveniences of the life of servicemen and women, instead of chilling out, they were willingly adopting an army-style discipline (Maoz 2004, 111). The "settlers," in turn, were to look for their new "spiritual center" (Cohen 1979) in India, since they were tired of living in a very demanding society with very strict rules (Maoz 2004, 117; more on the topic in Maoz 2005). In practice, how the scholars studying the phenomenon of the young Israeli backpacking pointed out, this is a dialectical situation: on the one hand, military patterns are transferred to traveling, on the other, people need to unwind the military regime (Noy and Cohen 2005, 12).[22]

The distinction made by Maoz does not map onto the everyday experience of the Israeli tourists living in and around Hampi and Dharamshala, who seem to fall into both categories at the same time.[23] Most of them arrived in India in their twenties after they had finished their compulsory—both for women

[22] Noy and Cohen (2005, 23–24) suggest that this may be related to the nature of the military service. Those who have served in combat units will be more inclined to look for fun and relaxation, while those who performed other backstage tasks treat the journey as a kind of compensatory test.

[23] See Uriely et al. (2002) on complex motivations of Israeli backpackers and different ways of experiencing depending on the chosen direction. Although India is a very popular destination for Israeli backpackers, it is not the only one. These tourists can be met also in South American countries (especially in places known for trekking opportunities), in other countries in South Asia (mainly in Nepal) and East Asia (mainly in Thailand), as well as in the African continent. In addition, Australia, New Zealand, and the United States are growing more popular, and young Israelis there often combine tourism with paid work enabling further travel (Noy and Cohen 2005, 21, 27).

and men—military service lasting several years.[24] Many researchers of Israeli backpacking see it as a rite of passage (van Gennep 1960) between the military service (that ended the period of youth) and entering adulthood (going to university, finding a job, starting a family), and thus reintegrating with the Israeli society (Noy and Cohen 2005, 1, 3, 255–257; Haviv 2005). India—a mandatory point on the route of such trips since the very beginning of Israeli backpacking history (Noy and Cohen 2005, 20)—seems to be an ideal choice. The cost of living is relatively low here, the climate is friendly (seasonality makes it possible to avoid winter throughout the year), the number of terrorist attacks is low, cannabis smoking is not fiercely pursued, and anti-Semitism is absent. Thanks to this, the holidays of the Israelis last for months, and some of them become "serial backpackers," returning to the same places year after year. But what is the most characteristic for this category of tourists is that they are engaged in grassroots enclavic tourism based on longer stays at places with infrastructure corresponding to their needs (Noy and Cohen 2005, 14–18). They prefer to live in the same guesthouses (where names and information boards are frequently only provided in Hebrew, which automatically eliminates most of the potential non-Israeli customers) and dine in restaurants offering the so-called Israeli food from the Hebrew menu. They rely exclusively on their own social networks, only rarely venturing out of them, passing on information about recommended places and services in a closed-circuit (Noy 2005).

I call this enclaves grassroots because, in the classic sense, enclavic tourism is controlled from the top. Cohen, writing about "tourist bubbles" (1972)—in reference to the concept of "environmental bubbles" by Daniel J. Boorstin (1962)—meant a conventional, institutionalized kind of tourism in which the bubble is created by tour operators, for example, in the form of resorts with all-inclusive offer. Such bubbles are physically separated spaces with limited access for people "from outside." Guarded by fences and guards, they are called "luxury prisons" offering all-inclusive exclusion (cf. Kaczmarek 2012 for Jamaica) and "leisure concentration camps" (Reid 1992, 75 about the Dominican Republic). Israeli enclaves do not have an institutional character—they keep a distance from others on their own (Cohen 2004, 48; Maoz 2004, 110), narrowing down their local service providers, who are

[24] There are also people who fled to India to avoid conscription, like my old friend who only saw his parents when they visited him in India. There were also "serial backpackers" who started out on a journey after their military service, and then embarked on subsequent expeditions (Uriely et al. 2002, 532–533). Either way, the vast majority of Israeli backpackers were both men and women, middle-class, of European descent, mainly from big cities and kibbutzim, mostly nonreligious, although this is also slowly changing (Noy and Cohen 2005, 9–11, 23).

able to tailor their offer to the needs and tastes of this specific group.[25] This can be interpreted as a form of a "conquest" or "colonization"—it is the Other who is expected to adapt here (Maoz 2004, 111–112). But closing oneself in enclaves can also be read as nostalgia for the sense of community that was supposed to mark the beginnings of Jewish settlement in Israel in the form of kibbutzim (Noy and Cohen 2005, 257).

The Israeli backpackers' chosen tactics of separation make them create their enclaves usually on the sidelines. They did not stop at Hampi itself but stop on the other side of the river in the former agricultural village, Virupapur Gaddi, where extensive farms had been converted into secluded backpacking centers for long-term tourists. The same is true in Dharamshala, where Israeli tourists lived mainly in the upper parts of the villages of Bhagsunath and Dharamkot,[26] both above McLeod Ganj, about half an hour's walk away. Both Hampi and Dharamshala were "too touristic" for this group of tourists, that is, too crowded and too expensive. Of course, the creation of "environmental bubbles," as already mentioned in this chapter, was also the case for the "European" backpackers. However, the distinguishing feature of the Israelis is that their enclaves were always mono-national—some researchers believe that one of the reasons for this is their limited knowledge of English (Noy and Cohen 2005, 15). Meanwhile, the enclavic character of backpacking tourism as such is, on the one hand, multinational[27] and, on the other, "permeable," in the sense that it is supposed to provide an opportunity of meeting the Other in the zone of familiar comfort. Such an "enclave" is located somewhere between a tourist space and a locality (Wilson and Richards 2008). In the case of backpackers from Israel, the tightness of the enclave was highly visible. In many places popular with this group, such as Manali in Himachal Pradesh or Anjuna in Goa, the Chabad Hasidic movement opened prayerhouses, trying to limit at least a little the hedonistic lifestyle of young Israelis, or to take advantage of the need for spiritual quest some of them had. Sometimes such houses were created as grassroots initiatives—one of them operated

[25] It is also increasingly common for Israeli tourists themselves to turn into migrant entrepreneurs catering for their countrymen by setting up guesthouses or restaurants as joint ventures with Indian citizens. However, these are usually small businesses, a way of keeping the Indian holiday going rather than for substantial profit.

[26] The example of the Dharamkot can be used to trace the development of tourism in similar places "discovered" by backpackers in India. It is an ordinary village where Gaddi people still keep buffaloes and hens and grow corn on terraced plots, while some are trying to benefit from the presence of tourists by running makeshift "cafeterias"—covered with sheets of plastic and a tin roof supported by a few poles, right by the dusty road, but profiled for tourists, that is, decorated with Bob Marley prints, equipped with mattresses for sitting (favored by Israelis, for prolonged lounging), offering "real coffee" (while a few years earlier in India, especially in the Northern states where there is no tradition of drinking coffee, one could at best hope for instant coffee) and "vegan burgers."

[27] In India, this is manifested by the inclusive use of English in backpacker enclaves, while Israelis speak Hebrew, which is pretty effective in excluding undesirable others.

in Virupapur Gaddi at a holiday resort run by an Israeli person who settled there. There was another one in Bhagsunath. The Jewish holidays organized in these "cultural centers" attracted many Israeli tourists, not for religious reasons but because of homeliness and the sense of being together that they offered. In this sense, these venues were another factor that reinforced the functioning of the Israeli enclaves (Haviv 2005, 69–70).[28]

For all the above reasons, these tourists were not perceived by workers and small entrepreneurs in the informal tourism sector in Hampi and Dharamshala as being open to local residents. Although some of them attended meditation courses in Hindu or Buddhist centers, they did so in an "experimental" rather than an "existential" mode (Cohen 1979), and many resigned in the process, mainly because of the restrictions that prevailed in such centers (Uriely et al. 2002, 530–532). In practice, Israeli backpackers spent most of their time eating, sleeping, chatting with one another, and partying (Noy and Cohen 2005, 4). This form of relaxation was associated with the consumption of large quantities of cannabis, most often in the form of charas (homemade hashish) and ganja (marijuana), and sometimes even stronger psychedelic drugs (see Mevorach 1997, after Noy and Cohen 2005, 26).[29] This explains why Hampi, as a holy place with restrictions on the sale and consumption of meat and alcohol and night silence, did not attract young Israelis.

The shopkeepers and service providers with whom I worked during fieldwork do not hold the best opinion about Israeli tourists for several reasons. First of all, Israelis had a reputation for being stingy. However—as I have shown using the example of the "European" tourist category—this could be forgivable and explained away by the need to budget a very long holiday, were it not for the lack of respect that Israelis demonstrated in relation to India and its people, according to my interlocutors. The terms that were used to describe them were "loud," "rude," "not very much in tune with the feelings of others" (cf. Maoz 2004, 109–110 and Hottola 1999, 77, who described Israeli backpackers in the wider context of South Asia as "aggressive, impatient, self-confident and self-centered as a group"). To support their claim, my

---

[28] Of course, there were also Israelis who avoided the Israeli enclaves, cocreating a heterogeneous group of international backpackers, which made them less visible for researchers, the latter often originating from the same Israeli backpacker background (Noy and Cohen 2005, 32–33; see, for example, Maoz 2008).

[29] Israeli backpackers used to identify those among themselves for whom drug use was an integral part of the journey by the term *haManalim*—"those of Manali" (Maoz 2005, 179–181). Manali is a hill station in the state of Himachal Pradesh, which was very popular among Israelis at the time, known mainly for the widespread availability of cannabis. Later on, many Israelis moved to nearby Kasol, for the same reasons that pushed Israelis to move from McLeod Ganj to the surrounding villages. This enclavic group of Israeli backpackers resembles the post-hippie participants of the "trance Goa" scene in terms of their lifestyle in India (some of the Israelis were indeed part of that scene, see D'Andrea 2007, 198–199 and Saldanha 2007, 40–41, 114–115).

interlocutors cited, among other things, frequent cancelations of orders in restaurants under any excuse—the dish is then put in the trash and the customer does not pay for it (the catering industry workers stress that "Europeans do not have such a habit"). Lack of respect for "Indian culture" was expressed in inappropriate clothing (especially during visits to temples—women's bare shoulders, men's shorts, see-through clothes), refusal to participate in rituals (which is usually strongly encouraged by Indians), ostentatious display of feelings (public kissing, embracing) and, generally, lack of interest in India-related things, so valued in "European" tourists (see Mevory 1997, after Moaz 2004, 111). According to Noy and Cohen (2005, 254), this may result from the lack of intercultural competence in such young people, who are brought up in the spirit of distrust toward the Other. Maoz (2004, 116), as well, claims that young Israelis, strongly indoctrinated in the army, perceive Indians in the same terms as Palestinians. The Hebrew word *mekomiyim*, literally meaning "local," is used both in military jargon for non-Jewish people living in the occupied territories and in the backpacker slang for "Third World" residents (Noy and Cohen 2005, 13).[30]

In Hampi, an additional factor in favor of negative assessment of this category of tourists is the context of displacement. It is the photographs of kissing Israelis and their bathing in the holy river in skimpy clothes (evoking the legendary stories of hippie nudism and taken as evidence of sexual permissiveness, if not promiscuity), public consumption of alcohol or all-night full moon parties (organized mainly in some resorts on the other side of the river[31]) that became an excuse for Hindu religious leaders to paint an Occidentalist image of Hampi. They called tourism a source of moral contamination, with tourists

[30] Arun Saldanha, who had contact with Israeli backpackers during his research on the Goa trance music in the village of Anjuna in Northern Goa, indicates their frequent involvement in quarrels and fights with the locals and the fact that some guesthouses decided not to serve the Israelis due to their bad reputation (2007, 41, 137). Saldanha directly claims that the community of Goa trance music fans—including many Israelis—is characterized by racism against Indians and aversion to the "charter flight" tourists and other backpackers. Their sticking together—which Saldanha calls "viscosity"—results in "racial segregation," manifested in not dancing until domestic tourists leave the stage; not inviting them into the circle of *chilam* smokers; trying to take over certain beach spaces; ostentatious ignorance of vendors, beggars, and service providers (2007, 3–6, 77, 82, 124, 184, 190). Saldanha, himself an Indian educated and living in Belgium, experienced these practices firsthand (2007, 125).

[31] Such open-air events encompassed electronic trance music and, often, psychoactive substances (e.g., LSD or ecstasy). However, they were limited to neither Israeli tourists nor Indian locations (see, for example, *Thailand Backpackers: Full Moon Party*, dir. Richard Parry 1998), although the Goa scene played a major role in the dissemination of this practice (Saldanha 2007; D'Andrea 2007, 175–221). The village of Virupapur Gaddi, situated across the river from Hampi, became popular in this regard after the police and the courts' clamp down on the trance music scene in Anjuna in Goa in the middle of the first decade of the twenty-first century. It was then that some Israelis moved to Hampi (Saldanha 2007, 178). It seems, however, that Goa trance scene started leaking into Hampi as early as in the 1990s, as the village residents remember techno parties from that time.

presented as hedonists and those who cater for them as greedy businessmen benefiting from economic exploitation of the holy place. These charges made up a file, which—submitted to the state court—made a major contribution to the "spatial cleansing" of Hampi (see Bloch 2017a).

The Tibetans have a slightly more ambivalent attitude toward Israelis. On the one hand, they see the Jewish diaspora as the living proof of the possibility of cultural survival after dispersal and the successful recovery of the homeland; Israelis thus personify the hope for a positive solution to the Tibetan cause. One of my old friends, a freedom activist and at one time a tenant of a tourist café operating at one of the NGOs in Dharamshala, once confessed to me, "We used to say to ourselves: 'Let's be like the Israelis!' Because they stick together, even when they go abroad to live and work, they have special bank accounts, on which they collect money, which then they transfer in support of their government. We should do like them." On the other hand, this idealized, martyrologic image does not fit the observed reality of socially unacceptable behavior. In the past, elder Tibetans feared that such "guests" would have bad influence on the Tibetan youth. Over the years, the venue for joint events held to the sounds of trance music and cannabis smoke was an open-air place called Shiva Café, located by a waterfall above the village of Bhagsunath. During the day, young Tibetans, dressed like their Israeli peers, would sit with guitars along Bhagsunath Road in McLeod Ganj. This example also shows that Israeli backpackers fit the Occidentalist optics much better than those for whom it was created, that is, the descendants of European colonizers. Tibetans reduced this cognitive dissonance, explaining the behavior of young Israelis by "mental problems," which according to them were the consequence of the trauma after the experience of violence, that is, service in the army in a country in permanent armed conflict. They were therefore referred to with a kind of compassion and tolerance.

Above all, however, although apparently ubiquitous and very visible, Israeli tourists rarely went beyond their own enclaves, hence most of my research partners working in the tourism sector did not know them as clients. According to Phurbu, the owner of a seasonal restaurant in Hampi and Kasol (the first one recommended by *Lonely Planet*), "Other tourists, having read in the guide book that the restaurant is good, will go to it once to check it out; if the food is OK, they will be coming regularly. But not the Israelis, they do not follow the guide book at all." This made them challenging customers: one needed to have a product that would arouse their interest ("European collections" of clothes did not qualify here), find an access channel, and gain their trust. Maoz claims that for Israeli tourists—especially of the "conqueror" type—"the locals are there for decor and service only and they are mostly used as porters and errant servants" (2004, 112). From the studies collected by Noy and Cohen it also appears that for Israeli backpackers, local people are a background, mere shadows on a stage dominated by an inbred Israeli group (2005, 32).

However, a few of my research partners managed to reach the Israelis with their offer and they tended to have a slightly different opinion about them. According to Dolkar, selling her goods from a stall in Dharamshala, "If you make friends with them, they are very good customers, because they bring other people along." A Kashmiri shopkeeper told me "in secret" that "once you've made friends with them, you have a guaranteed business because they will recommend you on some of their sites, and you won't even know you're there, but all their friends will come to you" (it would not bring much but regular income). A young Lambani, who runs the only T-shirt shop in Hampi, appreciated this group of customers very much and boasted that he gained their trust: "They like my designs, they think they're funny, take pictures of them.[32] I give them good discounts and then they bring others, they come in groups." Jewelry sellers from Rajasthan working in Hampi, including Udai and Mohan, who moved to Bhagsunath and Dharamkot for the summer season, had good relations with Israelis, who turned out to be the main participants of their jewelry-making workshops. Long-term stays in one place and the search for alternative sources of income made this group of tourists quite keen to acquire practical skills and fill up their spare time. In addition, Rajasthani vendors often assisted Israelis in buying more substantial amounts of jewelry, which they planned to sell at home or in Europe. As a result, information and recommendation letters written in Hebrew were ubiquitous in Rajasthani shops, and the shopkeepers themselves were able to conduct simple business talk in this language. Their relationships with regular customers from Israel seemed to be long-lasting and close—these tourists return to the shops they know for years, sit there for hours, often on the shopkeeper's seat, look through the goods on their own, clearly feeling at home. Nevertheless, my observations and conversations indicated that these relations involved not so much openness of tourists to the local inhabitants and their lives but a complete adaptation of the latter to the needs of the former.

Russians: New Challenges

In Hampi, Russians tended to be as unpopular as the Israelis, and my research partners definitely did not include them among the inhabitants of Europe. For

[32] Top-selling patterns were two elder indigenous women (Australian Aborigines) smoking cannabis; the inscription "Don't worry, be Hampi" (a paraphrase of Bobby McFerrin's tune *Don't worry, be happy*); a modern sketch of Ganesha, a Hindu deity with the head of an elephant; and varieties of the Sanskrit syllable "Om."

the workers and small entrepreneurs of the tourism sector in Hampi, contact with a Russian tourist was a new, big challenge, widely discussed in daily conversations. Over the past few years Russian citizens have taken over the northern part of Goa. Its hippie image, which in the early 1990s transformed into a trance culture with "full moon parties,"[33] was already partially undermined by British, Scandinavian, and German pensioners, and finally erased by the Indian police (cf. Wilson 1997a; Routledge 2001; D'Andrea 2007, 175–221; Saldanha 2007; Gupta 2014). The Russian massive presence was recorded on Goa beaches at the end of the first decade of the twenty-first century, when Russian travel agencies started to send charter flights to the tiny international airport of Dabolim, thus filling the gap left by the crisis-ridden tourists from Western Europe.[34] Unlike Israelis who tend to flock together on their own, package tourists from Russia are "looked after" by Russian travel agencies with their branches in Goa, Russian tour leaders and guides, and even Russian drivers. Even if they wanted to, it would be difficult for these tourists to escape from this bubble.

It was from Goa, more or less since 2011, that the Russians began arriving in Hampi as part of one- or two-day organized tours. As with the Israelis, my research partners had quite limited contact with these tourists, although for other reasons. Visits of Russian groups to Hampi were short, and therefore their schedule was very tight. This meant that there was no time for the ritual of bargaining and chitchat, and the groups spent nights in Hospet or large hotels outside the village. However, it seemed that the main obstacle to establishing

---

33 It was the electronic music played during these events that gained its own genre name "Goa trance" and spread around the world (D'Andrea 2007, 176; Saldanha 2007, 39–43). Anthony D'Andrea (2007, 54, 176–177) calls the participants of these events "post-hippie, post-punk 'freaks.'"

34 The Russian presence in Goa's tourist spaces is widely commented in the local and national press, but there is a lack of research studies on this group of tourists in India (they are only mentioned in Gupta 2014, 99–101). It is known that the first organized groups of tourists from Russia started to appear in Goa at the beginning of the twenty-first century, and already in the 2009–2010 Russian charter flights landed at the international airport in this smallest state of India. The state government made a strong contribution to this by promoting the region during the Moscow tourism fair, facilitating access to visas ("3,500 Russians to Descend on Goa's Shores Every Ten Days." *The Times of India*, October 6, 2010. http://timesofindia.indiatimes.com/city/goa/3500-Russians-to-descend-on-Goas-shores-every-ten-days/articleshow/6716963.cms?referral=PM). In 2013 and 2014, when the research for this book was conducted, the Russians were the fifth largest national group of tourists in India, and in Goa they accounted for more than half of all foreign tourists, with 250,000 visiting in 2013. This number decreased significantly in the following year due to a decrease in the value of the Russian ruble following the sanctions imposed on Russia by the European Union and the United States following the occupation of Crimea, as well as the growing prices in Goa and the removal of India from the list of safe countries. However, in 2016 the number of Russian visitors increased again due to the terrorist threat of other destinations popular among Russians, that is, Egypt and Turkey (Banerjee 2015; "Russian Tourists Attracted to India after Fall in Ruble Value." *Russia Beyond*, January 2016. https://in.rbth.com/economics/business/2016/01/26/russian-tourists-attracted-to-india-after-fall-in-ruble-value_562505). What attracts Russian tourists to Goa is the tourist season coinciding with winter in Russia, relatively low prices, and short flight distance.

relationships with this group of tourists was the lack of common language. Russian visitors to Goa usually hardly spoke English (with the exception of Russian backpackers, who, however, relatively rarely visited Hampi), and tourism workers from Hampi—with a very few exceptions—did not yet know Russian (unlike their colleagues in Northern Goa, where waiters—especially in beachside restaurants and bars—already spoke Russian and offered Russian-language menus; also the information boards in Goan shops were in Russian; and even some restaurant names were spelled in Cyrillic). The owner of a T-shirt shop, who got along "even" with the Israelis, said, "The Russians don't respect us." When I asked what he meant, he explained, "When I tell them about Buddha [when asked about a T-shirt with the motif of the Buddha Shakyamuni meditating under a Bodhi tree], the French or Spanish listen carefully, while the Russians just leave!"[35] However, they probably left because they did not understand what the seller was saying to them—they just wanted him to show them the price on a calculator or to write it down. It looks that the lack of communication was interpreted by the tourism service providers as lack of respect. As a result, contact between the parties was mediated by Russian tour leaders or Indian guides from Goa, which was generally not conducive to a friendly atmosphere, since both these categories of intermediaries usually worked on commission, which was quite unusual on the low-budget market in Hampi.[36] In addition, Russian tourists were perceived as loud and demanding, which was interpreted as aggressive attitude (the term "angry bulls" was often used to refer to Russians). Finally, there was the lack of intercultural competence interpreted as lack of respect for local culture. The owner of a family-run guesthouse in Hampi explained the "wrong" behavior of Russians in this way: "They only know Goa, where they don't have to dress properly. Sometimes they are not dressed at all, so it happens that the police have to stop them in front of the gate of the temple because they try to enter without any shirt on. They don't travel, they come here only because Goa is close to Hampi."[37]

As with the Israelis, Russian tourists were seen as reluctant to accept the Indian way of doing things and looking for recreating the familiar. They liked meat and alcohol and did not care about the ritual impurity of these (Phurbu laughed that for breakfast they added vodka—which they carried along—to

[35] Note the perception of Buddhism as an integral part of "Indian culture" expressed by a man who is a Hindu from the Lambani group.

[36] Russian tour leaders—often women—did not have a good reputation in Hampi either. According to my interlocutors, as soon as they learned "what was what" after two or three visits, they tended to economize on guides by offering tours themselves or taking one guide for three groups.

[37] There is one more interesting information that emerges from this statement, namely that Goa is not India. It is all the more interesting as conservative circles in Goa, invoking Catholic morality, also protested against female nudity on the Goa beaches (Saldanha 2007, 115–117).

black coffee and tea;[38] I myself witnessed how the vegetarian breakfast served in the restaurant was supplemented with sausages brought by tourists in a jar or dried sausage). In other words, Russian tourists in Hampi appeared to be a classic example of tourists in a "recreational mode" (Cohen 1979, 183–185; cf. D'Andrea 2007, 214). Their journey was not about self-fulfillment or the search for transformative experiences, it was not an identity project for which the authenticity of the experience as understood by MacCannell (1976) would be crucial. What they needed was a carnival regeneration of physical and mental powers, mainly through bodily pleasures, so that they could later return to their own society and their assigned roles. In this respect, they were perfect target of the institutionalized tourism industry, as they expected comprehensive service. As such, they were mostly lost as clients for the informal tourism sector, to which they had limited access.

At the same time, however, they were a temptation for small entrepreneurs of this sector. Between them and the Israelis, in the opinion of my research partners, there was one fundamental difference: they did have money. They had it because they most often belonged to the new Russian middle class, and also because their holidays were not long—often limited to a week, ten days, or maximum two-week stays. Consequently, they spent a lot in a short time. But they also haggled mercilessly (most often with a calculator in hand and/or through gestures, starting from 10 percent of the initial price; according to my interlocutors, they were trained to do so by Russian tour leaders in Goa) and only bought specific products (which did not include the Rajasthani "European collection," but rather gold jewelry or statues). Above all, however, Russian tourists did not happen to visit most of the shopping and service points in Hampi. First of all, they shopped mainly in Goa, where they had a lot of time between sunbathing and eating in restaurants. Secondly, as a rule, they were not even admitted to the main shopping area in Hampi by their tour leaders and guides, who led them along circular routes and back alleys, so that they "accidentally" ended up in specific shops on their way to an agreed restaurant. Still, there were stories circulating in Hampi which sparked the imagination of shopkeepers—about drunken Russians coming to the so-called night bazaars[39] in Goa and buying large quantities of goods (it was presumed that they were intended for trade in their home country[40]).

[38] The popularity of Goa among recreational tourists is also due to widespread availability and low prices of alcohol—in India individual states have different restrictions regarding the alcohol sale.

[39] Night markets—cyclical bazaars with goods for tourists taking place in the late afternoon and early night hours.

[40] Although the massive presence of organized Russian tourist groups is relatively new in India, their presence as businessmen—mainly involved in importing Indian goods—has been noticeable since Soviet times, especially in the north of the country. Traces of this presence are visible, for example, in the knowledge of Russian among Indian producers and wholesalers (see Suryanarayan 2017 writing about one of the bazaars in New Delhi, where the language of everyday communication is

Moreover, restaurant owners made good money on them, although the list of complaints was long: that they wasted a lot of food ("they order three juices per person, take one sip and that's it"; "they have no respect for food"), that the food must be exactly in line with their order (e.g., juice without water, although restaurant owners ensured that the drinks were made with filtered water "and it's impossible to do otherwise with papaya"), that they destroyed things ("they smash everything when they are drunk and say 'I'll pay,' but that's a lame excuse").

If this group's presence in Hampi increased, it would surely change the local tourism market. The asking prices, which so far were not particularly high in Hampi, would likely go up in order to ensure profits from the new haggling style. In addition, Russian tour leaders started to introduce to Hampi practices typical for mass tourism which had been previously absent, like bringing customers only where the tour leader was paid a commission or offered discounts or free food. Several of my research partners started to consider learning Russian.[41] In the only big restaurant that already had a deal with Russian tour leaders, one of the waiters already spoke some Russian. The young boys, whose job was to bring in customers, greeted these new tourists with a cheerful "*Paydyiom rebyata*!" (Russian for "let's go") and women selling fruit encouraged "*Pasmatree*" (i.e., "have a look"). There was already a Russian-language guidebook to Hampi published, which was the result of a business idea of a Russian tour leader working in Goa, who persuaded the local bookshop owner to join this publishing joint venture. Not surprisingly, the bookshop—located, of course, on the route frequented by Russian tourists—soon mainly sold the statues (bought back from one of the expelled Kashmiri shopkeepers). A Tibetan friend who worked seasonally in Goa as a masseur, once admitted, "Sometimes I feel guilty for not speaking Russian. It's crazy, but they really expect people in India to speak to them in Russian!"

The above expectations and perceived offensiveness sometimes provoked resistance, directed not so much against this group of tourists as against attempts by its leaders to take over the local market. In the winter of 2014, a strike of local taxi drivers "against the Russians and the police" broke out in Northern Goa, paralyzing the entire region for three days. The protest concerned Russians taking over transport services—they rented cars and then drove Russian groups around, depriving Indian taxi drivers of their jobs (the police turned a blind eye to this illegal practice because it was to receive

---

still a pidgin based on Russian). So, perhaps the former business custom of imposing a framework of contact has become a pattern in tourism.

41 Some wanted me to teach them basic expressions and vocabulary.

a share).[42] Phurbu, in his commentary on this event stated quite seriously, "Soon Russian tourists will be banned from entering India."[43] In addition, the local press regularly reported news of the Russian mafia on the Goan real estate market, money laundering, gambling, drug trafficking and prostitution (Datta 2007; Barretto 2010),[44] and the offensive presence of Russian tourists in Goa started to be described by the term "Russian invasion" (Gupta 2014, 100). Such reports systematically reinforced the negative image of Russians, even though some of the Russian migrant tourism businesses were run just like any other businesses set up by tourists from the Global North countries who decided to settle in India, and many Russians traveled around India as backpackers, often as spiritual practitioners (see below for the situation in McLeod Ganj). This is how Russian tourists from Goa have become synonymous with mass package tourism and all its negative impact described in the introduction to this book.

In Dharamshala, the Russian language was increasingly common. The presence of Russian-speaking tourists was noticed to such an extent that a pharmacist friend, who decided to fight the phenomenon of begging in the town, made an English leaflet for tourists, which he then translated using an online translation tool and printed out in Russian as well. However, the approach to Russian-speaking tourists in Dharamshala was fundamentally different from that in Hampi or Goa, simply because they were different tourists, mainly Buddhists from the Siberian republics of the Russian Federation (such as Tuva and Buryatia[45]) as well as "seekers of spirituality" from the European part of Russia, that is, adepts of Tibetan Buddhism. The owner of a bookstore in McLeod Ganj, specializing in books related to Buddhism, was more and more often asked about relevant publications in Russian language. The growing presence of Russian-speaking tourists was also reflected in the

42 Indian taxi drivers, on the other hand, are accused by tourists of price-fixing.

43 This finally became the bone of contention between Russian tourists and residents of the places they visited—there were attacks on tourists, who entered temples half-naked or touched sacred objects. A few of these high-profile cases in the 2014–2015 season provided the basis for Russian authorities' decision to remove India from the list of safe destinations ("Blow to Goa as Russia Strikes India Off Its Safe Travel Destinations List." *The Times of India*, November 29, 2015. http://timesofindia.indiatimes.com/india/Blow-to-Goa-as-Russia-strikes-India-off-its-safe-travel-destinations-list/articleshow/49965574.cms). In practice, however, according to commentators, this was a result of deteriorating mutual relations after the change of government in India in 2014, and the attacks on Russian citizens were not so much the result of cultural misunderstandings, but a reaction to long-term attempts by Russians to take over part of the tourism services market in Goa ("Are Russian Tourists Losing Their Taste for India? Dwindling Numbers Suggest So." *First Spot*, June 1, 2016. http://www.firstpost.com/india/are-russian-tourists-losing-their-taste-for-india-dwindling-numbers-suggest-so-2810052.html).

44 Tourists from countries of Western Europe in resorts in Turkey and Egypt shared the same image of Russians as "dangerous savages" (Ageenko et al. 2013).

45 And from Kalmykia, located in the European part of the Federation, where people traditionally followed Tibetan Buddhism.

offers of various forms of self-development and therapy. In Dharamshala, one could come across notices in Russian advertising classes of "honey massage—a traditional Siberian massage therapy." In nearby Dharamkot there were daily yoga classes taught in Russian.

Less Visible Ones

Less noticeable categories of tourists in Hampi and Dharamshala included Japanese and South Korean visitors. Like the Israelis, tourists from these countries tended to stick together and stay out of the way, spontaneously creating their own national "environmental bubbles," even though they often traveled solo (cf. Prideaux and Shiga 2007, 50). The difference, however, was that Japanese and Korean tourists were not perceived as offensive, and their withdrawal—spending a large part of their time in guesthouses, having in-group interactions, long lonely trips with a camera, avoiding parties, and so on—often made them invisible to tourism sector workers and thus they rarely appeared in their conversations.[46] They were mainly spoken about by those small entrepreneurs who specialized in services tailored for such tourists. Bhavani, the owner of two rooms for rent in Hampi, whose brother runs a small restaurant with Korean food and referred his customers to her for accommodation, complained that these tourists were reserved, rarely stroke a conversation, and—she made a gesture of putting her palms over her eyes—were not interested in reality around them. When I suggested that maybe the reason was their insufficient knowledge of English (Prideaux and Shiga 2007, 48), Bhavani retorted that "even without a language you can get along, but they don't want to, because they don't trust anyone." Both Japanese

[46] The same applies to the studies focusing on these groups of tourists in India. While Japanese conventional tourists who go on package holidays are interesting for researchers (e.g., Chon and Inagaki 2000), the ones that opt for less organized travel are still rarely the focus of researchers' scrutiny. "Asian backpackers"—initially mainly Japanese—as a new "phenomenon" challenging the "Anglo-American backpacking tourism monopoly," have been broadly researched for several years now (see, for example, Hampton 2010, 18). There are single reports on their presence mainly in Thailand, Australia, and Europe. Peggy Teo and Sandra Leong (2006), researchers from Singapore, adopted the postcolonial approach based on decentralization of the "Western" perspective on backpacking while conducting a short-term research among "Asian backpackers" in the Khao San district of Bangkok. Hamzah Muzaini (2006), a Malaysian researcher from Singapore, did an autoethnography in Thailand, Cambodia, Laos, and Myanmar. In both cases, the authors focused on the racial aspect in the context of the absence of a visible difference between the local population and tourists (Muzaini calls it "looking local"—which sometimes is not beneficial for tourists, as they become subject of discrimination in the same way as local residents do, for example, Asian female backpackers are mistaken for Thai prostitutes). Tourism-led migration of these groups to Australia and New Zealand was examined by Cooper 2002, and for studies of Japanese backpackers in Australia see Prideaux and Shiga 2007 (the last study was based exclusively on quantitative methods). Japanese backpackers in Europe were described by Anderson, Prentice, and Watanabe 2000. I have been unable to locate any studies focusing on Japanese or Korean backpackers in India (although such studies may have been published in their languages).

and Koreans were considered very frugal, even obsessed that they are being cheated. In addition, they were supposed to stick to their habits, especially in terms of food, so they rarely immersed themselves in the local culture, in contrast to "Europeans" who frequently did so, and which local people appreciated (one of the shopkeepers in Hampi laughed that "they won't even have a decent [i.e., Indian] breakfast, only two or three bananas").

Again, however, as with the Russians, this group of tourists in Dharamshala was perceived differently than in Hampi, since they were mostly Buddhists (apart from Japan and South Korea, also originating from Taiwan). They opted for volunteer tourism, were students of Tibetan spiritual masters and often sponsored monasteries. Their presence in the tourist space of Dharamshala was reflected in a public audience with the Dalai Lama in June 2014. The audience organizers decided to offer tourists visiting Dharamshala the possibility to take a joint photo with His Holiness. However, since over a thousand tourists from McLeod Ganj and surrounding villages filled the square in front of the main temple to the brim, it was necessary to divide them into smaller groups. In the proposed division "Asia," represented mainly by Koreans, Japanese, and Taiwanese, was a separate category next to "Russia," "Europe" (by far the most numerous), "Israel," "America" (which I elaborate below), and "India," to which I devote the last part of this chapter.

However, both in Dharamshala and Hampi, backpackers from Japan and South Korea tended to stick together. They took accommodation in the same guesthouses, often run by their fellow nationals who had settled in India, dined in restaurants with Japanese/Korean food, and attended courses in their own languages. They used Japanese- and Korean-speaking online forums and blogs to share reviews of places and service providers. At the same time, however, there were also tourists who, like Tamiko whose story I recalled in chapter 1, often traveled for months "around the world" as part of the gap year,[47] combining tourism with casual work (see Cooper 2002). Moreover, contrary to their enclavic tendencies, they often settled down in India, forming relationships with workers or entrepreneurs in the tourism sector and trying to make a living in the sector themselves—like Tamiko—often by skillfully designing offers addressed to their fellow countrymen. They developed migrant entrepreneurship, stimulated by tourism and guided by the choice of a lifestyle, rejecting the draconian capitalist regime and resulting from it a sense of spiritual emptiness and social alienation (Kajiwara 1997), that is, all those ills which, according to researchers, afflict the inhabitants of

---

47 Young Japanese were encouraged to do so by their own government, as part of a program promoting long-distance travel to improve language competences that was initiated in the early 1990s (Anderson et al. 2000).

postmodern societies of the so-called Global North who decide to settle in countries perceived as not yet contaminated by modernity.[48]

Other less visible categories of tourists in Dharamshala included Americans from the United States, Canadians, and Australians, which did not mean at all that there were few of them but that they were usually included into a category more capacious than "Europeans," that is, *Inji*. *Inji* is the term used by Tibetan refugees to refer to representatives of the widely understood "West." It is probably a phonetic record of the word "English," as it was the British who were the first whites to be encountered by Tibetans back in Tibet—on the occasion of the armed invasion in the early twentieth century. Tourists from North America and Australia were less likely to come to the south of India, to the extent that they were not even perceived by workers and small entrepreneurs in the tourism sector from Hampi as a separate group with some specific characteristics. If a North American tourist appears in Hampi, he or she was simply treated as "an European."

Of course, social reality was far much complex. Take, for example, Natasha, an artist from Ukraine: while studying dance in Israel, she found out that there is a fantastic butoh school near Dharamshala; so she came to study Japanese dance theater in a Tibetan settlement in India. Another case in point is Suzie, a belly dance instructor from the UK, permanently commuting between Thailand, Turkey, and Israel: a friend in Thailand told her about a Tibetan woman who founded the first Tibetan dance studio in Dharamshala, where one could learn swing, salsa, and Bollywood-style dance; so she came and ran a course for Tibetan, Indian, and *Inji* women from all over the world. There are plenty of such stories, but in this chapter I wished to focus on the way in which tourists and their expectations are imagined by the workers and small entrepreneurs in the informal tourism sector, as it has a direct impact on the tactics adopted toward visitors, which is the topic of chapter 3.

(UN)OBVIOUS ECONOMIC DISPARITIES

In the introduction, I have demonstrated that tourism into postcolonial countries is often interpreted both as an effect of global economic inequalities and as a practice that reproduces these inequalities (mainly due to the phenomenon of leakage). The disparities in economic development on a global scale

---

[48] In Hampi, a Japanese woman, who married the auto-rickshaw driver, opened a guesthouse consisting of two rooms. Another Japanese woman came there with her three daughters following her husband, an itinerant Indian salesman, with whom they run two shops. In Dharamshala, a Korean woman who married a Tibetan run a popular Korean restaurant (and also an NGO), and another woman from Korea who was in a relationship with a Tibetan man managed an Italian-Tibetan restaurant. I come back to these stories in the following chapters.

are, in turn, revealed as the legacy of colonialism. As Fanon wrote, Europe is literally a creation of the Third World, as "the well-being and the progress of Europe have been built up with the sweat and the dead bodies of Negroes, Arabs, Indians, and the yellow races" (1963, 95). Many postcolonial critics argue that the deep differences in economic development that exist today between the so-called Global South and Global North were initiated and perpetuated in the era of colonialism (Ferro 1997). Those who work in tourism in the countries of the Global South are aware of this imbalance. At the same time, however, activities within the informal sector and direct access to tourists offer them many opportunities to overcome these discrepancies. In the course of my research, I noticed, for example, that pointing at economic inequalities between India and the tourists' countries was a rhetorical tactic often used by workers and small entrepreneurs in the process of negotiating prices. This is illustrated by a scene from my mobile ethnography field notes:

> Ewelina [a tramping tourist] is trying to bargain on my behalf and reduce the price of trousers I wish to buy from 350 rupees to 200. The "final deal" is 250. Ewelina proposes a "compromise," i.e., 220. Sangeeta, a young Lambani saleswoman, points out to her: "You're arguing over two bottles of water" (suggesting that we are wealthy as none of them drinks bottled water). Sangeeta's younger sister talks about the same 30 rupees: "This is one meal for us." (the message between the lines being: you just ate a fried marlin for 250 rupees!). I am convinced, embarrassed, and pay 250 without any more word.

A twelve-year-old girl from Hampi, a daughter of an auto-rickshaw driver, believed that tourists, although relatively wealthy, had no empathy for the poorer. To support that claim, she told me about a scene she observed with two elder tourists who bought something from a street stall and got ten rupees small change. When a beggar woman came up to them with a child in her arms, they were to say, "Go away, stupid woman!" The girl, shocked by this behavior, gave the woman five rupees, which her father gave her so that she could buy some sweets. At the same time, she saw nothing wrong with her father charging tourists more than local customers, stating the simple fact, "You have more money."

I must admit that I personally considered this logic to be legitimate and accepted double rates. Anyway, they were legally sanctioned—access to all Indian monuments from the UNESCO World Heritage list, including some of the sites in Hampi, was a dozen to several dozen times more expensive for foreign tourists.[49] Kristen Ghodsee, an anthropologist who conducted research

[49] As a rule, citizens of countries belonging to the South Asian Association for Regional Cooperation, namely Sri Lanka, Bangladesh, Nepal, Bhutan, Pakistan, Afghanistan, the Maldives (and

among employees of one of the Bulgarian resorts, also drew attention to the dual pricing system—for "own people" and tourists—as one of the ways of thinking about how to overcome global inequalities. According to Ghodsee, this post-socialist legacy explained in part how her research partners justified their own actions, which they would otherwise consider to be quite dishonest: taking the money left at the bar by drunk customers, minimal but regular overcharging in the restaurant bills, or inexact small change operation, because tourists were not proficient in handling unfamiliar banknotes and prices, or introducing informal "fees" and financial "penalties," for example, for a glass broken by a tourist (Ghodsee 2005, 72–75).

The small entrepreneurs in the informal tourism sector with whom I worked in Hampi and Dharamshala often complained about the myth shared by the tourists depicting India as an incredibly cheap country, which they actively tried to deconstruct. One of the saleswomen explained to a tourist who tried to buy some of her goods for as little as 100 rupees (just over 2 dollars), "These are not the days when everything was so cheap in India. Now the prices here are getting closer to those in your country. In the past, when I used to buy such a sheet for 50 rupees, I could sell it for 100. Now the wholesaler wants 150." The T-shirt seller from Hampi made his own comparative calculation: "I know you have money. For one euro you can have breakfast, you can have a cappuccino or a Coke. And what can we get for one rupee? Nothing." An employee of one of the NGOs in Dharamshala, pondering over whether 100 rupees for a ticket to a folk show was too much, said to himself, "It's only two dollars." The owner of a small restaurant in Hampi believed that people from the West led a "good, comfortable life" in India because they could afford to do nothing, just living off the difference in the purchasing power of money: "How much rupee can you get for one pound?"—he asked rhetorically.

The conviction about the systemic and widespread economic divide between tourists from the so-called Global North and tourism workers in the so-called Global South is prevalent in the literature on the subject. Saldanha, quoted many times in this chapter, on the basis of research carried out in Goa at the turn of the twentieth and twenty-first century stated the following clearly:

> Crucial to contact zones like Anjuna [one of tourist destinations in Northern Goa, popular especially because of the trance scene] is the *economic* [emphasis in original] disparities that constitute them. The Anjunkars are more or less compelled to tolerate insolent behavior, while tourists realize they can get away with it—as the guests with money and freedom. . . . On the whole, especially

sometimes Thailand and Myanmar under other regional cooperation—BIMSTEC) are excluded from the category of foreign tourists.

> since the early nineties, white travelers in India are well fed, financially secure, and can fly home at any time. Even without any education, any Indian from the age of three knows this. It is apparent in all financial transactions between freaks/backpackers and Indians. . . . Being white in public in the third world demands almost uninterrupted engagement with one's body—tourists are "walking dollar signs," perceivable from afar. (Saldanha 2007, 137, 168, 169)

It is possible that Saldanha's decisive and generalizing interpretations—just like the alleged racism of Goa trance music fans—stemmed from his own positioning as an Indian educated in the country of the so-called Global North. It is also possible that these inequalities were more obvious before the crisis that began in 2008. However, my research has revealed that the binary division between rich tourists and poor service providers in the informal tourism sector in India is not as obvious as it seems.

During my fieldwork and hours spent in shops, stalls, or restaurants, I witnessed seemingly paradoxical situations in which a tourist seemed to be the poorer party in relation to a small entrepreneur. This reversal of global positions can be explained in two ways. Firstly, the perception of haggling as an integral part of "Indian culture" results in tourists coming to India being told that this practice is the norm and skipping it will amount to lack of intercultural competence. Each tourist guidebook features instructions on how to haggle (e.g., divide by three or start at half the proposed price),[50] and many foreign tour leaders also offer such advice. Moreover, tourists, especially at the beginning of their stay in India or if they come for a short time, are completely unaware of the prices and afraid that they will be cheated. The following excerpt from the mobile ethnography field notes showcases such dilemmas:

> Tourists are serious about haggling. Marek assists his pal Roman in purchasing postcards in Hampi. After the transaction, Marek states: "I wouldn't give that much." Roman tries to save his face: "It's not for me," and Marek retorts: "He figured out that you don't speak the language [that's why]." Marek himself haggles about every single rupee, even when he buys bananas, and constantly compares prices. Whenever the group discusses cheating, Marek repeats the slogan he associates with the Arab countries: "It is OK to cheat the infidels," which shows his conviction that everyone wants to rip him off. Magda, a tour leader, tries to challenge this harmful generalization with a self-reflexive joke: "Back at home, the 'fidels' cheat the 'fidels,' not sure what is worse."

---

[50] Lonely Planet's guide to India, in a special box entitled "The Art of Haggling," recommended, "If you have absolutely no idea of what something should really cost, start by slashing the price by half (even more in touristy spots)" (Singh et al. 2005, 1059).

The result may look like this: a tourist brings some clothes to a local tailor in Hampi asking for minor alterations and mending. The price is twenty rupees (less than half a dollar) per piece, but the tourist tries to negotiate a lower price of ten rupees. The tailor bursts out laughing derisively, but finally he waves his hand and says, "Ok." He is annoyed that tourists haggle for such small amounts, not even knowing what the real prices are. "What can you buy for 10 rupees?"—the tailor asks me after the client leaves.[51] The owner of a clothing shop, a young Lambani, made similar comments on the behavior of the tourist couple who visited his shop day after day to just look at the goods, try them on for more than half an hour, and then leave without buying anything, deciding that they need to think about it and that they would come back the next day. The shopkeeper was astonished to see them ponder so hard over a blouse for 250 rupees (less than $5). The Rajasthani painter, who runs a small painting studio in Hampi, stated, "Tourists care more about the discount than the price."

The obsession with not paying more than the locals, and generally spending as little money as possible, is an integral feature of backpacking, as many researchers have already noted. Hazel Tucker, who conducted research among backpackers in Turkey, pointed out that although they are usually not less well-off than other tourists, backpackers tend to hold on to the principle of "spending less," which is a kind of marker of their way of travel (2003a, 59–60). Not paying more than the local price—just like haggling—are also seen as expertise and competence in local reality (Saldanha 2007, 168). On the one hand, such attitudes can be interpreted as a manifestation of inability to acknowledge global economic inequalities—if so, it is more lucrative for a small business in the tourism sector to cater not for a backpacker seemingly well versed in local reality but for a tourist who is not so keen to express their "expertise" in local conditions, one who comes to India for a two-week trip and wants to have fun, spend the money one saved for this short holiday.[52] At the same time, especially in the aftermath of the economic crisis in Europe and the rise of prices in India, traveling for longer periods of time is actually much more costly and cutting down one's expenses then becomes necessary, which many workers and small entrepreneurs in the informal tourism sector—as I have already shown in this chapter—seemed to understand.

Moreover, the reversal of economic positions in the relationship between a tourist from the so-called Global North and a service provider from the

[51] Actually, not much: a small packet of chips, a few onions or one glass of chai (milk tea) from a street stall.

[52] This paradox has already been noticed by the circles of independent tourists, who have been increasingly vocal about backpackers' haggling habits being unethical.

Global South occurs in the case of those tourists who "have gone native"[53] and settled in India. These are the "nomads from affluence" (Cohen 1973), the contestants of the political and cultural domination of their part of the globe, disdainfully rejecting the tourist infrastructure and wishing to "become locals," whom, at the same time, they both exoticize and idealize. They expect the Other, who should not have consumer needs, to engage in non-commodified exchange, as illustrated by an excerpt from my field notes from Hampi:

> The tenant of a tiny massage parlor was approached by a young Finn wearing just *lungi*,[54] with a naked torso, and asked for a massage for 50 rupees [about 1 dollar], while the price for an hour-long massage is between 300 and 1000 rupees. He stated that "it is all I have," that his daily budget is 150 rupees [about 3 dollars], that he sleeps on the rooftop of one of the guest houses and eats food from street stalls. The masseur laughed and explained to the boy that "nothing is free, I need to pay the rent and massage oils cost money too," but in the end he offered the man a 15-minute massage, because he then did not have any other clients at this moment.[55]

Similarly, an elderly American woman, who had been living in India like a *sadhvi*[56] for fourteen years, was staying on the rooftop of a restaurant rented by a Nepalese man in Hampi. She claimed that she worked as a volunteer teaching English to monks somewhere in the northern states. It was unclear why she came to Hampi, she had no money at all. The owner of the restaurant let her sleep on mattresses used as seats during the day and agreed for her to eat free of charge the same dishes as the tourists whom she would bring to the restaurant.

In Dharamshala there were many people who seemed to "have gone native": they drank hot water instead of Coca-Cola (as recommended by Tibetan doctors), learned Tibetan from Tibetan friends, hanged out in monasteries, took Tibetan names, wore *chubas*[57] or monk robes. Erik Cohen called

---

[53] By this term I mean attempts to imitate local practices and customs, although often quite unsuccessful (Laxson 1991).

[54] *Luṅgī* (Tamil)—a men's outfit worn in the south of India instead of trousers. It consists of a strip of fabric wrapped around the hips, reaching to the ankles or rolled up to the knees. Nowadays, men wear it with a button-down collar shirt, or at least a T-shirt.

[55] One of the operators of a small, round, plaited boat used to react in the same way—in response to clients offering him embarrassingly low rates he did not refuse to provide the service, but instead of making a whole circle around the lake, he used to row into the lake a bit, turn the boat around a few times, and then bring the customers back to the shore.

[56] *Sādhvī* (Sanskr.)—itinerant ascetic in Hindu and Jain tradition (male form: *sadhu*, Sanskr. *sādhu*).

[57] *Phyu ba* (Tib.)—a kind of coat or dress worn by both Tibetan men and women, although in exile in India mainly in official situations or for festivals. The refugee version of *chuba* is—because of

them "experimenters"—trying different things, visiting other people's worlds in search of themselves, but still unable to convert (1979, 189). They could be met in an Indian ashram,[58] in an Israeli kibbutz, or in a hippie commune. Sometimes the "experimenters" overdid mysticism and/or hashish or other drugs—that the Tibetans called them "freaks" (see McGuckin 1997, 365), the same term that Goa trance fans used to call themselves (Saldanha 2007, 54), but also one used by the Indians to refer to white hippies staying long term in India before the term "backpacker" was introduced (Hampton 2010, 9). Those tourists brought an inconsistency in the generally positive image of *Inji* held by Tibetan refugees. I must add, however, that Tibetans were surprisingly tolerant and flexible when dealing with "freaks," whose odd behavior was usually explained—in line with Occidentalism—by confusion resulting from living in a world of alienating, unbridled consumerism.[59] At the same time, Tibetan refugees made a clear distinction between these "wretches" and those who have found a sense of life in Tibetan Buddhism. Cohen called the latter "existential" tourists, that is, those who, having found a new spiritual center away from home, abandoned the old ways that were assigned to them by birth (1979, 189–192). These people often led a double life, working in their country of origin (or another country of the Global North) and spending their holidays in a monastery or a Buddhist center, where they recharged their spiritual batteries for the whole following year. Some of them decided to move permanently to their "elective" center.

Recognizing the heterogeneity of the tourist category and the nonobvious economic inequalities, many small entrepreneurs in the informal tourism sector in Hampi and Dharamshala differentiated their offer according to the customers with whom they were dealing. They set different rates for the same services and goods, depending not only on their assessment of the tourist's economic status based on the country of origin but also on tastes and preferences in relation to specific offers and styles of traveling. For instance, the

---

the climate—significantly modified and standardized in relation to the varieties worn in different parts of Tibet. On a daily basis, most young Tibetans in McLeod Ganj wear jeans and T-shirts.

58 *Āsrama* (Sanskr.)—a meditation center, a place of retreat, spiritual teachings, and practice in Hindu tradition. Since the 1960s, many ashrams in India have also enabled spiritual development of the inhabitants of the so-called Global North, that is, "spiritual tourists" (Norman 2011) or "ashram tourists" (Sharpley and Sundaram 2005). Tibetan refugees also offer the *Inji* modern Buddhism, stripped of dogmas but shrouded in exotic mysticism; on why and how they do that, I elaborate in chapter 3. Siv Ellen Kraft (2007) in her analysis of the *Lonely Planet* guide showed that according to its authors experimenting with different spiritual traditions was an integral part of the journey through India.

59 Saldanha (2007, 81) cites opinions of Indian psychiatrists working in Goa, according to whom backpackers are particularly vulnerable to psychosis. As reasons, they point to long-term absence from home, being on a tight budget, often traveling alone, having close contact with poverty and bureaucracy, functioning outside the disciplinary regimes of family and work (for Israelis they also listed trauma related to military service), partying and using drugs.

masseur I have mentioned before charges "Europeans" 1000 rupees for an hour-long massage (even though the financial crisis undermined their financial standing, Ayurvedic massage, being part of the "Indian culture," is highly desired by this group of culturally sensitive tourists), Russians pay 500 rupees (not because they are poorer, but as a "special offer," because they rarely use his services due to the fast pace of their visits to Hampi), Japanese and Koreans are charged 300 rupees (because they are frugal and restrained), and others pay 700 to 800 rupees. However, the greatest dissonance in the perception of global economic inequalities as understood by workers and owners of small businesses in the informal tourism sector in Dharamshala and Hampi was caused by Indian tourists.

DOMESTIC TOURIST: OUR OWN OTHER

According to the owner of two large stores and a Tibetan handicraft center in McLeod Ganj, "Western tourists are finished." Since 2008, his clients were mainly Indians. He did not complain, because "they freely spend their money." A cloth salesman stated the same thing: "The Indians are now richer than foreigners. They still haggle a little, but it's out of habit." In Dharamshala, a "white tourist from the West"—affected by the crisis, with a limited budget, economizing—is no longer a "walking wallet." This role was taken over by representatives of the fast-growing urban Indian middle class, speaking English, dressed in "Western fashion," wearing sports shoes, men in jeans and polo shirts, and women in leggings and tops, only lightly stylized as Indian, and no longer in sari or *salvar kameez*.[60]

Studies of tourism often fail to see—due to the denial of the privilege of cosmopolitan mobility to the Others from the so-called Global South, already discussed in chapter 1—that those who are collectively perceived as tourist service providers or just elements of the exotic landscape can also be tourists. In both Hampi and Dharamshala, Indian tourists were an important group of visitors equipped with economic, social, and cultural capital. They provided an important context for the perception of global inequalities and categories of Otherness by workers and small entrepreneurs in the informal tourism sector. In the context of the Global South, domestic tourists introduced an intriguing dissonance in the Orientalist dichotomy between a mobile cosmopolitan "Westerner" and a sedentary, traditionalist "Easterner" (see Said 1979). The share of domestic tourism in direct income from "travel and tourism" sector

[60] The attire traditionally worn by women in India. Sari (*sāṛī*) is, in a nutshell, a long piece of cloth wrapped around the body as a dress, worn mainly in the south, while *salwar kameez* (*śalvār qamīz*), that is, trousers with tunic, is more popular in the north of the Indian subcontinent.

in India's GDP in 2016 was as high as 88 percent (compared with just 12% spent by foreign tourists), and the income generated by Indian tourists was still growing (WTTC 2017, 6, 12). The number of domestic tourists in 2015 was sixteen times bigger than the number of international tourists (Ministry of Tourism 2017, 86–87).

At the turn of the century, Shalini Singh noted that 94 percent of Indian domestic tourists were those traveling for religious reasons (2001, 139). The changing profile of tourists in Hampi and Dharamshala over the past few years indicates that this statement is no longer valid. A secular, domestic tourist from urban middle class has emerged as a new important actor within the Indian tourism market. Just as Singh predicted, there is an urgent need to study Indian middle-class travel habits (2001, 140). In 1989, Richter—the researcher of the tourism sector in India, already quoted in the introduction—wrote that domestic tourism had gone beyond religious pilgrimage, and that holiday travel was becoming increasingly important for the middle and upper classes (1989, 106).[61] Until recently, however, both the central and state authorities responsible for tourism policies neglected domestic tourists (Wilson 1997a, 71). Only a few years ago, domestic tourism had just begun to be treated on a par with international tourism. The Indian Ministry of Tourism noted that "the Indian economy is developing at a steady pace, the middle class is growing, and has growing surplus income that could be spent on recreation" (Ministry of Tourism 2017, 25; see also Bhargava 2017).[62]

In Hampi, small careers in tourism began with Indian pilgrims—the villagers catered for their needs long before foreign tourists arrived in the 1990s. Over time, villagers started to specialize and the services market was divided into those serving tourists and those targeting pilgrims. Although both of these types of travelers rub shoulders in the narrow streets of Hampi, sit on the stairs leading to the river and meet in the temple, their paths become

[61] At the same time, the number of Indian tourists traveling abroad is also growing rapidly—in 2015 it reached over twenty million (Ministry of Tourism 2017, 87). According to the World Tourism Organization, India was one of the fastest growing countries sending tourists around the world already a decade ago, so much so that UNWTO and the European Travel Commission commissioned a special study on the perception of Europe as travel destination among citizens of India, which focused on Indian tourists' behavior patterns—their choice of holiday setup, ways to spend money, activities, use of media and Internet. The research results also included a list of recommendations on the best ways to promote specific tourism branches on the Indian market (UNWTO 2009). The second most popular destination among the Indian middle-class tourists was the United States (Singh 2001, 141–142).

[62] The problem seems to be further-reaching and concerns non-Western tourists in countries of the Global South, ignored both by governments of these countries and by researchers of tourism (see Cohen and Cohen 2015; rare examples are Cohen and Neal 2012 or the already mentioned research on Asian backpackers). The Indian Ministry of Tourism is now also changing its approach, opening up to tourists from the countries of the Global South, as evidenced by, among others, cooperation agreements on tourism development signed in 2016 with the government of Qatar, South Africa, or Kyrgyzstan (Ministry of Tourism 2017, 66–67).

separate on the level of infrastructure. Indians sleep in the pilgrim houses run by the local Brahmins[63] who also provide them with spiritual services (celebrating pujas[64] for their deceased whose cremated ashes are thrown into the holy Tungabhadra river, which is also a place of ablutions, blessings, etc.), while foreign tourists stay in guesthouses. Pilgrims eat food served by Brahmins in makeshift canteens, while tourists are the clients of restaurants. Secular tourism workers in Hampi used to say ironically that Brahmins are involved in "karma business."[65] Only few of my research partners provided services to both these groups. One of the villagers run a guesthouse which only hosted pilgrims sent over to him by his Brahmin friend, whenever the Brahmin's own pilgrim accommodation was full. At the same time, he rented the rooftop of this guesthouse to a Nepalese who set up a restaurant for foreign tourists there. It should be noted that pilgrims were clients coveted by some people renting rooms in Hampi—hence the ironic comments about Brahmins who tended to monopolize them. Pilgrims were considered to be "easy to handle," as—in the opinion of my interlocutors—they did not need to be lured either by prices (as one of the local entrepreneurs explained, "Tourists, that is backpackers, go here and there, check, compare prices. A tourist will pay 250 rupees [over 5 dollars] for a room, for which an Indian will pay 400–500 rupees, even if it is a low season, otherwise a few people will sleep in a double room and you will earn enough anyway") or by esthetics (rooms for them did not have to be repainted every year, and the lack of attached bathroom or window view was not a problem[66]), or by marketing (such rooms did not have to be recommended by *Lonely Planet*).[67]

Indian pilgrims who came to Hampi exclusively for ritual purposes, in principle did not challenge the opposition of the religious "Easterner" versus the materialistic and hedonistic "Westerner." However, as I have already written in chapter 1, many Indian tourists combined religious travels with sightseeing and recreation—they came to Hampi because it was not only a sacred Hindu place but also a UNESCO World Heritage site, a source of national pride, located not far from Goa. These tourists chose accommodation

---

[63] The highest of the four Vedic social strata (*varna*).

[64] *Pūjā* (Sanskr.)—offering of worship ceremony.

[65] *Karman* (Sanskr.)—the sum of constructive and destructive acts committed through thought, word, and action, guided by will, from all previous incarnations, that determine our future incarnations. The rituals performed by Brahmins are to reduce the negative effects of destructive acts, that is, improve the karma.

[66] Actually, attached toilet might be a disadvantage, to the point of deterring clients from this category of tourists, since it involves a ritual impurity which should be kept out of the place where you live.

[67] On the other hand, some guesthouse owners in Hampi did not want to take in pilgrims precisely because of the ritual impurity—before the Brahmin sets up the date for the puja for the deceased family members, their ashes are stored in the rooms along with the pilgrims.

in hotels in the surrounding towns and used the same restaurants and shops that were frequented by foreign tourists.

However, a serious challenge to the perception of the above dichotomy in Hampi was posed by the presence of another group of Indian visitors—weekend tourists. These were mostly young, middle-class Indians, often having corporate jobs, mainly in the IT sector (less often students), from metropolises such as Mumbai, Bengaluru, Pune, or Hyderabad. Just like European and North American hippies in the past, and now Israeli long-term backpackers, these hardworking professionals, often doing long office hours, under great pressure, but very well-paid, discovered Hampi—a peaceful, beautiful location, only a few hours away from their workplaces. Their way of taking leisure was not particularly different from how the Israelis relaxed, except that they had less time. During the day they rested in secluded resorts across the river, smoking hashish or marijuana, and in the evenings they partied. Unlike the Israelis who needed to save money for their long-term holidays, these tourists had a lot of money and were willing to spend it. All this made them an even greater conceptual challenge for workers and small entrepreneurs in the informal tourism sector in Hampi—precisely because they were Indians. Indian young women, who dressed and behaved in line with the Occidentalist notion of "Western" women, generated very mixed feelings: they wore shorts (or at best tight jeans) and skimpy tops displaying fashionable tattoos, and, moreover, they often smoked cigarettes in public and (when lounging across the river) drunk beer and smoked charas. These practices, transferred from big city clubs to the Indian province, caused great cognitive dissonance among tourism workers who were only accustomed to seeing them but only among foreigners. It seemed that, contrary to what the Hindu leaders wanted to promote in their Occidentalist rhetoric, it was not the foreign tourists who were desecrating Hampi (Bloch 2017a), but the domestic tourists from the metropolitan middle class—the familiar Other—whom pious Indian pilgrims were to consider sinful.

The same kind of visitors appeared within the tourist space of Dharamshala, transforming the image of this hilly town within the last few years. Prior to 2011, there were no such tourists at all, while today Indians have become the dominant group, especially in the hot period preceding the monsoon, from April to June, and, just like in Hampi, on the weekends. *The Times of India*, in an article, "Weekend Rains Bring a Rush of Tourists," published at the beginning of June 2013, reported that in just one weekend 100,000 Indian tourists—mainly from Delhi, Punjab, Haryana, and Chandigarh—visited Dharamshala and the nearby towns of Kangra and Pathankot.[68] McLeod Ganj,

[68] "Weekend Rains Bring a Rush of Tourists." *The Times of India*, June 7–9, 2013: 4.

which is a small hill station, on Fridays and Saturdays began to resemble a busy city, with vibrant nightlife, crowds in the main square, and stunning noise of people and honking cars.

This tourist boom brought about changes in infrastructure. Like mushrooms after the rain, there emerged new streets with guesthouses and hotels, many of them advertised as "deluxe," featuring men in uniforms guarding the entrance. Businessmen from other states, like Punjab, already entered the local hotel market. Many of these investments were unprecedented in McLeod Ganj, built in violation of the seismic-damage prevention rules of the area—multistory, large hotels with standardized rooms. Unlike small Tibetan guesthouses offering a variety of rooms, the big hotels could provide accommodation for big groups of tourists. In addition to hotels, there were also Indian restaurants offering a completely vegetarian menu. However, what was most striking when one compares McLeod Ganj before and after 2011 was the appearance of fashionable clubs and discos. Previously, there were a few places where one could dance or drink beer on the weekends, run by both local Indians and Tibetans (e.g., the famous McLlo Restaurant Beer Bar, which has been operating on the main square since 1990, the Mount View Bar, or the former government-run Hotel Tibet, still preferred by the elder Tibetan elite, served by old-time waiters). The new joints, however, had different characteristics: very loud music played until late spilled out into public space, and the entrance was guarded by security guards, doing door selection, which looked quite bizarre in a small Himachali town, as in India this practice has so far been associated with several-star hotels in metropolises (the perfect example is a club called Black Magic, located in the heart of the old part of McLeod Ganj).

I had the opportunity to experience this change of tourist profile personally. On the first day after my arrival to the field, right on the main square of McLeod Ganj, I was chatted up by an Indian man of about fifty. He introduced himself as a real estate specialist from Mumbai, whose grandfather bought land in Dharamshala to build an ashram where everyone would be provided free of charge, "not what people do here, just business." The man said that since we had such a good talk, we should go somewhere to have tea. When I refused, he lost all interest in me. On the same day, on my way to the neighboring village of Bhagsunath, I was accosted by two young Indian men. In the evening, a man in his early forties offered to open a chakra in my body—I politely thanked him, saying that I was not interested. On the second day, in the afternoon, an Indian man of my age approached me, complaining that tourists did not want to talk to him, that they were "not nice." He claimed he was from Rajasthan where his family ran many restaurants and a large tourist resort and that he came here to buy land in order to open a catering establishment. After this introduction he suggested we should go have tea together. When I refused, explaining that I had an appointment, my new friend was clearly annoyed. On the same day, in the evening, another man of around forty-five asked me if I liked dogs, as he noticed that I was watching a monk playing with two puppies. After this introduction, he asked if I had attended Dalai Lama's teachings—he himself had. Then, inevitably, he suggested tea. There followed many more such episodes.

TV broadcasts of the national cricket games in the stadium in Lower Dharamshala depicted not only idyllic images of a hill station surrounded by greenery but also spots from McLeod Ganj showing scantily dressed blond female tourists. It is this image of exotic sexuality in close proximity that became a magnet for weekend male tourism, especially from the nearby state of Punjab. The images created the demand for discos and Black Magic–style clubs, where Indian men hoped to befriend foreign women whom they had seen in the cricket spots on TV. Many of them, affected by a patriarchal puritanism, were quite convinced that women from "the West" were sexually permissive (Saldanha 2007, 116). The origin of such conviction can be traced back to American pornographic cinema, but they are also caused by a misreading of cultural codes, such as revealing clothing, solo travels, or the lack of restraint in everyday contacts between men and women (e.g., women speaking to men). This made Indian men who came to Dharamshala quite direct in dealing with female tourists, which was interpreted by the latter as pushy and resulted in "European" tourists moving from McLeod Ganj to more secluded surrounding villages.[69]

---

[69] All Goa tourism researchers, whom I know, write about the phenomenon of "weekend bus tours" practiced by Indian men as an example of foreign tourism being an attraction for domestic tourism: "On Sundays, buses full of eager young men shout and whistle at any white female as they drive into Goa" (Saldanha 2007, 116). Gupta stated that domestic tourists go to Goa to "gaze upon Western-ness in all its excessive forms . . . and where they allow themselves to indulge in and experiment with ideals of romance, sexuality and alterity in an affordable manner, particularly if travel to the West is not a feasible destination option in the near future" (2014, 106, see also Gupta 2009). D'Andrea recalls that "during the New Year's Eve [peak season in Goa], main tourist resorts organized 'raves' that attracted thousands of Indian males. They had to wait in long lines, and pay a hefty entrance fee (Rs. 500 (US$10)), whereas foreign 'hippies' were immediately allowed in for free. This was an attempt to render such events more 'authentic'; after all, 'hippies' are a tourist attraction for Indian tourists" (2007, 196). The context of the beach, of course, fuels—as Saldanha calls it—"racialized sexism," through the persisting myth of hippie nudism and bikinis worn by Russian tourists (2007, 116). In this sense, Hampi and Dharamsala are less affected, although the fact that adverts for cricket matches show white female bodies suggests that this context is also relevant there. Researchers think that even without sexual overtones, Western tourists are an attraction for Indian tourists and that Western tourists in Goa are looked at not only by young Indian men but also by couples on honeymoon trips (Gupta 2014, 106). Based on my observations, this is less true in case of middle-class Indians who have repeatedly had the opportunity to be foreign tourists themselves, traveling in Europe or North America, and more true in case of young men from small towns and villages who—especially on the feast of Holi, celebrated by dancing in the streets and throwing colored powders at each other—come to Hampi to be able to experience the presence of foreign female bodies. I elaborate on this issue in chapter 3. However, it would be too much of a simplification to reduce these mutual perceptions solely to race. Saldanha himself points

Tibetans found it difficult to accept that it was no longer Buddhism and the Dalai Lama that were the main tourist attractions in Dharamshala. Opinions that "*Inji* tourists in Dharamshala are over" were quite common, and so were comments that "everyone fled to Bhagsu[70] because of the Indians," since McLeod became "too crowded and loud, just like India." At the same time, it was said that Indian tourists who started arriving to McLeod Ganj had money and were willing to spend it, as I already noticed in the previous section. However, their tastes and the nature of tourism they practiced meant that only few types of businesses benefited from it. Small entrepreneurs in the informal tourism sector complained that only restaurants (they said "Indians eat a lot") and upmarket hotels (they were supposed to pay 5,000 rupees, that is, more than 100 dollars a night without blinking an eye) made money on these new tourists. The owners of small Tibetan restaurants worried that even though tasting *momo*[71] was an important part of Indian tourists' itinerary in Dharamshala, they preferred to eat in Indian restaurants. They also liked pizza, but it had to be served quickly—big restaurants which employed a lot of staff and were able to provide such service remained in the game, while others lost. Shopkeepers were devastated, especially since their offer was mainly tailored to "Europeans," including, in the case of Tibetan shopkeepers and stalls, those *Inji* who sympathized with Buddhism and/or Tibet.[72]

However, Tibetans were not the only ones not satisfied with the change of a tourist profile in Dharamshala. I noticed, for example, that young Indians who run a guesthouse in McLeod Ganj—my hosts during one of my field trips—did not go to the bus station on the main square to search for customers among the interstate bus passengers, who used to arrive every morning. When asked for the reason, they replied that foreigners did not like to be solicited, so they did not want to accidentally "catch" the Punjabi tourists, whom they clearly did not like, claiming that they drink, are loud, and party at night, disturbing other guests. This group seemed to have a strong influence on the general perception of Indian tourists in Dharamshala, even though another large category were families and young couples from Indian metropolises—New Delhi, Mumbai, Hyderabad, Bengaluru, or Chennai—about whom tourism service providers said that they were not very different from "Europeans"; they were, as Phurbu said, "nice, polite, cultured." This is certainly due to the

out that for charter flight tourists from the UK, Germany, or Scandinavia it was the white freaks who constituted an attraction (2007, 141).

70 Short for Bhagsunath.

71 *Mog mog* (Tib.)—steamed dumplings, considered as "traditional Tibetan dish."

72 The Tibetan interpretation of the differences in the economic status of Indian and *Inji* tourists is interesting, as it reflected the preference given by Tibetans to "European" tourists, despite their current relatively smaller purchasing power: the wealth of the Indians was supposed to result from having "dirty money," while "Europeans" had no money, "because the government takes everything from them in taxes."

specificity of tourism from Punjab to McLeod Ganj—young men who hopped to the neighboring state for the weekend in groups of several people, looking for entertainment, strongly based on racialized sexism, which in practice meant club-crawling in the hope of establishing intimate relationship with foreign female tourist.[73]

However, Indian tourists were also diverse in terms of economic status. In the stories about theft told by stall owners and shopkeepers, it was the Indians who appeared as the main culprits. This may be due to the fact that Tibetans in particular did not have much confidence in the Indian police, so they hardly ever reported cases of theft, which could encourage this kind of practices that exploited the disadvantaged position of refugees. Some Tibetan guesthouse owners also complained that less well-off men who traveled as part of weekend tourism rented double rooms but stayed there in bigger groups, assuming they did not come to Dharamshala for two days to sleep. The room served them mainly as a place to have drinks before going to a club, which saved them money, as drinks in clubs were quite expensive.

To sum up, the presence of domestic and foreign tourists in Dharamshala and Hampi was very differently valued. While in Hampi the presence of tourists from "the West" was interpreted by the representatives of Hindu nationalism in terms of contamination, in Dharamshala it is the Indian tourists who were accused of desecrating the Buddhist atmosphere of the place. In both cases, however, the overlap between domestic and foreign tourism largely canceled out the colonial character of the latter, as most tourist spaces there were heterogeneous, that is, available to both Indians and foreigners, thus avoiding the impression that enclaves of luxury and leisure were created only for foreigners (Richter 1989, 105). Of course, the class factor played a major role here, because Indians who could enjoy these "luxuries" were, as Richter pointed out, the wealthy local elite (1989, 106). At the same time, however, many white tourists traveling in India were economically disadvantaged due to the crisis, which, to some extent, undermined both global economic inequalities and India's highly developed backpacking tourism. As a result, for many workers and small entrepreneurs in the informal tourism sector Indian tourists appeared to be richer than "Europeans," who shared a lot of spaces, such as guesthouses or restaurants, with less affluent Indian tourists. Finally, the small size of Hampi and Dharamshala meant that the local social networks overlapped with the economic sphere, so that even those local residents or small entrepreneurs who could not afford to be customers of a café, could sit there as guests of the owners or other employees.

---

73 I address representations of gender and race and their effects on mutual relations in tourism in chapter 5.

Chapter 3

Not-So-Empty Meeting Grounds

Self-Representations and Relationships

IMBUING GOODS AND SERVICES WITH AUTHENTICITY

Anthropologists involved in projects to protect traditional cultures have quite unanimously condemned tourism as a source of undesirable cultural change due to commodification (Crick 1989a, 311–312).[1] In anthropology of tourism one of the first and canonical texts using this rhetoric was an article by Davydd J. Greenwood (1977) entitled "Culture by Pound: An Anthropological Perspective on Tourism as Cultural Commodification," in which the author, using the example of festivities held in a Basque town and included in the tourist offer, demonstrated that the commercialization of culture leads to a loss of authenticity.[2] On this basis, there occurred a hierarchical distinction between ethnic and tourist art (i.e., souvenirs produced for the tourist market), the former supposedly spontaneous and intended for local needs, that is, authentic, and the latter created purposefully, tailored for tourists' tastes, and thus artificial and less valuable (Hume 2014; see critique by Wieczorkiewicz 2003). Such logic is also based on the premise that ethnic art is indigenous and traditional and has never been subject to any external influence, which blatantly overlooks the fact of cultural exchanges with other, that is, neighboring-, trading-, invading-communities, as well as the effects

---

1 See, for example, the special issue of *Cultural Survival Quarterly* 6 (3) from 1982, "The Tourist Trap—Who's Getting Caught?," or Haunani-Kay Trask (2000) in the same journal, about "the prostitution of Hawaiian culture" through tourism.

2 Although Greenwood's approach became less radical over time, at that time his attitude toward the impact of tourism on local cultures was clearly negative. He wrote that "the commoditization of local culture in the tourism industry is so fundamentally destructive" (Greenwood 1977, 131).

of colonialism and the "imperial gaze," to which the Other had to adapt to varying degrees (Pratt 1992).

Such distinctions construct the opposition between tourism and culture, in which tourism always entails lack of authenticity due to its market-oriented character (Shepherd 2002). It is precisely the assumption about an ontological division between the sphere of the "backstage" (i.e., the real space, where real life takes place) and the sphere of the "stage" (i.e., staged tourist space), along with the underlying concept of authenticity as the basic category organizing the tourist experience, that was often criticized in Dean MacCannell's theory (1976). Researchers also pointed out that, certainly, not every tourist is guided by a desire for authenticity (see, for example, Cohen 1979), for example, people motivated by the pursuit of pleasure or diversion are satisfied with the carnival-like reversal of "normality" offered by mass tourism. For them, a tourist trip serves the purpose of regenerating physical and mental forces, so that upon their return, they can better function in their own centers. It is a means of *recreation*, but not *re-creation* within a specific identity project, which often takes on a form of travel (Cohen 1979, 183–185). Russian tourists from Goa who just pass by Hampi provide a perfect example.

On the other hand, authenticity is indicated as an important part of tourist experience of travelers most frequently encountered in Hampi and Dharamshala, that is, backpackers, tramping tourists, and travelers engaging in more profiled forms of tourism, such as volunteer tourism. It is them, people whom Cohen calls experiences seekers (1979, 186–188), who are MacCannell's modern pilgrims, yearning for authenticity. Such desire is said to be caused by a sense of alienation born out of modernity, industrialization, spatial mobility from rural to urban areas, secularization, and progressive individualization. There is a narrative of loss of the natural relationship that linked humans and their birthplace in the premodern world, embedded them within a local community, a professional or kin group—the lost feeling of belonging to a place and a community, belonging to a larger, holistic order. Alienated and confused, they look for truth, naturalness, authenticity in the lives of other people—at other times and in other places—whom they perceive as traditional, and thus non-modern.[3] They look for a genuine "native"—not the one envisaged by Thomas Hobbes (1968 [1651]), living in the natural condition of mankind, being "solitary, poor, nasty, brutish, and short" (Hobbes 1968, 186), but a being of nature taken from the idealistic, romantic image by John Dryden (1672): noble, free, equal, happy in his simplicity, constructed as a binary opposition to ourselves—who are lost in civilization, detached from our roots. In the postcolonial context, this desire

[3] See more about the need for authenticity among tramping tourists in Bloch (2016a).

is further reinforced by the "imperialist nostalgia" by which people mourn for what they have destroyed themselves (Rosaldo 1993, 69–70). Based on the discourses about the Other that originate in their world, tourists develop expectations toward the communities they come to visit and their cultures. These representations result from multiple processes—of generalization, exotization, essentialization, ahistorization, idealization, and pathologization, all happening often at the same time (Bloch 2013b), as well as of homogenization, de-contextualization, and mystification (Albers and James 1983, 1988). All of them bring about the discourse of Orientalism, which was an ideological basis of colonial power (Said 1979).

However, when we realize that what we call cultures are not, in fact, entities sealed off within geographical borders (Gupta and Ferguson 1992), we realize that the way each community exists is not a given state but an ongoing process of constant, diverse appropriations, more or less violent. We can thus look at tourism as yet another sequence of influences, and see the goods and services that it generates as "manifestations of the local cultural activity," even though they may display elements that outside observers consider to be characteristic of a given region or which are borrowed from other groups, if they happen to be more desired (Wieczorkiewicz 2008, 53–54). The critique of commodification, fueled by nostalgia, is an attempt to freeze the Other in a timeless ethnographic present (Fabian 1983; Thomas 1996). It not only means paternalistic refusal of another people's right to change. If commodification amounts to ascribing monetary value to different cultural elements through their inclusion within market economy, such refusal suggests that the Other should not have the right to manage their own cultural capital (by, for example, monetizing it) and the right to have the same needs as the so-called Western people (e.g., consumption of goods). Obviously, we need to remember that commodification often stems from global economic inequalities—the Other has no other resources to monetize apart from culture. However, this could also be interpreted as reclaiming the right to culture that used to be appropriated by colonialism and anthropology: "Culture, now, is also intellectual property, displaced from the 'museum' and the 'anthropological' gaze, no longer 'naked' nor available to just anyone pro bono" (Comaroff and Comaroff 2009, 3). This always begs a fundamental question of who benefits from it, that is, about the relationship between the capital and labor. Nevertheless, a clearly negative assessment and nostalgic refusal of this process does not allow us to see its complexity and its less than unobvious consequences.

Many researchers started to point out that "the interest of outsiders in local culture and the demand for craft and artistic products can make people appreciate what they thought was no longer worth much" (Wieczorkiewicz 2008, 70–71). This is especially important in the context of postcolonial countries, where colonization policies have led to the depreciation and marginalization of the native cultures as allegedly less valuable than the European ones. In this sense, a positive value ascribed to local cultures by tourism provides a counterbalance to the mimicry of the colonial period as understood by Homi Bhabha (1984), when the colonizers—mainly through the educational system and Christianization—sought to control the Others by attempts to shape them in their likeness, while at the same time maintaining the irreducible difference between the original and the imitation. The colonized wanted to imitate the colonizers, hoping that by accepting their culture and rejecting their own—which was perceived as inferior—they would become equal to their "masters." Tourism seems to wield the power to reverse these value-judgments. Michel Picard (1990, 1996) revealed this process using the example of the Balinese culture reinvigorated by tourism. Walter E. A. van Beek (2003) also observed that the interest shown by tourists in visiting the Dogon country, Mali, strengthened the local sense of cultural pride.[4] The same occurred among Tibetan refugees. One of my research partners from Dharamshala, the founder of a modern Tibetan theater, who also runs a café, told me that the fact that tourists were so interested in Tibetan culture proved its great value: "Sometimes we even tell our children: 'see, these people from abroad come here from so far away to learn our language, and you're just bumming around.'"

Tourism can also provide an impulse and a scaffolding for representations of one's own culture—as was shown by Jeremy Boissevain (1996) (in his article, describing celebrations in Malta, he directly critiqued Greenwood's approach) or Eric Silverman (1999) (writing about the production of tourist art in Papua New Guinea). The presence of tourists can encourage cultural exchange and increase the value of local products (Yu and Littrell 2003). This, in turn, can serve as a means of raising economic capital, which will then be used to maintain or revitalize local traditions that would otherwise be at risk of extinction, as demonstrated by Erik Cohen in his landmark text about Bali, where he challenged the binary opposition of authenticity versus commodification and introduced the concept of an "emerging authenticity" (Cohen 1988; see also Picard 2008). Moreover, the producer/performer and the consumer/viewer can satisfy their different needs: something that is a spectacle for the tourist does not have to cancel out local meanings, as performers are also viewers (Comaroff and Comaroff 2009). Ethnic entrepreneurship, as John L. and Jean Comaroff called it in their influential book *Ethnicity, Inc.* (2009), makes it possible to profit from the tourists' encounters with Otherness, which is at

[4] However, Van Beek showed that in Northern Cameroon, Kapsika people had completely different view on tourism—encounters with tourists made them feel inferior regarding the position they attributed to tourists. According to the researcher, the differences stemmed from the type of tourism practiced in these two places. In the first case, tourists visited the villages inhabited by the Dogon, whereas the other consisted of nature sightseeing expeditions where local people were marginal. The impact of tourism on the local community should thus be seen in a broader context.

the same time "authentic, albeit [served] in consumable form" (2009, 140). Thus, commodification of ethnicity does not necessarily mean its destruction, as it allows people to tell their story anew and to create or reinforce a sense of identity[5]—"the survival of culture has, therefore, been replaced by the idea of survival *through* culture" (Golinowska 2013, 108, emphasis added).

To sum up, commercialization of culture (its products, practices, knowledge) can provide both a means of self-creation and a source of livelihood for marginalized communities deprived of other resources (Comaroff and Comaroff 2009; see, for example, Esman 1984; Hillman 2003). In the following section, I look at the impact of tourism on the refugee version of Tibetan culture and the tactics used by various employees and petty entrepreneurs in the tourism sector—not only Tibetans—to create their self-representations. I analyze practices serving the purpose of authenticating goods and services and look for their underlying causes. I also focus on the competition over the right to cultural heritage and benefits from its commodification.

Tibetanness as a Tourist Attraction: Tibetanization Practices

Unlike in Hampi—whose contemporary inhabitants "inherited" the tangible cultural heritage in the form of the remnants of the Vijayanagara kingdom (which ultimately turned out to be their curse)—Tibetan refugees had to invent their heritage. They have achieved spectacular success in this regard, turning a refugee settlement into a popular tourist attraction and making exile version of Tibetanness a widely recognized and desirable cultural product (McGuckin 1997). To this end, they skillfully used the existing imaginations of Tibet. After coming into exile, Tibetans discovered that Tibet—imagined in the spirit of Said's Orientalism (1979)—had already been there (Bishop 1989; see, for example, Hilton 1933[6]). And it was largely reduced to Buddhism and personified in the figure of the 14th Dalai Lama—the very aspects that were to become the main tourist attractions of the Tibetan diaspora.

Today, it attracts both tourists seeking spirituality, disgusted with the consumerism and materialism of the disenchanted Western world, and those for whom exotic Tibetanness carries the promise of "cool authenticity." The term

[5] Comaroff and Comaroff point out that ethnicity is not based on the rationality of classic market economy. Commodified Otherness, despite its mass and commercial character, does not have to lose its primary value, since "its 'raw material' [i.e., ethnicity] is not depleted by mass circulation. To the contrary, mass circulation reaffirms ethnicity. . . . Greater supply, in other words, entails greater demand" (2009, 20). Ethnic products, therefore, have the potential to become "authentic reproductions" (Bruner 1999; Edward M. Bruner created this category to overcome the unhelpful dichotomy of authenticity versus commercialization; see also Steiner 1999).

[6] James Hilton, the author of the novel *Lost Horizon* (1933), which was later turned into a movie by Frank Capra, contributed to the spread of the myth of Tibet as a land of eternal happiness, spiritual wisdom, levitating monks, and third eye opening.

was first introduced by Tom Selwyn (1996a) to describe those characteristics of the world sight-visited by tourists, that they identified as truly different from their own (post)modern reality (in this sense, "cool authenticity" is close to the category of exoticism). Tramping tourists expressed their desire for this form of authenticity when wishing "to get to know such an exotic and culturally diverse country," to "experience adventures in a very different culture, see a different 'world,' how people live there," to participate in "a very exotic journey" and "a completely different culture," to "finally see a piece of a completely different world, and what world is more distant from ours than this [i.e., India]?" The experience of exoticism, Otherness, and cultural diversity were the reasons most frequently given by these tourists for selecting India as their destination, immediately after their interest in culture, religion, or philosophy[7] (see Bloch 2016a). Within this arrangement, characteristic of an "imaginary paradise" (Dann 1996, 68–71),[8] local inhabitants play the role of a "scenery" and of cultural markers (see MacCannell 1976), like, for instance, a "typical" Tibetan monk meditating in a monastery, or a "typical" Indian snake charmer. Thus, the difference here is a prerequisite for satisfying the tourist need for "cool authenticity."

"The Western gaze . . . does not simply see but also produces that which it wants to see," writes Vincanne Adams in relation to the Sherpas, who also faced a powerful set of expectations from the so-called West (1996, 24).[9] However, this does not mean that people automatically adapt to external representations, but that multiple processes come into play here, including self-Orientalization (Lie 1996),[10] that is, producing evidence of one's own cultural

---

7 The first of these reasons was indicated by 27.8 percent of tourists who participated in the questionnaire, second by 17.2 percent—in their answers to the following open question: "What were the reasons you decided to go to India?"

8 Graham Dann analyzed 5,172 photographs in the brochures of leading UK travel agencies (focusing on the representations of people they created) and distinguished four types of "tourist paradise" that these images constructed. A "paradise contrived," based on romantic gaze (images of lonely beaches, uninhabited islands), was set up outside social relationships and even if local people appeared in it, they were exotic extras. Two other kinds of "paradise" were based on the need for familiarity: in a "paradise confined," the tourist community lived in exclusive, closed, safe enclaves, while in a "paradise controlled" local people appeared in strictly defined service roles, as waitresses, bartenders, or animators employed by all-inclusive hotels. In a "paradise confused" the local and tourist spaces overlapped, and guests and hosts could interact and mingle.

9 Sherpas—an ethnic group of Tibetan origin living in Nepal near Chomolungma (Mount Everest); they entered Western imaginary through contacts with international Himalayan mountaineering teams. Nowadays, Sherpas play an important role in Nepalese tourism. It is interesting that Sherpas' allure has been largely built on their Tibetanness—Buddhism and shamanism—and not the fact that they are Nepalese.

10 John Lie created this category based on Said's Orientalism to describe how the contemporary image of a culturally unique Japanese has been constructed. In this way, especially in the era of globalization, the community can create its own distinctiveness and use it as a source of pride. This is the case of San of South Africa as described by Comaroff and Comaroff, who by "enacting them repeatedly for tourist-consumers . . . were *not* [emphasis in original] just shanty dwellers in

uniqueness based on already existing representations. Such intentional self-essentialism may in turn serve groups to achieve their goals—something that Gayatri Chakravorty Spivak (1997) called "strategic essentialism." For Tibetans living in exile, such constructed exoticism is a strategy used to gain international support for the Tibetan cause, as well as moral and economic support for the refugees themselves. In this way, the discourse of Orientalism has been appropriated by "contriving others" (Ryang 1997, 28), becoming a source of their agency. The Tibetan elites skillfully used the Orientalized ideas about Tibet in their identity politics, in order to create an idealized and romanticized image of a Tibetan as a compassionate Buddhist, combining mystical exoticism with safe modernity.

Creating the self-representation desired by tourists, this kind of seduction strategy (Adams 1996, 24, 217), is not a matter of passive acceptance of imposed images, or a calculated manipulation of the audience, but one of an active consent expressed for a given image, whose positively valued elements have been internalized. This is how the self-representations created for the needs of the tourism industry can be in fact affirmative, as the basis of positive self-identification. As such, they are more a source of "self-*affirmation* rather than self-subjection" (Kanemasu 2008, 120; italics original). Yoko Kanemasu described this process using the example of amiability and friendliness, that is, the attributes inscribed by the employees of the hotel industry in Fiji into the indigenous Fijian identity, even though they appeared to be a continuation of colonial representations, imposed on the hospitality sector workers, who were supposed to do badly paid jobs with a smile. These workers—maids, waiters, bartenders—interpreted the meanings of the key elements of their touristic image in a different way than an analysis of visual representations of Fijians in touristic brochures might suggest. Amiability, for instance, was understood by them not as an act of serving—and thus a manifestation of voluntary subordination—but as a practice set in the context of reciprocity, with reciprocal amiability being indicated as the main source of satisfaction of tourism workers (Kanemasu 2008, 119).[11] This is, therefore, not just an "image management" technique, done in the hope of financial gain, but—as Adams points out for the Sherpas—the "culture work" which abolishes the difference between what is materially viable and what is beneficial symbolically (Adams 1996, 219). I write more about this in the second part of the present chapter and in chapter 5.

Western rags, desperately immiserated though they might have been. They were seen, and reciprocally were able to see themselves, as a named *people* [emphasis in original] with a 'tradition and a way of life.' In other words, a culture" (2009, 11).

11 The amiability was also used as a way to resist subjection by tourists—the more indifferent they were, the more the workers tried to imbue the host-guest relationship with a friendly dimension, forcing reciprocal amiability (Kanemasu 2008, 126).

The same process can be observed among Tibetans, for whom the identity politics, consistently implemented by the refugee elites, has led to a positive revaluation of the refugee status. In other words, being a Tibetan has become attractive, also in the field of tourism. This explains why people whose ethnic identity is much more complex tend to emphasize its Tibetan aspect in their self-representation created within tourism. The example of Norbu and Dolma, the shopkeepers from Hampi, the "Indian Khampas" (*gyakar Khampa*) I described in chapter 1, is a case in point. Tourists are usually unaware of the internal divisions within each group, so they classified Norbu and Dolma—on the basis of their anthropomorphic characteristics and, above all, the assortment offered in their shop—as Tibetans. Indeed, they traded in goods that suggested connection not only to Tibetanness but also to Buddhism and refugeehood: Buddhist prayer flags, wall hangings with quotes from the Dalai Lama, Buddhist statues, notebooks, and bags with the label "Product of Tibetan Refugees." Dolma often drew tourists' attention to the latter, even though her own attitude to Tibetan refugees was very ambivalent. The *gyakar Khampa* are, at best, operating on the outskirts of the Tibetan diaspora, being perceived as "second grade," that is, Indianized Tibetans. This is visible, for example, in competition for customers—"Indian Khampa" businesses are not usually allowed on the Tibetan bazaars scattered all over India. Refugees assume that the right to trade within such markets should be reserved to the disadvantaged noncitizens, and *gyakar Khampa* do not belong to that group. Indeed, Norbu and Dolma were asked to leave one of such bazaars in Goa.[12]

Reacting to such Tibetanized spaces, tourists—especially if they belong to "the European" category discussed in chapter 2—activate their acquired cultural competences. They say "Tashi Delek"[13] when entering, and then initiate talks about the Dalai Lama (how an extraordinary person is he), about Dharamshala (they have been there and liked it very much), about mantras (and their meanings), about *tsampa* (that Tibetans eat it for breakfast, lunch, and dinner[14]). Dolma tried to joke that they did not eat *tsampa*, because temperatures in Hampi can reach over forty degrees Celsius in the shade, and nobody there grows barley.[15] When tourists asked Norbu about where they lived, he usually answered that "now close to the Tibetan border," by which

[12] Dolma's younger sister successfully runs a stall there, but her husband was a Tibetan refugee.

[13] Tibetan greeting (Wylie: *bkra shis bde legs*).

[14] *Tsampa* (Wylie: *rtsam pa*)—roasted barley flour, most often mixed with salt tea with butter, barley beer, or just water. In Tibetan national discourse *tsampa* is presented as one of the symbols of national unity, constructed in opposition to the Chinese "rice eaters" (Shakya 1993). In Dharamshala, *Tsampa Eater* T-shirts are up for sale, developed and reproduced as a national symbol for refugees themselves, although it is also one of self-representations for the outside world, including tourists. I elaborate on the subject in the next chapter, in the section devoted to political tourism.

[15] But I do not think they ate it even when in Manali, as is the case of Tibetans in Dharamshala who eat *tsampa* occasionally, maybe except the elderly.

he meant the culturally Tibetan region of Spiti in Himachal Pradesh, where the family spent the summer season. This is how this peculiar communication kept going, which was less about hosts adjusting to the expectations of the guests and more about keeping a certain vagueness. During just one conversation with a client, Norbu used terms, "our Spiti," "our northern India," "our Tibetan [something]" and "our Indian [something]," which he felt was not contradictory at all. From among these self-identifications the tourists selected those which were most readable and consistent with the assortment displayed in Norbu and Dolma's shop.

The restaurant on the opposite side of the street, run by a Tibetan from Nepal, was managed in a similar fashion. Norbu and Dolma often sent their customers there: "Tourists know we are Tibetans"—Dolma said, acknowledging the fact of external identification—"so they ask us where in Hampi one can eat good *momo*.[16] Then we send them there." The restaurant was called "Little Tibet Kitchen"—with a branch, a bakery called "Tibet Quality Bakery"—and there were many Tibetan cultural references there: Buddhist prayer flags, a large framed picture of the Dalai Lama, and the national Tibetan flag bearing the inscription "Tibet for Tibetans," although the menu was no different than in the restaurant run by the Nepalese, Phurbu and Padma (who also served *momo*, had a portrait of the Dalai Lama, and had wall hangings depicting Buddha). The owner observed that "people from the West like Tibet, all those Tibetan refugees, they often say that they had met the Dalai Lama in Europe, they ask if I am a Tibetan." The young man who runs this restaurant did not seem to have a simple answer to that question. He was born in Nepal, left his home village when he was ten to take up a job in Kathmandu. His mother died shortly afterward and his relationship with his father had always been difficult. Thus, neither did his parents tell him a story of escape into exile nor did they plant in him a picture of his "imagined homeland," that is, all those narratives that constitute the identity of "born refugees" from India (see Bloch 2011). He did not know where his parents came from in Tibet or when they had escaped. He was not included in the exile schooling system, so he was not "made" into a Tibetan. Neither did he know the arguments in favor of the independence of Tibet nor did he know any important figures within the Tibetan freedom movement in exile. He barely spoke Tibetan—once, when Tibetans came to his restaurant, he had to explain himself by saying that it was "because I was brought up in Nepal." He used to joke that when he got old, he would go to Tibet, shave his head, and wear monks' robes—evidently, for him, like for tourists, Tibet was associated with Buddhist spirituality. Like Phurbu and Padma, he freely oscillated

[16] *Momo*—steamed dumplings—are another culinary symbol of Tibetan identity in exile (although appropriated from the Chinese cuisine).

between Buddhism and Hinduism, having the image of Buddha Shakyamuni tattooed on one arm and the Hindu deity Ganesha[17] on the other, a vivid illustration of the Nepalese religious syncretism. At the same time, although one might think that he treated Tibetanness only as a way to attract customers, he was also engaged in bidding for authenticity so characteristic of Tibetan refugees in India (Bloch 2014a), for instance, when he talked about his *gyakar Khampa* friends, Norbu and Dolma, as "not exactly Tibetans."

In terms of goods and services offered to tourists, Tibetanness is also claimed by people coming from outside of the Tibetan cultural area. A salesman from Delhi, an ardent follower of Shirdi Sai Baba,[18] seasonally trading in Hampi, sold not only Hindu-related artifacts but also Tibetan handicraft (statues, *thangka* paintings[19]), along with "Save Tibet" stickers. Just like the restaurant owner from Nepal, described above, he admitted that "people like Tibet, they want to buy this stuff." The scale of this phenomenon in Dharamshala was even greater. Tenzin, who runs a travel agency, noticed, "When tourists come here, they are under the impression that all of McLeod Ganj is Tibetan, which is not true." New Indian-owned hotels and guest houses that mushroomed around McLeod Ganj in recent years exhibited grand, gold-painted statuettes of Buddha in their reception areas, along with big pictures of the Dalai Lama. Most tourists were neither well versed in ownership matters nor nuanced self-representations. They could be unaware that they were accommodated in an Indian—and not a Tibetan—hotel, especially since one of the popular authenticating practices was hiring Tibetan-looking employees as front-desk staff, especially at the reception.[20]

Even though it seems that Tibetanness is an attraction tailored mainly for non-Indian tourists motivated by their desire for authenticity, for domestic tourists it also represents something of a value that adds to the stunning views and climate of Dharamshala, and the new cricket stadium. Contrary to common complaints regarding this category of tourists (see chapter 2), domestic tourists are—perhaps with the exception of young men from Punjab coming for weekend club crawls—the consumers of Tibetan material culture, although clearly different ones than "the Europeans" and Buddhist tourists. I spent long hours in one of the bigger Tibetan shops in McLeod Ganj and noticed that Indians were shopping there a lot, buying all sorts of goods,

---

[17] An important figure in the Hindu pantheon, the personification of wisdom, and patron of scholars and science.

[18] An Indian spiritual teacher, active mainly in the second half of the nineteenth century, who combined teachings of Hinduism and Islam.

[19] *Thangka* (Wylie: *thang ka*)—Buddhist images painted (less often embroidered) on scrolls. Handmade and easy to transport, they are very popular among tourists.

[20] Tibetan entrepreneurs also did that, employing Tibetans as waiters or receptionists, while most of the "invisible" work (Lozanski 2010, 754–755)—cooking, dish washing, cleaning—was done by workers from Bihar (see chapter 1).

including patriotic T-shirts displaying the Tibetan flag and proclaiming "Tibet needs you." Mostly, however, they did not treat it as an expression of their support for the Tibetan cause but treat it as a way to satisfy their need for "cool authenticity." Tibetans reminded them of the indigenous "tribal" groups of the Northeast India—the "Wild East" and its Others versus the Hindu majority. What clearly sets them apart, however, is the figure of their spiritual leader. Many Indians, who are used to show respect to all spiritual teachers, regardless of the specific religion they represent, treat the Dalai Lama with reverence. On one occasion, during prayers held in the main temple of McLeod Ganj (for the nun who had just self-immolated), I could observe Indian tourists who happened to visit the temple bowing to the Dalai Lama. Indian entrepreneurs and workers in the tourism sector in Dharamshala were aware of that, and tried to make a reference to Tibetan cultural capital in the practices of authenticating their own offer.

Whose Heritage Is It?

These practices were seen by some Tibetan refugees as appropriation and led to competition over the right to cultural heritage (and economic benefits from it). This rivalry was particularly evident in Dharamshala, which is heterogeneous in terms of its more and less permanent inhabitants, as opposed to Tibetan camps or aforementioned bazaars, where the only traders are Tibetans. In the camps, refugees generally live on land leased to them by the state authorities, with quite a lot of freedom in terms of self-governing. However, Dharamshala is one of the scattered settlements, where only individual properties are leased to the refugee authorities. Tibetans, who have deliberately maintained their statelessness over subsequent generations born in exile and who therefore are mostly non-Indian citizens, enjoy limited rights to acquire land and real estate (see chapter 1). Local regulations in force in Himachal Pradesh make the situation even more complex by limiting the possibility for non-native residents to trade in land, just like in other mountainous states of India, where agricultural land resources are scarce but attractive for other types of investments, such as tourist infrastructure. Of course, there are ways to circumvent these regulations, but they are more accessible to people who have considerable financial resources and extensive social networks.[21] As a result, even in places that tourists commonly associate with Tibetanness, Tibetans more often than Indian citizens find themselves trading from street stalls, while Indians—in McLeod Ganj these are in particular Kashmiris—run

[21] One of the most popular is *benami*, that is, making an informal arrangement with a local resident in whose name the land and/or property is bought, although in practice it is used by another person. Of course, such an arrangement does not provide any legal protection for the actual buyer.

large and more elegant "antique shops." The same is true for other hill stations in Himachal Pradesh, such as Shimla or Manali, as well as Leh in Ladakh.

This contrast was noticed by Tibetans and caused quiet dissatisfaction among some of them. The main accusations of appropriation of Tibetan cultural heritage and engaging in "Buddha business" were directed at the Kashmiris. The allegations made in this context were of a moral nature: "They have the most fashionable shops, while Tibetans have to trade in the dust, rain, are pushed by cars, and bullied by the police."[22] In this narrative, the Kashmiris were portrayed as the ungrateful beneficiaries of Tibetan culture, who did not show due respect for both the culture "from which they make a living" and its rightful owners, the Tibetans. One of my research partners asked a rhetorical: "Why don't they sell their own culture? They can sell Allah or something!"

The second group in Dharamshala that was targeted for the same reason were the Nepalese. They, in turn, were accused of desecrating Tibetan culture by mass production of cultural artifacts without respecting traditional rules. I often heard the opinions of my Tibetan interlocutors that the Nepalese "learned everything from them [Tibetan refugees]," when the latter settled and set up their handicraft centers, *thangka* painting workshops, and carpet factories in Nepal, employing local population as workers. During my fieldwork, the Nepalese run numerous tiny *thangka* studios in McLeod Ganj (Kashmiris also sold *thangkas*, but along with many other items that referred to Hindu and Buddhist traditions). Manufactured by Nepalese craftsmen for the tourist market, their *thangkas* were relatively cheap, hence the ones painted under the watchful eye of Tibetan masters—often for months, in line with the sacred painting ritual practice and following strict rules at each stage of the process—had no chance to get through to the consumers in a competitive "free" market. Dharamshala was constantly gossiping about *thangka* paintings displaying letters of the Tibetan alphabet arranged in a meaningless way, which of course tourists were not able to detect. These were, therefore, classic accusations which appear at every instance of commodification, but were all the more acute in this case, as the benefits went not to those who considered themselves to be legitimate heirs to tradition.[23]

---

22 From time to time, there were police operations conducted in McLeod Ganj—stalls were removed, as they were supposed to negatively affect the safety and aesthetics of the place. Theoretically, there were plans to relocate stalls to a large retail complex, but it was not clear where to find a suitable location for such a structure in a hill town. These operations did not target Tibetans in particular, but since Tibetan traders run the vast majority of stalls, that was how they were interpreted. In fact, this may relate to the practice of "spatial cleansing" (Herzfeld 2006) and the postcolonial ideas held by Indian government about modern public space arrangements (see Bloch 2017b).

23 All this bidding for "cool authenticity" with regard to Tibetan culture and its presence on the tourist market was also taking place in the wider context of Tibetan exile and political rivalry for legitimate representations of Tibetan culture created in the diaspora and in Tibet, where the PRC

Above all, however, I noticed that there was a fear of the "made in China" label among the sellers on the Indian tourism market—and not only among Tibetans. They all were aware that customers from Europe are reluctant and suspicious of goods from the People's Republic of China. Firstly, because of the need for "cool authenticity," which is certainly not met by an external manufacturer famous for counterfeit goods, and secondly, for moral reasons, because of the growing popularity of campaigns calling to boycott products manufactured in violation of labor (and human) rights. For Tibetan refugees, this is even more significant for political reasons. Local activists are aware that trading in products made in China by Tibetan entrepreneurs presents a contradiction of the refugee self-representation and causes cognitive dissonance among tourists supporting the Tibetan cause, especially tourist-activists (sponsors, volunteers, activists of pro-Tibetan organizations). Already in the past there were campaigns organized in Dharamshala—stickers proclaiming "We do not sell products 'Made in China'" (in three languages: English, Hindi, and Tibetan) were distributed among sellers and service providers, who were supposed to place them on their premises. The boycott of Chinese products has been also enforced during numerous political demonstrations organized by Tibetan refugees in India—hundreds of "made in China" hot water flasks were smashed to pieces and clothes were burned. The shop run in McLeod Ganj by the Indian branch of the international NGO Students for a Free Tibet used to sell a collection of T-shirts saying "Made in Tibet" with an added barcode as an ornament. On the other hand, those who openly trade in Chinese products (although it is not an offer addressed to tourists but to local clients, these stalls are clearly visible on the main streets of McLeod Ganj) are mainly newcomers from Tibet. Without a network of contacts which would enable them to obtain goods from Indian sources, they use those known to them from their country of origin, that is, Chinese Tibet, brought via Nepal. Tibetan activists have already realized that it would be morally ambiguous to expect these refugees to boycott products "made in China." At this most difficult stage of adaptation to a life in a new country, they simply have to make a living.

As a result, most traders avoid answering the question of where their goods come from—they usually pretend they do not understand the question or answer by stating their own place of origin. In practice, as I have observed, they often simply do not know the answer. They obtain goods either from large wholesalers, primarily in New Delhi or Jaipur (in the state of Rajasthan), or from middlemen—itinerant wholesalers who tour Indian tourist destinations, often offering items on credit and collecting debts at a

authorities pursue their own cultural politics (see Calkowski 1997 on Tibetan opera and Harris 1999 on the art of painting).

later date.[24] The traders do not know where the goods provided by Indian wholesalers are produced, and I assume that, due to the tourists' need for "cool authenticity" of which they are well aware, they probably prefer not to know. Instead, they may suspect that a large part is manufactured by global capital-ventures somewhere in Thailand or Singapore—that is why the same "traditional" headscarves can be bought in India and in Morocco (see Cohen 1993a; Hume 2009). At the same time, many of the Tibetan items destined for the tourism market—for example, those sold in Kashmiri shops—actually come from small factories and handicraft centers in Nepal, where labor costs are lower than in India.

Tibetan handicraft workshops—producing carpets, woolen clothing, objects carved in wood, metal statues, or *thangkas*—have been shut down, even though they were supposed to be the pillars of refugee economic self-sufficiency, especially in the northern states of India[25] (Phuntso 2004, 140–141). Their failure was due to not only the mechanisms of global capitalism (insufficient financial resources, which prevented more mass-scale and thus cheaper production, combined with insufficient know-how about ways to market more luxury products, and international campaigns which affected the flagship Tibetan product, that is, carpets[26]) but also a significant increase in the level of education among Tibetans born in India, for whom handicraft work was no longer attractive. They were replaced by Nepalese workers, who became the main suppliers of Tibetan crafts to the Indian tourism market. Some of the Tibetan handicraft products are made in India, such as the popular "100% yak wool Tibet shawls," which come from Ludhiana in Punjab, the country's major textile center. Finally, most of the various types of goods are produced in China. Dolkar, a street seller from Dharamshala, defended the Tibetans who sell products "made in China" by stating, "Today everything is

[24] Those small-scale wholesalers offer a limited range of products and at higher prices, but in this way retailers do not incur the cost of time-consuming expeditions to procure goods (often requiring travel to several states). These services are mainly used by smaller shop and stall keepers, such as Dolkar, but also those who want to restock. Those itinerant wholesalers are another group of hypermobile workers in the informal tourism sector in India. They spend half of their lives on trains. They usually set out on Monday—traveling, for instance, from New Delhi to Goa—and reach on Tuesday afternoon, do business for a day and a half, and come back on Thursday morning, on Friday they are on site, on Saturday they restock, to be back on the road on Monday. They carry all their assortment in handbags and know how to avoid tax inspection at railway stations (often they even have agreements with individual inspectors, who are part of the business).

[25] In southern states (especially Karnataka) and eastern states (Orissa) large agricultural settlements were set up, but even these included handicraft centers and carpet factories.

[26] Here I refer to the fair trade campaign launched in the 1990s, which led to a reduction of Western markets demand for carpets from South Asia (especially from Pakistan, India, and Nepal) and Morocco, because their production was said to involve child labor. Tibetans tried to save this industry by making their carpets "Child Labor Free," but it did not have the desired effect and sales dropped dramatically. See the campaign against child labor in the carpet industry by the Anti-Slavery Society: http://www.anti-slaverysociety.org/.

made in China. Recently I got a gift from a friend from the United States and it was also made in China."[27]

Cool-Hot: A Game of Authenticity

It is increasingly difficult for tourism workers to appeal to the "cool authenticity" of the offered products within the context of global capitalism. It is much easier to authenticate services and create a sense of "hot authenticity." This concept, also introduced by Selwyn (1996a), refers to tourists experiencing reality when they cross the barrier created by the practice of sightseeing and the "tourist gaze" (Urry 2002). Mary Conran refers to it as authenticity through intimacy (2006, 275). Most often it takes the form of sensual experience—that is, tasting, listening, touching, incorporating Otherness by letting it into one's own body (Wieczorkiewicz 2008),[28] instead of objectifying, distancing looking at—and fraternizing with the natives (MacCannell 1976), that is, establishing a close relationship, not mediated by the tourist industry, with people living in the world into which the tourist arrived (Lozanski 2010, 745). This need—and the promise to satisfy it—is strongly emphasized in the marketing of tramping tourism. Both visual and textual representations of India created by Polish tramping agencies online suggest the possibility to enter a "backstage" of a tourist spectacle (MacCannell 1976). In other words, they offer hope of escaping from the tourist enclaves—"luxury prisons" mentioned in the previous chapter (Kaczmarek 2012)—and arriving in a "paradise confused" (Dann 1996), where "true natives" live. The latter are supposed to be friendly and hospitable toward travelers. One can experience closeness by eating together, participating in local festivals, or by mixing in the crowd at a local bazaar (Dann 1996, 78). Tramping offers such visions of communion with the Others in their "natural" environment: at "the fruit, vegetable and floral market" in Mysuru[29] or at the floating "vegetable and floral market" in Srinagar; during a visit to "a healer, who makes herbal healing oils"; or in "the Tibetan refugee camp," during "a ride on the ordinary bus"; or an "overnight stay in the temple dormitory," which is supposed to be an excellent opportunity to get to know the pilgrims."[30]

The promises of dining in local restaurants and traveling by public means of transport translate directly into the perception of tramping tours by their

[27] In 2008 a scandal broke out in the People's Republic of China when it turned out that one of its factories manufactured Tibetan national flags—the workers explained that they did not know what they were producing and the orders from abroad were good ("'Free Tibet' Flags Made in China." *BBC News*, April 28, 2008. http://news.bbc.co.uk/2/hi/asia-pacific/7370903.stm).

[28] More about the multisensory experience of India by tourists in Bloch (2016a).

[29] Formerly, Mysore, a city in the southern state of Karnataka.

[30] All quotes come from the tramping trips' itineraries posted on the agencies' websites.

participants, who stress feelings like, "for a while you can feel like an inhabitant of the country you are visiting," you can get to know it "not from behind the window of the tour bus, but in its natural form," "to encounter the authentic everyday life of people." One of the tourists described the need for authenticity and the possibility to satisfy it through tramping in such a way: "We may get invited to someone's house, to a local or family celebration. We visit places that are less frequented, usually inaccessible for mass tourism, unaffected by the tourist industry." The last statement reveals yet another condition for experiencing authenticity, namely being open to the unexpected, described in literature by the term "serendipity"[31] (Tucker 2003a, 43–68). Hazel Tucker points to the search for serendipity as an integral part of being a backpacker: it involves avoidance of planning, declared deviation from the usual routes, longer stays in low-cost locations, which is supposed to create opportunities for adventure. The same is true for tramping, which promises its participants a high level of autonomy, freedom, and flexibility (Bloch 2016a, 13–14).

The distinction made by Selwyn was critiqued for operating the etic (in relation to "cool authenticity") versus emic (in relation to "hot authenticity") dichotomy, that is, the opposition of what is objective (authentic objects) and what is subjective (authentic experience) (Cohen and Cohen 2012, 4). Meanwhile, it is worth noting that both these authenticities are socially constructed and interpreted from the bottom-up perspective, and that we should treat authenticity itself as one of the epistemological categories which people use to organize their world. Employing it in the field of social sciences does not mean making us experts in what is or is not authentic (Wieczorkiewicz 2008, 90–93), but rather studying the processes of "authentication," that is, social production of the conviction about authenticity. In other words, it is about exploring how certain views and experiences become authentic for some people (Cohen and Cohen 2012; Fillitz and Saris 2013).

MacCannell claimed that authenticity, including the authenticity of the "backstage," is staged by the tourist industry. Tourism sector employees are supposed to be aware of tourist needs and expectations and in response to them produce a sense of authenticity (MacCannell 1976). Tim Edensor (2001) distinguished three groups of social actors involved in this practice: "directors" and "stage-managers" (owners of travel agencies who create scenarios of tourist experience; travel guides; and tour leaders who tell tourists where and how to look, what souvenirs to buy, and what food to eat); those who "perform otherness" (i.e., performers of "traditional" music, producers of "traditional" crafts, etc.); and "cultural intermediaries" acting as mediators

[31] See footnote 21 in the introduction for the origin of this concept.

between tourists and the local community. The problem with such conceptualization, however, lies in the fact that it presupposes the inevitable inauthenticity of the interpersonal contact that takes place within the framework of tourism: its staged performativity. However, participant observation conducted for this book proved that the relationship between guests and hosts in tourism is much more complex and dynamic,[32] and that using different tactics to develop such relationships does not mean that the parties do not experience them as authentic.

During my field research, I often had the opportunity to observe practices of engaging tourists by involving them in daily and professional activities. Udai and Mohan, for instance, like other Rajasthani jewelry retailers, offered crafting workshops. In part, it was of course a tactic to somehow cope with stuckedness in shops and long working hours during the season (see chapter 1). Surely, making jewelry in front of customers also served to authenticate products in the "cool authenticity" sense by enhancing the trader's authority as a professional and guaranteeing that the goods are not "made in China" (although only a small part of the assortment was usually produced in-house). Above all, however, these mini-workshops offered tourists the possibility of direct contact with the manufacturers and with their material culture by participating in the process of its production, and not just watching, as is usually the case during "staged backstage" tours (see MacCannell 1976).

The most disadvantaged workers in the informal tourism sector also used their own cultural capital. One example is Geeta, who, as a result of the "spatial cleansing" of the bazaar, lost the premises previously leased for several years to Phurbu who runs a restaurant there. This allowed her to support herself and her two daughters, one of which was disabled, after her husband's death. After the demolition of the bazaar, she was taken in by her mother-in-law and brother-in-law but was still looking for a way to ensure her economic independence. She started an informal business offering tourists cooking classes of "traditional Indian food" in the kitchen of her temporary home. Her take on the "traditional cuisine" actually involved a north-south fusion: *aloo gobi*,[33] *sambar*,[34] *dosa*,[35] *idli*,[36] and *roti*.[37] Since her mother-in-law started renting the roof of her house to Phurbu to run a restaurant there, customers who wanted to use the toilet had to go downstairs and enter the

[32] And often going beyond economic motivations, as I explain in the next section.
[33] A potato and cauliflower dish, popular in the northern states of India.
[34] Spicy vegetable soup/sauce based on lentils and tamarind fruit, served in the south of India, usually with *dosa* or *idli*.
[35] Stuffed cornflower pancake, common in the southern states.
[36] The already mentioned in chapter 2 steam-cooked, round, flattened dumplings with no filling, popular in the southern states.
[37] Unleavened bread prepared on a griddle, an integral part of many meals in India.

house—this is how Geeta recruited tourists interested in learning how to cook (one dish cost 100 rupees, that is, about 2.5 dollars, including ingredients and consumption). Sitting with Geeta's daughters in the adjacent room, I had the opportunity to eavesdrop on these encounters. While cooking, Geeta talked about her life, her widowhood, her daughters, and the demolition, thus giving tourists the opportunity to establish a close relationship with an "ordinary local," in addition to granting them access to the private space of her (temporary) house. This also provided her with an opportunity to advertise her other services: painting hands and feet with henna, that is, *mehendi*[38] (which was done mainly by her younger daughter), or selling sari.

Ravi, a self-taught guide, also offered cooking classes. He was motivated by a growing competition from licensed guides. He learned how to cook while working in a restaurant at the bazaar, where he was a waiter, but he also helped in the kitchen. His offer was a package: 500 rupees for five dishes. It all started when, responding to the need for "hot authenticity," he invited tourists for refreshments to his rented room (he was a permanent immigrant in Hampi) and, over time, he created a demand to which he himself responded. He also offers massages—when asked what kind, he answered, "special, relaxing, my own idea." He saw the guarantee of "cool authenticity" coming with ethnic associations: if someone was an Indian, they could do "Indian" massage (and cook "Indian" food). He had no massage parlor, and could not afford to rent a separate space, so he worked by the river, which he claimed to be a "natural place" for this type of activity. Since he did not incur rental costs, he could offer lower prices than other masseurs, which gave him competitive advantage. This line of business developed when he started offering massages to tourists weary from sightseeing. After some time, he decided to train in it and do it commercially.

The examples of Geeta and Ravi demonstrate how flexible the workers of the informal tourism sector are, and how they are able to respond to existing—or self-generated—demand. As they are rarely able to make a living from just one activity, they combine different kinds of work, constantly reshaping themselves and improving their skills. The key here is access to tourists. Services such as cooking classes in one's own—or rented—home are a great opportunity to establish close relationship and build trust, which can then pay off in many ways (about which I write in chapter 5). The culinary sphere is ideal for this. On the one hand, it satisfies the need for "cool authenticity"—in each tourist guidebook one can find whole sections devoted to local food, with descriptions of more exotic dishes, recipes, and a mandatory list of places to go to in order to "eat like the natives" (Wieczorkiewicz 2004, 274). On

[38] *Mehandī*—a dye paste made of henna shrub and used for decoration of hands and feet (and sometimes forearms and calves). A custom once reserved for brides.

the other hand, "tasting the world" is a way of satisfying the need for "hot authenticity," as shown by Anna Wieczorkiewicz (2004, 2008) through her concept of "a tourist taste." This literal consumption of Otherness is an expression of openness to another culture, symbolized by food and by letting it in. Wieczorkiewicz uses the opposition of interior and exterior as understood by Mary Douglas (1977), with the mouth representing the borderline, a specific checkpoint (Wieczorkiewicz 2004, 279). When I still worked as a tour leader, I had many opportunities to observe the inner struggle of tourists buying street snacks, as I was encouraging them to taste the food. Above all, however, food also allows people to "fraternize with the natives" through the myth of feasting together: "The images of a wanderer being admitted to a tribal feast by a bonfire, a tourist being offered a bowl of rice or a traveler being invited into the home for dinner, depict the communal sharing of a meal, evoking the sense of returning to an unmediated, 'pure' human relationship" (Wieczorkiewicz 2008, 303).

RELATIONSHIPS: INTIMACY, RECIPROCITY, RESPONSIBILITY

According to Kristin Lozanski, the natives whom a tourist wants to "befriend" cannot be just any natives—relationship with the tourism service providers will be seen as commercial (2010, 754–755). Since modernity has been founded on capitalist money exchange, an authentic thing will denote something that is not commodified. Tourists want to have an insight into a reality that has not been prearranged for their needs, they want to believe that "the world was like this before they have arrived and will be like this when they leave" (Wieczorkiewicz 2008, 78). For this purpose, "ordinary people" are needed—the "picturesque Others," as Deborah P. Bhattacharyya calls them in her critique of human representations contained in the *Lonely Planet* guide to India (1997, 383). However, my research has not confirmed these observations. First of all, I have noticed that the workers of the informal tourism sector—despite the context of the service provider–customer framework, which seemed unfavorable for fulfilling the need for authenticity—have been very successful in building significant relationships with tourists. Secondly, this context did not seem to bother the tourists themselves. According to Lozanski, it is not only tourism workers who do not satisfy the tourist desire for authenticity but also those "natives" who resemble the tourists too much. Tourists are supposed to crave contact with Indians who are different from them in terms of class, mobility, and cosmopolitanism (2010, 754). My research did not, however, confirm Lozanski's claim. Familiarity with the global world and high level of intercultural competence of workers and small entrepreneurs did not affect their authenticity, but rather facilitated the establishment of interpersonal relations.

Numerous anthropological studies show that even in a "paradise controlled" (Dann 1996), that is, in all-inclusive resorts, where the division of roles between providers and recipients of services is strictly defined and formalized, tourism workers are able to overcome these limitations and build close relationships with tourists (Kanemasu 2008; Ghodsee 2005, 43–75). However, the informal nature of the sector, as is the case in Hampi and Dharamshala, offers greater opportunities for this to happen (cf. Simoni 2017), not only because it is a more democratic tourist space and that it allows for flexible role changes. It is also encouraged by the fact that work in the informal sector, especially in small businesses, is a combination of attachment to a workplace and boredom while waiting for customers. Seasonality forces tourism workers to be at work all the time: with no days off, from morning to evening, or even longer, because in the evenings the tourists—busy with sightseeing during the days—finally have some time for leisurely shopping, which is additionally fostered by more bearable temperatures. Moreover, labor mobility often leads to blurring of the boundaries between private and public spheres. Many migrants—especially those who move without families—do not rent separate accommodation for themselves to reduce costs, so the workplace becomes their home. However, even in the case of family businesses, the family members spend several hours a day in a shop: eating, watching TV or videos on smartphones, playing with children, taking a nap. The life of the "natives" thus becomes fully accessible to tourists (see Krog 2007, 95), while interactions with tourists offer the informal tourism workers a form of entertainment during long working hours. Norbu used to say, "This work is pleasant because you can sit and chat with customers."

Julio Aramberri (2001), in an article with the provocative title "The Host Should Get Lost"—referring to the metaphor of hosts and guests introduced by Valene L. Smith to describe both sides of a tourist contact—suggested that these categories are inadequate to describe encounters in the area of tourism. In his opinion, protection (extended by the host to the guest), reciprocity (i.e., a mutual bond), and duties for both sides (e.g., respect for the host manifested by obedience) that characterize the traditional hospitality are absent in the commercial hospitality based on money exchange and service provision. However, Aramberri's conclusions are based on an analysis of mass tourism and the formal sector, while anthropological studies show that many minor forms of tourism fall somewhere in between these categories and exhibit certain characteristics of the traditional hospitality (see, for example, Heuman 2005). The following section provides exactly that kind of contribution. It shows how my research partners—the hosts and their guests—positioned themselves in a partnership relation and how they activated the principle of reciprocity in mutual relations, proving that tourism has a potential to create not-so-empty meeting grounds.[39] I discuss the roles of "cosmopolitan natives" and "caring guardians" assumed by the employees and small entrepreneurs of tourism sector, that is, emphasizing their own intercultural competences and taking responsibility for tourists. Finally, I look at their attempts to distance themselves from the capitalist nature of tourism through the category of "tranquility."

"A Friend, Not a Servant": The Need for Closeness and Partnership

One of the jewelry sellers in Hampi, when considering moving his shop to the other side of the river due to the precarious situation caused by the demolition, indicated the dehumanizing nature of the service provider–customer arrangement as a justification for his decision:

> Rich people have recently started arriving in Hampi, staying in [upmarket] hotels in Hospet. But they hand us money as if we were beggars [as an illustration my interlocutor made a gesture of throwing money at the counter]. On the other side [of the river] the tourists are different, nice, they sit down with you, talk to you, smoke with you. This is real tourism!

Distancing oneself from the capitalist nature of tourism does not only involve tourists treating locals in a friendly manner but also involve the same kind of positioning of self in these relations. My housekeeper Bhavani's brother put it this way: "Those people in Hospet do not know how to deal with tourists, they serve tourists as if they are their servants. Nobody will talk to you in those [expensive] hotels like a friend, like we do, nobody will advise you, give you support." Research among hotel workers in Fiji, to which I already referred in this chapter, has shown that while they have largely internalized the image of a friendly native Fijian serving tourists with a smile, they have not accepted some of its elements, including those of servility, suggesting their self-subordination. Their resistance was aroused by photographs from tourist brochures, such as the one depicting a fully dressed waiter, knee-deep into a sea to serve drinks to a couple of white tourists wearing swimsuits and floating on mattresses (Kanemasu 2008, 119, 121–123).[40] At the same time, Kanemasu noticed that they valued the reciprocal courtesy of the tourists,

---

[39] Here I refer to Dean MacCannell's critical essays, *Empty Meeting Grounds* (1992).

[40] Charlotte M. Echtner and Pushkala Prasad point out that such performances resurrect asymmetrical relations between former colonizers and colonized, often characterized by master-servant power relations (2003, 674).

which was the starting point for friendships sustained over the years through successive visits, gift exchange, or keeping in touch by phone. Seeing oneself as helpful friends allows people working in the tourism sector to build a sense of self-esteem (one of the waiters in a Nepalese restaurant in Hampi completely gave up this service role by claiming that he "only helps his friend in running the restaurant").

I often witnessed tourists lingering in shops, stalls, travel agencies, restaurants, and reception desks of guesthouses, chatting with owners and employees. A tourist from France, for example, discussed with Norbu the superiority of Ayurveda—the ancient Indian medicine—over conventional treatments, Paris being expensive, Chinese buying vineyards in the south of France, and how many hours a day French children spend at school as compared to Indian ones (Norbu's son, Tsering, had just started his school education, so his father was vitally interested in this topic); an Irish tourist, as another example, chatted about the Om syllable,[41] about what *karman* is (Norbu recommended him to read the book by Kelsang Gyatso[42]) and about the symbolism of swastika, but also about business and life; another tourist from France initiated a conversation about the construction of the road from Lhasa to Kathmandu via Mustang (the Frenchman made a comparison with the roads built in Europe by Hitler for strategic purposes; both him and Tenzin—the owner of the travel agency in Dharamshala, where this conversation took place—were nodding their heads, gloomily), about the Return March to Tibet organized in 2008 by Tibetan refugees (which route coincided with the itinerary of the Frenchman), about India being too "soft" for China, and so on. Apparently, small entrepreneurs in the tourism sector did not seem particularly disappointed when such talks did not result in a transaction. Arun Saldanha, a researcher focusing on the informal tourism sector in Goa, also pointed this out: "Being bored, the vendors [on the beach] are happy to chat even if they know it won't lead to purchase. The supposedly economic motive then becomes a source of irony" (2007, 120).

One of the tramping tour leaders with whom I worked during mobile ethnography tried to translate the cultural practice of haggling precisely in the context of building direct relations with the hosts. She told tourists, "Treat it as an opportunity to chat, to make contact." Reacting to an essentializing comment of one of the tourists, who stated that "they [Asian salesmen] do not respect people who do not haggle," the tour leader replied, "It's not that someone wants to fool you. It is more for fun, not hard-core negotiations,"

---

[41] The most important syllable in Hinduism, considered to be the beginning of the universe. It has often been used in products meant for tourist, for example, as part of jewelry.

[42] Founder of the controversial New Kadampa school, a branch of the Tibetan Gelugpa tradition (Wylie: *dge lugs*), which came into conflict with the Dalai Lama (see Dreyfus 1998).

and to another tourist's comment that "this is a waste of time!," she explained, "This is how it is in our country, because we don't have time at all. Here it is a form of interaction, and not, as in our case, just keying in the codes and asking for 60 zloty."[43] In this statement, chatting was represented as an objection to commercialized, dehumanized interpersonal relations, reduced solely to a monetary transaction; in other words, it appeared to be an opportunity to satisfy the need for "hot authenticity."

On the one hand, such conversations allowed tourism workers to act as experts in their own—very broadly understood—culture, a kind of cultural interpreters (see Nash 1977). This is particularly important in the case of Tibetan Buddhism, since its modernized, exile version attracts tourists who are "seekers of spirituality" (as described in the previous chapter). They expect every Tibetan they meet to be not only a Buddhist but also a Buddhist scholar. In the Tibetan shop in Dharamshala, where I used to spend long hours drinking tea, the owner often explained to his clients, students of Buddhism, how to make an evening puja or how to read Buddhist texts. Acting as an expert-interpreter may be motivated by economic gains (increasing the "cool authenticity" of goods and services), but it may also serve to build a sense of self-esteem and dignity (being an authority on the culture desired by tourists). On the other hand, for workers of the tourism sector, such talks are an opportunity to get firsthand information on life in other countries and learn how to deal with foreigners (Kim 2001), strengthening their own intercultural competences, which they can then use in building relationships with other tourists. The tourists themselves, driven by the need to "fraternize with the natives," seem to be keen to enter into such relationships. Marek, an entrepreneur and a regular client of tramping tours, pointed them out as one of the main advantages of his preferred form of travel: "There is better connection with local people, because there [during tramping] we do not just get into the bus and step off the bus—we have conversations, [time] to talk to people, make contact." Roman, an electrician, talking about a long tiresome journey on the Indian train, stated, "You can totally get along with these people. You can talk, sit, laugh with them." A student named Ela recalled her stay in Dharamshala:

> I felt such a cool atmosphere, that you can join in, talk to everyone, take a stroll, it is a small town, . . . such an atmosphere, I would say relaxed and calm. . . . I had a nice conversation with a boy who showed me *thangkas* and *mandalas*, explained everything. I sat there [in the shop] for half an hour. We chatted, and this man was so nice that I really enjoyed sitting there.

---

[43] Polish currency.

Speaking about Sangmo, a Tibetan woman running a small bakery and one of the protagonists of the last chapter, Ela said the following:

> She is an amazing person. She does not just greet you, but asks about your day, wishes you to "have a good day," she includes you in an interaction . . . you feel like you want to talk to her. . . . And although you are aware that she may talk like this to a hundred or two hundred other people, you feel [like someone] special, there is something in her voice, such an authentic care, it is not just about having more customers . . . you feel that it is for real. Not just professional but having a deep wish to answer your questions.

The above statements indicate that the business-related context of these encounters did not prevent tourists from experiencing them as authentic. Staszek, who also likes tramping, stated unequivocally:

> If such a person is a shop owner, then you know that it is in their best interest to somehow encourage you. I take it with a pinch of salt, but I do benefit from it to talk about their family or work, or to ask how the business is going, whether they are happy. For example, when I was buying this plate, I asked how many people there were in this family, etc.

Research by Magdalena Banaszkiewicz (2012), conducted among the participants of few-day trips from Poland to St. Petersburg, confirmed the above observations. The tourists, in spite of such an extremely standardized format of traveling, and, in addition, clearly focusing on material culture, were hungry for encounters with the locals and considered such interactions to be very valuable moments during the trip. Moreover, they did not mind at all that these contacts mostly included people who served them (receptionists, waitresses, trainmasters).

I myself often saw tourists inviting guides or traders for meals to a restaurant, showing each other family pictures, and exchanging contacts. A retired couple, who after bringing the children up, realized that they were very fond of tramping, reminisced, "One day, we were sitting there and these women were selling vegetables—we even took their address because they asked us to send them pictures. . . . [W]e promised to send them. . . . [T]hey were happy, they were so happy, they hugged me." Such encounters often happened without verbal communication, which did not prevent both sides from experiencing them as meaningful. Mary Conran calls them "affective interactions" (2011, 1463), while Rebecca Ogden writes about "affective capital" involved in establishing relationships in tourism (2015, 177–193).

On the other "side" of the contact, the needs were similar. A painter from Rajasthan, who run his small studio in Hampi, was convinced, "Tourists are an inspiration for me. When they come and say: 'I like your paintings,' I feel like I want to continue working." When asked what kind of tourists he liked the most, he didn't use a geographical or national classification, but started talking about meetings that proved important to him: with John from the United States, who, when he entered the shop, had tears in his eyes and the painter hugged him; with Nicolas from Spain, who asked him to put his signature on the picture, which he does not usually do, because the paintings are pretty mass-produced, but in the end he agreed, because the Spaniard "was an artist too" and, as proof of his close relationship, he added, "With love." In one of the letters that the painter received, the tourist wrote the following:

> I learned from u a lot about art and heart. Never give up art! You are kind and gentle, and [you are] the worst businessman I ever met. I hope you stay like that, loyal to yourself. I wanted you to remember me every day, so my gift for you is a Ganesha statue. I wish God will hear your prayers and give you all you ask for. It is a real pleasure and honor to meet you. In true love, [signature].

These encounters are often followed by "excursions into the ordinary" (Joy 2010, 61; see Horolets 2012) so desired by tourists, because they satisfy their need for "hot authenticity." Raj—the owner of a guesthouse in Hampi—used to invite tourists to a banana tree plantation that belonged to his family. He even took several clients to the funeral ceremony of his deceased uncle. My housekeeper Bhavani invited her long-term tenants for special festive meals at home during important celebrations. Sharing meals—what I discussed in the previous section—was an experience that tourists often pointed out as significant. When during interviews I asked what impressed them the most during their trip to the north of India, many tourists indicated that it was the meal to which they were invited in the Sikh temple in Amritsar. I recorded, for example, the following statement: "Before this trip, if someone asked me a few years ago if I would sit with pilgrims and eat with them, from one bowl, I would never have imagined it would be a reality."

Of course, such closeness can be interpreted in terms of "distanced friendship" and "apparent participation" (Wieczorkiewicz 2008, 207), as it is often limited to one-off meeting, enabling the tourist to maintain a safe emotional distance from the economic and social inequalities in the backdrop of these relations (Horolets 2012, 124). However, this is not always the case, as I describe in chapter 5 of this book, referring to specific stories of such encounters and their follow-up. Moreover, the availability of relatively cheap smartphones in India and the prevalence of instant messenger platforms, such as Facebook and WhatsApp, allow both sides to stay in touch after the tourists return home, which fundamentally changes the ephemeral nature of these relationships, extending them beyond the holiday.

It also happens that the need for closeness does not really benefit the workers in the tourism sector. John was supposed to go for lunch with the Rajasthani painter, but never got back to him to set the date, and Nicolas did not buy the picture—the artist tried to contact him through me to send the painting as a gift, but the email address written on a piece of paper was illegible. I was surprised to learn how often my research partners showed a far-reaching openness in their relations with tourists, something bordering on naivety. The story of a jewelry vendor from Rajasthan is just one example. The man sold a tourist from the United States goods worth over 700 dollars agreeing to a later payment. There was no ATM in slowly vanishing Hampi, and in the nearby shops, where one could withdraw cash and where they went together, the American could not procure the money ("because it was a new card"), but he showed the seller with his smartphone that he had money in his account. The next day he had to continue his journey, but he promised to pay everything once he was in Dharamshala next month (where the Rajasthani trader was also heading for a summer season). When I met him there three months later, the American still texted from time to time that he was already on his way, and then his Indian cell number stopped working altogether. The salesman, who I thought was a victim of a fraud, seemed unaffected and clearly did not intend to treat this experience as a lesson to limit trust he placed in his relations with tourists: "One in ten people can be like this, the rest is fine," he said.

Nelli was an interesting victim of tourists who wished to establish close contact with the Other. In fact, she attracted them as their own Other. Her visibility was not only due to her European origin but also due to the arrangement of the shop, which was so tiny that it was virtually impossible to sit inside, so Nelli was constantly visible. As a result, she often had customers who would not buy anything, but would sit for an hour or longer, talk about their problems or ask about her affairs. The conversation typically started with a question, asked in complete disbelief: "How did you end up here?" Nelli was particularly annoyed by the latter as she knew that in Europe people would not ask such personal questions, but the need to "fraternize with the natives" pushed tourists to trespass on her personal space.

"Sharing Is Caring": The Principle of Reciprocity

Mohan, a jewelry salesman from Rajasthan, said once, "I like this job because it is about taking and giving, sharing things and thoughts, meeting people." Such an exchange can involve a variety of "goods" (in addition to the know-how serving to develop intercultural competences). One of the most popular are photographs. On the one hand, these are photographs of the hosts taken and gifted by the guests, which I have already mentioned in the example of a tramping couple and vegetable vendors. After the forced displacement from the bazaar, in his miserable shack erected among other temporary housing structures, the man who looked after the temple elephant held large prints of photographs taken by an Australian journalist who regularly visited Hampi, depicting him proudly sitting on a festively decorated Lakshmi elephant during a procession. My housekeeper Bhavani received a photograph of her whole family taken on the roof of a Korean restaurant run by her brother. Tourists kept sending her pictures from the restaurant she lost in the demolition of the bazaar, which made her very moved because it brought back memories. On the other hand, the family albums of tourism workers were full of photographs of tourists—most often referred to as "friends"—offered during their subsequent visits or sent on various occasions (weddings, birth of a child, daughter's performance as an angel in school Christmas play, or holidays spent elsewhere). However, photographs of both parties posing together were the most common. In Geeta's album one could see, among others: a British woman, "a good friend for twenty years," with both women posing with their children; further on, the British's six-year-old daughter, now a teenager, sitting on a bus surrounded by the children from Hampi; then, the picture from Geeta's wedding shows a tourist dressed in sari; there were also photographs of her already deceased husband in a restaurant, posing with tourists who embraced him; there was a picture taken a year ago showing Geeta's niece in the company of two European children; and so on.

Today, these albums are usually digital: Facebook and other communicators, especially WhatsApp, make it possible to send photos in real time at no cost, thanks to the availability of wireless networks in tourist destinations. A tourist who takes a selfie with a new friend can immediately send the photo to his or her phone and then, once back home, send other pictures from family celebrations or simply from everyday life. Many times, while hanging out with tourism sector workers, I witnessed them receiving photo messages from tourist friends, for example, a photo of a snow-covered street in Amsterdam (at the time when the temperature in Hampi reached thirty degrees Celsius). Photographs were also sent the other way, so tourists could keep up with events happening in the life of the community and their befriended families. This virtual-visual proximity is very inclusive because it overcomes the limitations of communication resulting from a lack or poor knowledge of English. Moreover, the audio messages allow for communication with completely illiterate people, such as young Lambani shopkeepers.

The act of taking a photograph itself can also be a form of communication via a digital camera display or a smartphone screen. This is how Staszek used to initiate contact: he took pictures and then showed them to the people he photographed, who in this way became simultaneously the recipients. In his case, the camera was not a barrier between the photographer and the outside world; it was not a shield with which he was protecting himself, in line with the common interpretation of how tourist photography practice works (e.g., Bruner 1989; Wieczorkiewicz 2008).[44] Until recently it was a contact initiated almost exclusively unilaterally, in which power relations resulting from the technological and economic superiority of tourists from the so-called Global North were acutely expressed—it was the tourist, as the owner of the camera, who chose the object, measured it, and pressed the shutter button. Today, however, thanks to the availability of smartphones and the growing number of domestic tourists, foreign tourists themselves are becoming the subject of the gaze (what I already discussed in the previous chapter) and the photographed objects. Although for many of them this reversal of roles caused surprise and even annoyance (when having to pose for photographs with the whole class of students made it impossible to "properly" visit the Taj Mahal), others saw this change through the prism of exchange and reciprocal relations. Ela, in a conversation with her trip companions, said, "It's a bit intrusive, but we're taking pictures of them too, so it's cool." Ewelina, who, as a blonde, was frequently the object of these practices, said, "It's actually touching—we not only take from them, but also give something." The popularity of selfie, in particular, makes it possible to move away from the classic approach to photography as a distancing and objectifying practice because it blurs the division of roles between the photographer and the photographed. Ewelina willingly agreed to be photographed—with young boys or with a child in her arms handed over by a proud father—and she initiated such joint photography sessions, for example, with Tibetan monks.[45] By the same token, those who previously acted as objects became codirectors. Not only posing and watching became a collective endeavor, but also the shooting itself: the joint review of the results, the request to take the picture again or delete it (see Haldrup and Larsen 2010, 122–153). If we move away from the interpretation of tourism photography through the lens of its representations of Otherness, as proposed by Michael Haldrup and Jonas Larsen in the chapter of their book devoted to the performative aspect of the social practice of photography in the context of tourism, we will see that "photography can now also be an end in itself . . . a source of pleasure, creativity and sociability" (2010, 126).

Visual evidence of a close relationship with tourists, one that is based on reciprocity, can be inscribed in the body. The auto-rikshaw driver from

---

[44] Wieczorkiewicz wrote, "By aiming the lens [of our camera] at something we want to include in the set of our trophies, we get separated from a reality that may somehow threaten us" (2008, 179), while Bruner argued that "the camera serves as a protective device for the tourist-photographers, socially isolating them so that they do not have to relate directly to the New Guineans, face to face, eye to eye. They can hide behind the camera lens" (1989, 441).

[45] Moreover, this variability of roles allowed tourists to become aware of the relativity of the category of the Other, when they played this role in the eyes of Indian tourists.

Hampi had a deal with an Italian tourist friend, a professional tattoo artist, who tattooed him in exchange for a tour. He commented on this exchange: "Sharing is caring." When the Italian visited him once again while attending a tattooing tournament in New Delhi (he had to travel almost 2,000 kilometers to do so), the driver gained a new tattoo of which he was very proud because, as he stressed, "All my tattoos are European[46] and all were done for free!" Ganesha and the head of Buddha Shakyamuni tattooed on both arms of a Tibetan restaurant owner from Nepal, mentioned in the previous section, were also the result of such an exchange: tattoos for meals. Many tourists, like the driver's Italian friend, return to Hampi and Dharamshala regularly, often over several years, becoming the "veterans" mentioned in chapter 2. Guided by the principle of loyalty, they usually stay in the same places, becoming someone oscillating between a regular customer providing a secure income (all the more so as they usually stay longer) and a family friend. They call on holidays, ask about health, about business, about friends. They bring gifts, send parcels, especially clothes for children—the toddlers then parade dressed in fashionable, branded clothes along the village dusty streets. Once, a friend who made a living by guiding tourists to a nearby viewpoint, appeared dressed in militant-type trousers. The temperature already exceeded forty-two degrees Celsius in the shade, and everybody were boiling wearing thin lungi, but he claimed that he was not hot at all: "It's from a friend from Germany," he explained proudly, "you won't get a pair like this in India." Sometimes workers and small entrepreneurs in the tourism sector visit their foreign friends and stay in their homes and vice versa, which was the case of an Australian woman, who used to bring groups of fellow Australians to India as part of spiritual tourism and thus regularly visited Hampi, where she has made friends with Norbu and Dolma—at least a few times she spent her holidays in their empty house in Manali.

In terms of reciprocity, such relations are based on an exchange of gifts and favors which, by binding both parties within a system of commitments, also enrich them (Mauss 1966). Every gift must be reciprocated, that is, the recipient of a gift is obliged to reciprocate. However, this is not a barter—the exchange has a "moral dimension," as Alfred Radcliffe-Brown wrote in relation to the Andaman Islands in India stating that "these exchanges were less important economically than as maintain relations of friendship" (1922, 174). Though, this does not mean that the value of the gifts did not matter, as the reciprocated gift should be of equivalent value. As far as tourism is concerned, this principle, classically in anthropology referring to the social

46 There was one tattoo studio in Hampi on that time, run by a young Lambani from a nearby village (tattooing is a common practice in this group), but the auto-rikshaw driver said that it "reproduce[s] patterns from the Internet, while those [made by an Italian] are original and unique."

norm regulating the relations between different groups in premodern societies, was studied and reinterpreted by Adams (1992). She demonstrated that the relationship of reciprocity can be developed between tourists and workers in the tourism sector, thus widening the definition of the sector as based on market-oriented service provider–consumer relationship (see, for example, Sahlins 1972, 149–183). It often turns out to be a reconstructed form of traditional relationships—for Sherpas, who Adams worked with, it was a system of reciprocal work embedded in kinship, mapped onto new realities (see also Heuman 2005). Exchange, in the sense of the principle of reciprocity, must be extended over time—as opposed to one when payment for goods or services automatically ends the relationship.

My research partners valued relationships and the capital—social, cultural, but also economic—that they had the opportunity to accumulate through tourism, in exchange for what they themselves offered to the visitors. These relationships, which were often built over the years, are difficult to classify as they bring various benefits to both sides, both material (earnings, discounts, good quality goods, gifts) and not so easily measurable. In the last two chapters of this book I describe how workers and small entrepreneurs in the informal tourism sector use these benefits to strengthen their own, often disadvantaged position and overcome the constraints imposed on them by their political and social position.

Cosmopolitan Natives: Cross-Cultural Competences

As I have already mentioned several times in this chapter, contacts with tourists allowed employees and small entrepreneurs in the informal tourism sector to acquire intercultural competences. These competences included not only the knowledge of foreign languages, especially the lingua franca of the tourism sector in India, that is, English, which is particularly important for people who entered labor market as children without any formal education.[47] Above all, it is about gaining "practical" knowledge through contacts with other people, who have different cultural and situational backgrounds, ways of perceiving reality, and of communicating. This requires openness to others, a desire to understand the rationale that guides their opinions and actions (see Gudykunst 2003). In Kristen Ghodsee's research conducted in a tourist resort in post-socialist Bulgaria, the opportunity to meet other people, and through them "the world," as well as daily, free language training with

[47] Previously, this was only interpreted by researchers as a manifestation of power relations, because, as a rule, less educated staff learned the language of usually more educated tourists (Nuñez 1977; see also Cohen and Cooper 1986).

native speakers[48] were perceived by the resort's employees as significant nonfinancial benefits of working in tourism, which allowed them to construct "international" identities (2005, 64–65).

Tucker, who carried out ethnographic research on backpacking in Turkey, uses the term "cosmopolitan" in relation to the competences acquired by those dealing with tourists in the informal sector (2011, 34). Cosmopolitanism here is understood as "a willingness to engage with the Other," "a state of readiness, a personal ability to make one's way into other cultures, through listening, looking, intuiting, and reflecting," and "a built-up skill in maneuvering more or less expertly with a particular system of meanings" (Hannerz 1996, 103; see also Vertovec and Cohen 2002). The tourism employees described by Ghodsee claimed that these benefits were comparable to traveling (2005, 64–65). It is, therefore, a kind of cultural exchange, which allows employees in the tourism sector to avoid seeing their relationship with tourists as one-sided, and themselves as inferior. Working in tourism can thus be interpreted as traveling-in-dwelling (Clifford 1992, 108, 1997), that is, enjoying the cultural diversity of the world without having to move physically.[49] Tourism sector workers, usually perceived in opposition to cosmopolitan tourists as settled and having a clearly defined identity, also become participants in the world of mobile meanings and ideas, thus transcending the binary division between rootedness (in territory/culture) and cosmopolitan openness (see Glick Schiller and Irving 2014).

This, in turn, makes them "highly competent in playing to those imaginings, but they also seem able to play with them, frequently engaging in ironic performances and caricatures of the tourists' images of themselves as cave-dwellers" (Tucker 2011, 33). Tucker describes how young men from Göreme—a Cappadocian village of rock-carved houses—played with the tourist image of themselves as cave-dwellers, reenacting scenes from Flintstone family comedy cartoon in front of tourists (2003a, 169–180). This was to undermine the power relations underlying such representations, in line with the logic: I am aware of your expectations, I reflect on them and show that I can distance myself from them, because—in fact—I am just like you. On the one hand, then, the cultural translation that occurs in the field of tourism, necessarily simplified and comparable to a pidgin (Graburn 1984), may strengthen the existing mutual representations. On the other hand, direct

[48] One of the bartenders said, "I can learn new vocabulary and improve my pronunciation. I can learn jargon and slang, things I cannot learn from books" (Ghodsee 2005, 64).

[49] We travel this way when going out for sushi to a nearby restaurant, making green tea from Sichuan at home or watching programs on National Geographic channel. One can also live by dwelling-in-traveling—in truck cabs, on cargo ships, or in chain hotel rooms during endless business trips. These concepts indicate that spatial mobility does not necessarily entail the experience of Otherness, just like this experience does not necessarily have to involve physical departure.

contact that takes place in the informal tourist sector, the intercultural competences of its employees, and their high self-awareness can serve to deconstruct these representations.

All of this allows workers and small entrepreneurs in the informal tourism sector to position themselves as cosmopolitan natives, especially in contrast to people who have no experience in working in tourism. I often heard references to such knowledge, gained from interactions with tourists[50] or from tourism workers' own experiences as tourists, in actual conversations with tourists. For instance, I had an opportunity to listen to Raj, an owner of a guesthouse, telling tourists from the Netherlands how much he loved the modernist architecture of their country, for which reason he was to go to Rotterdam. He traveled couch surfing, which he often mentioned in the presence of foreign tourists, his prime motive declaring to be the desire to "learn from other people's cultures." Bhavani and her brother running a Korean restaurant, while sharing a home-cooked meal with a young Korean man occupying a room adjacent to mine, named numerous Korean dishes (Bhavani was famous for making four different kinds of kimchi[51]), expressed opinions about the power of Korean liquors, and freely used Korean names. The tourist in question spent his time in Hampi on lone photographic excursions and sitting in a room (he was there longer than me), and he dined almost exclusively at Bhavani's brother. Bhavani herself, on the other hand, when I told her one day that I was going to a jam session organized by a bunch of ex-tourists settled in the Hampi area, exclaimed, "I love Miles Davis!" and she added, "and classical music too, and African music! In my restaurant you could always hear the music from all over the world." It turned out that tourists shared MP3 music files from their phones with her and told her about the music they liked. Even children in tourist destinations accumulate cultural capital: they volunteer to speak English with tourists and are familiar with new technologies (my laptop was constantly used for games, other tourists' tablets served for watching cartoons). India is obviously a technological tycoon, but one must remember that here we are talking about the country areas.

I myself was not only a source of information for my research partners but also served as an excuse to demonstrate their own cosmopolitanism resulting from familiarity with tourists. This was the case during my visit paid with

---

[50] Including those with myself. During long, informal conversations with my research partners, classic research relationships—the researcher as the one who wants to learn something and the "subject of research," who is supposed to act as the source of this knowledge—was often reversed. Like other tourists who spent a lot of leisure time in shops, workshops, or restaurants, I served as an expert on my own culture, answering numerous questions about cuisine, politics, weather, economy, and so on.

[51] A popular Korean dish based on pickled vegetables, served as appetizer or as part of other dishes. Bhavani's daughter recalled that the whole house had smelled so much of fermented cabbage that their neighbors and relatives stopped visiting them.

Raj to seasonal agricultural laborers from the Lambani group, who worked on the surrounding sugarcane and banana plantations. We sat in their temporary camp, drinking chai, and Raj acted as a cultural broker. The Lambani women in particular were shocked that I traveled alone—they said I was leading a "strange, crazy life." They asked about my husband, children, and finally about my income. Raj tried to protect me from having to answer ("I know it's a personal question"), and when I finally told them, and the Lambani could not help being surprised that I earned so much, he decided to perform cultural translation. He turned to the woman who was clearly the leader of this group and said, "You're rich, not her! You have a house and some land, and she probably has to rent a flat. You think she makes a lot of money and can afford a lot, but in her country, she can't afford a hotel room, oh no! Do you know how much is one *idli* in her country? And how much did she have to pay for the flight to come here?!"

The role of cosmopolitan natives required a benchmark. For Raj in the above scene there were Lambanis—backward villagers. A Kashmiri manufacturer of clothes considered "pushy" sales strategies used by other Kashmiri entrepreneurs to be unprofessional, meant to confuse customers: "I never do this, I only indicate the prices quoted on the clothes, inform about the availability of colors and sizes, and leave the tourists alone." The fact of doing business with his wife, a French woman, made the man position himself at a higher level of "civilized" trade as compared to other Kashmiris. The inhabitants of Hampi very often manifested their cosmopolitanism in contrast to the employees of the formal tourism sector from the city of Hospet. One of the restaurant owners from Hampi stated, "These people [managers and employees of hotel restaurants in Hospet] cook everything very spicy [hot], they don't serve espresso, don't know what Lavazza[52] is. They don't speak languages, so they can't explain what's on the menu. We not only speak English and French, but now also Russian." He was the first person in Hampi who decided to cater for Russian tourists and is now an expert in this category of clients: "They need at least half an hour before they place an order, because you have to explain everything to them. They have no idea what *thali*[53] is." Tibetan entrepreneurs, on the other hand, explained that the difference in their attitude toward clients—that is, "They [Indians] are pushy, we are restrained, not imposing ourselves"—stemmed from their familiarity with *inji*.

---

[52] An Italian coffee producer, one of the world leaders in selling coffee to restaurants. In 2011, Lavazza opened its factory in India, in the state of Andhra Pradesh, neighboring with Karnataka.

[53] A popular way of serving meals, especially in Northern India, where food is served on a special plate with compartments for different dishes, the combination of which makes up the whole. Being unfamiliar with the idea makes it difficult to place an order.

Caring Guardians: Responsibility for Tourists

The owner of a restaurant in Hampi serving Korean food, while talking about the attractiveness of Hampi for international tourists, referred to the "living" heritage founded on "hot authenticity," which he contrasted with the "cool authenticity" of "lifeless" monuments: "People come here not for monuments, not for [tangible] heritage, not for UNESCO, but because they feel good here, because we create a good atmosphere." This involves making self-representations of Hampi's tourism service providers as experts in catering for tourists' needs. The narrative of responsibility for tourists—their well-being and safety—often went hand in hand with emphasizing high competences in the field of "tourist culture." This was accompanied by "buttering-up," also described by Adams in the context of the Sherpas who were spending a lot of time hosting tourists and making them feel welcome in Nepal, making them feel that the trip was worth the money they invested in it. This did not only apply to their own clients but apply to tourists in general, and the extensive networks of reciprocity that they constructed in this way also included Nepalese and other Sherpas (Adams 1996, 216–217).

Since I stayed with Raj for a few weeks, in his family guesthouse adjoining the restaurant, I had the opportunity to observe how he was building his own representations of a caring host in his relations with tourists. Raj always paid much attention to all new clients. He talked to them at breakfast: "Is the food all right?," "Is everything OK?," "Happy in Hampi?[54]"; when tourists were paying the bill, he animatedly recounted them the history of Hampi as if he was narrating a documentary, claiming, "I'm so lucky I can work here. And you are lucky too, that you can be here. Each stone here tells you a different story. So slow down, take your time, because some people stay here for a few months [here he used to tell them stories of long-term residents]."[55] The trust that Raj gained in this way paid off. Although he did not officially run a travel agency, he was the first person that tourists addressed to as a trusted cultural broker, whom they asked to arrange trips to the surrounding area. He himself subcontracted these services, asking, for instance, Santosh to act as a guide, because—as Raj explained to tourists—"Santosh means 'satisfaction.'" Then he joked, "Make sure to watch him as he pours two liters [of gasoline] into

[54] It was also the password for wireless Internet in Raj's restaurant.

[55] Then he would come back to my table—my mobile office, where I worked on my field notes—and said he was going to sleep because he was tired and (in response to my incessant questions) that the Internet did not work, "because it doesn't work." Such scenes were a sign that my status in the field had clearly changed—I had been promoted to a buddy whom one could jokingly disregard. The same was true in my temporary home in Dharamshala, where I once expressed my concern about a key to my room that might have got lost and the response was "Well, then you'll have to sleep outside, *ji* [honorific, here used ironically]." The owners, young Indians, usually quite restrained in relations with customers, were often teasing the guests they liked.

the gas tank, and then take the key," thus laughing off tourist' fears of being cheated.

Workers and small entrepreneurs in the informal tourism sector are aware of the common perception of tourism as an immoral sphere. Tourist guidebooks, travel blogs, and tour leaders warn about the high probability of being "ripped off" at every step and offer plenty of advice on how to protect oneself against it (haggling instructions, use of various techniques to verify the authenticity of the goods,[56] limited trust, etc.). Malcolm Crick, in his street-corner anthropology, wrote—referring to street vendors and people whose job is to bring tourists to selected shops and restaurants—that "no matter how clear the evidence is at the overall systems level as to who benefits most from international tourism . . . , at street level, the tourist is frequently the victim. I certainly experienced the tourist arena very much as a predatory niche" (1992, 178).

This led to absurd situations when tourists, convinced that not haggling will be a proof of their lack of cultural competences and will increase the risk of them being cheated, forced sellers to inflate prices—in this way, the feared practices seemed to be self-induced. Small entrepreneurs, my research partners, often anticipated these fears, emphasizing their own honesty through such practices as introducing fixed prices on offered goods (often advertised with a board suspended at the entrance) or fixed rates (Udai and Mohan set prices for jewelry products by adding a 20 percent markup to the price of silver as regulated by the government—"no more, no less"); encouraging tourists to compare prices with those asked by other sellers;[57] "rewarding" non-haggling customers with unsolicited discounts; not pressing (Udai: "If they dream of a ring, they will come and buy it anyway"); openly informing tourists about the quality of their goods (Norbu and Dolma pointed out to the display cases with silver jewelry and those where it was made of other, less precious metals; when asked about a blanket with the label "100% yak wool Tibet shawls," they immediately admitted that it is not woolen[58]). Mohan claimed that they were tired of constant suspicions: "I've been in this business for thirteen years. I am tired of making up stories [about the origin of offered items], haggling, arguing. I never put pressure on customers. If they

[56] For example, using a cigarette lighter to set fire to threads from scarves was a method to check if they were made of silk.

[57] Longer stays in Hampi and Dharamshala allowed tourists to get a sense of prices and types of goods, especially as the various sales groups specialize in the same assortment. In more mass and anonymous spaces, through which groups of tourists pass quickly rushed on by tour operators, there are greater opportunities for overpricing and offering lower quality without damaging one's good reputation.

[58] This tactic seems to affect the tourist need for "cool authenticity." In practice, however, it serves to satisfy it—the indication of goods which are "non-authentic" confirms the authenticity of those with whom they are contrasted.

want to, they'll buy, if they don't, they won't. I earn enough. I don't need more. I offer reasonable prices." Many of my research partners were aware of the fact that excessive encouragement to buy goods or services was perceived by tourists as pushy: "In Hampi, everyone keeps on trying to attract tourists and to push something: bananas, T-shirts, rooms, postcards," complained a guide offering tours in French, "so I'm not looking for clients in front of the temple anymore; I only work when I get a call from the travel agency or from the guest house." He complained about unprofessional, overly pushy colleagues, who "spoil the market" and make other guides feel "like prostitutes" accosting tourists.

Building a sense of security does not only concern financial issues but also concern personal safety, especially in case of female tourists. My research took place at a time when the globally constructed image of India as a dangerous place for women, because of cases of gang rapes, resulted in a significant drop in the number of women traveling alone. One of the tactics used by my research partners to mitigate this fear was to consciously keep a distance when dealing with women traveling solo. A novice guide to Hampi told me about the advices he got from more experienced colleagues. First of all, he was to always require payment for his services, otherwise "she will think that you have some agenda" (a tip he got when he was first asked by a female tourist to guide her to a local viewpoint). Secondly, he was not allowed to send to the tourists text messages like "I'm glad you're coming, because we will be able to spend some time with each other," since it could be interpreted as having romantic undertones. A friend of mine working as an auto-rickshaw driver in Hampi told me a story about how he defended a female tourist, whom he was driving around the local sightseeing sites. During a stopover, a local man, on the pretext of taking a photo with the woman, suddenly kissed her for which the driver punched him in the face: "She was my client, I was responsible for her," he explained to me. The story was set within a narrative of competence in the "tourist culture," where competent men were contrasted with men from villages who "do not know how to behave towards women. They watch [Western] films and think that all Westerners behave like this [i.e., kiss women in public], so they can do it too. Fools!" Unlike them, they, or "Hampi boys," never do that, because they "grew up with tourists and know how to behave."

To sum up, workers and small entrepreneurs in the informal tourism sector acted as cultural brokers, equipped with the intercultural competence local experts, enabling tourists to have a safe contact with the sight-visited reality (see Smith and Brent 2001, 275–282[59]). These were not just guides (Cohen

[59] A separate section devoted to cultural brokers was only added to the third edition of *Hosts and Guests*, along with an introduction by Valene L. Smith.

1985) or travel agents (Adams 1984). A characteristic feature of the informal sector is that, as I have already pointed out in the introduction to this book, it allows a broad, democratic access to tourists, which in turn ensures the accumulation of intercultural competences necessary in dealings with tourists for a wide variety of actors: auto-rickshaw drivers, waiters, postcard sellers, shopkeepers, stall owners, and so on (cf. Rybus and Wieczorkiewicz 2014, 151). As noted by Crick, who worked in a similar field in Sri Lanka, "In the touristic arena, nearly everyone is some kind of stranger handler" (1992, 178), which makes the high position of intermediaries resulting from access to tourists and their capitals (cultural, social, and economic) more accessible. Crick refers here to the category of "professional stranger-handlers"—that is, people who specialize in guiding strangers around their world—a category with which anthropologists are very much familiar (Agar 1980, 85–86, 117). Just like ethnographers, tourists, especially those looking for "hot authenticity" and in the informal sector in particular, need the support of those "gatekeepers" and other cultural brokers who would enable them to relate to the reality they came to visit.

Shanti: Saying No to Capitalist Profitability

I was often surprised that workers and small entrepreneurs in the sector, which in essence should be governed solely by the capitalist rules, were often guided by other motives and did not treat profit maximization as the main determinant of their decisions and strategies. Norbu used to say that in Hampi—unlike in Goa—"you can relax." I was puzzled by this category, which seemed contrary to the economic nature of his migratory practices. Just as profitability was not the key preference criterion—with postcrisis "Europeans" as the favorite tourists—it was neither key to choosing the place where people run their businesses. All my research partners agreed that more money was to be made in Goa—and most of them had experience of working there—but they chose Hampi, because "Hampi is *shanti*,[60]" that is, tranquil. This sentence reappeared in statements of my interlocutors, recited like a mantra, and was a bit nostalgic in the context of the atmosphere of uncertainty caused by the ongoing displacement. In other words, the tranquility of doing business was clearly preferred to high income and was an important criterion of satisfaction. The protagonists of Ghodsee's book from the Bulgarian tourist resort also paid great attention to the fact that tourism gave them the opportunity to work in a nice place (and not in an office), in addition to being surrounded by "happy people who are on holiday," as one of the waitresses

[60] Sanskr. *śāntiḥ*.

stated, "If I were a doctor, I would have to work with sick and worried people all the time" (Ghodsee 2005, 64).

Of course, in the case of Hampi there were also other arguments which come into play: the costs of renting premises were lower (although, in fact, they had been so in the past, as demolitions generally substantially reduced the supply), and the competition was smaller, with the majority of small businesses,[61] which did not aspire to become big. It seems that the capitalist jungle my research partners recalled in their Goa stories clearly frightened them. They were aware that they would have to pay a lot to survive as businessmen in Goa. Firstly, it was about commissions. With more mass, more organized and more expensive forms of tourism in Goa, paying tour leaders and guides to bring customers to shops or restaurants is common practice, while in Hampi—at least until the arrival of Russian tourists and their "guardians"—the practice was pretty nonexistent. The same is true for Dharamshala, where individual tourists staying in India for longer visits make their own choices as to where and what they buy.

A second source of reluctance to doing business in Goa was the presence of a "mafia," who charged small entrepreneurs for "protection," that is, for the right to do business, even if it were just scarfs or cheap, artificial jewelry sold by Lambani women on the beach.[62] In Hampi and Dharamshala, where the division between local and incoming populations is fluid, such practices, based on the superiority of sedentarism over mobility—in line with the logic "this is our place, if you want to trade here, you have to pay us"—were not present. Moreover, the itinerant traders in Goa were trapped between the local "mafia" and the police, who organized "raids" on the beaches under the pretext of combating the practices of touting tourists. Lambani saleswomen from Hampi told me that they often had to run away and that if they failed, their goods were confiscated. The fact that this type of trade is carried out mainly by young women exposes them to various forms of violence, including sexual one (see Singh et al. 2012). In Hampi, the Lambani could afford to rent small premises and run more stable, secure businesses.[63]

[61] Businesses in Hampi are often the first independent economic ventures run by former employees of the tourism sector.

[62] Goa was a natural point of reference for people in Hampi as it situated relatively near, but similar opinions were expressed about other leading tourist destinations in India. When I asked Mohan why they did not run their store in their native Rajasthan, which is a popular tourist destination, he said that the high rents in Jaipur combined with the presence of the local "mafia" were forcing shopkeepers to overprice their goods, which he did not wish to do.

[63] Since the demolition of the bazaar, the street sales in Hampi were limited to teenagers from the nearby villages (mostly school students in the afternoons and on weekends) who sold postcards and several young men (migrant traders) who offered simple musical instruments. In addition, the women displaced by the demolition sold coconuts, flowers, and incense to pilgrims (used as offerings in the temple), and bananas bought by tourists. Around thirty people, mainly women, traded in plastic toys made in China displayed in trolleys set up at the bus stop belonging to the temple,

In my conversations with workers and small entrepreneurs in the tourism sector in Hampi, people often used the phrase "good energy." Mohan, for example, used to say that he "likes temple sites because they have a good energy." Norbu assumed that if Hampi were to eventually disappear—although he really did not want to leave—he and his wife would choose a village in Tamil Nadu,[64] "near some peaceful Indian temple or ashram." Phurbu initially worked on the other side of the river, but decided to move to Hampi, "because here is more *shanti*, close to the temple." The related restrictions, which are especially important when running restaurants, as the ban on selling nonvegetarian dishes and alcohol and the obligation to close the premises at 8:00 p.m. did not seem to bother him. Interestingly, combining the category of tranquility with the proximity of the Virupaksha temple—and the generally sacred nature of the site—appeared in the narratives of all workers in the tourism sector, regardless of their religion; they were articulated by Buddhists and Muslims alike. Similar comments were made in Dharamshala, where the Dalai Lama acted not only as the main magnet for tourists (at least until the cricket stadium was built) but also guaranteed peaceful character of the site. It is striking how inconsistent this perception is with the pathologizing image of tourism workers constructed by religious leaders, especially in Hampi, where the representations of tourism as an immoral economic exploitation of a holy place, promoted since the 1990s, significantly contributed to forced displacement and demolition of tourist-related infrastructure (see Bloch 2017a). None of those working in Hampi considered moving their business to Hospet, located twelve kilometers away—there was a general consensus that this was a large Indian city, "ugly, crowded and noisy." Noise, congestion, and growing pollution were also the most frequently indicated aspects of the negative change occurring in McLeod Ganj, noted by both long-time workers of the tourism sector and foreign tourists, especially those from "the Europeans" category, who moved to the quieter surrounding villages, especially Bhagsunath and Dharamkot, where the locals still farm fields and graze their cattle.

The category of tranquility also included relations with the local population. In this regard, again, the point of reference was Goa, whose inhabitants were considered by my research partners from Hampi to be "spoiled," that is, "drinking and smoking [hashish] a lot and chasing for money." There were numerous stories about overpriced accommodation, extortion, harassment, and generally bad treatment of seasonal migrants. In contrast, people living in Hampi appeared to be "good people who do not cause troubles to

---

which was an offer mostly for Indian tourists. Finally, several Lambani women from a nearby village used to sell petty handicrafts, also displayed in trolleys.

[64] The southernmost state of India.

the outsiders," "not very wealthy, and thus honest," "friendly and not jealous," who do not take advantage of their position (of being a local), and do not exalt themselves. Although the demolitions have seriously damaged this relationship—uncertain of the future, the inhabitants of Hampi began to raise rental prices in an attempt to maximize their profits, here and now—the local residents and migrants enjoyed a symbiotic relationship that had worked well for years. The tenants and landlords were usually connected to each other over many seasons which created a certain sense of closeness (for instance, a local woman, who rented premises where Phurbu run his restaurant, assisted in the birth of both of his daughters and supported his wife Padma after delivery). I would even say that these relationships were much better than those between the Hampi residents themselves, just as the image of them in the eyes of the migrants was much more positive than in their own opinion.

Chapter 4

Recovering the Subaltern Voices

Tourism and Engagement

POLITICIZING TOURISM: TOURIST AS AN ALLY

Culture can be used by marginalized groups as a political resource in their struggle for recognition, the right to land or economic survival, as shown by numerous anthropological studies (see, for example, Conklin and Graham 1995 on the indigenous Amazonian populations). Skillfully reproduced uniqueness—due to, for example, the processes of self-Orientalization and strategic essentialism discussed in the previous chapter—"make these groups visible on the international scene and thus attractive for tourists" (Macleod and Carrier 2010, 13). In this perspective, tourism has the potential to become a space for the recovery of voice by subaltern communities lacking political representation.[1] Skillfully politicized tourism can make tourists allies in the struggle in which these communities are involved, the issue that I wish to address in this section.

In both locations where I did my research the situation of subalternity does not only result from the fact that the inhabitants of the postcolonial country play a service role in relation to the privileged visitors from the so-called Global North but—as I wrote in chapter 1—is also a consequence of exile and statelessness (in the case of Dharamshala) and displacement and demolition (in the case of Hampi). Therefore, the tourist is offered a "two in one" special deal: the package of the cultural offer (i.e., satisfying one's need for "cool authenticity") contains a political message. This gives the tourist a new, engaging role: "the ambassador of the cause" (in the case of Dharamshala, fighting for free Tibet, and in the case of Hampi, fighting against displacement).

[1] See the category of subalternity discussed in the introduction.

The efforts of postcolonial critics to reclaim the subalterns' voice (Spivak 1993) are close to the anthropological search for silenced narratives that began in the 1970s. The anthropologist Edwin Ardener (1975a, 1975b), for example, referring to research carried out among the Bakweri in Cameroon, laid the foundations of a "muted group" theory, indicating that certain groups are silenced by the "dominant structures" of the society, which have the power to identify, that is, the power to state what the world and the people living in it "really" are. In the original sense, this concept was concerned with women and the criticism of anthropological descriptions of the world created exclusively from a male perspective—both of the researchers and of the informants. The greater presence of men in the public sphere makes their models of reality form a binding discourse—it is them who talk about the world of men and women, women being merely subject to observation (like Nuer's "cows"—as Ardener ironically compares [1975a, 4]). Women are "mute" because, although they speak, they do so in a language different from the male one, a language that is largely ignored—they are dummies in the theater of men (Ardener 1975a, 2). All they can do is to learn the dominant language (concepts and norms) to communicate with their environment, but this does not allow them to express their own perspective and vision of the world. Spivak, who has been reflecting on the (im)possibility of regaining the subalterns' voice, did so using the example of women. I return to this issue in the next chapter and its part devoted to the emancipatory opportunities offered to women through tourism.

Over time, the "muted group" theory was extended to other groups (ethnic minorities, lower social strata, non-heteronormative people, children, etc.) for which others spoke (nation-state, elites, heterosexual people, adults, etc.). One way to overcome silencing is to educate the dominant group about the situation and values of the subaltern group, which is to be done by the subalterns themselves (Obre 1995), that is, through creating their own self-representations. Spivak, like Ardener, was concerned about whether, in the face of the power of the current discourse, there are ways in which subaltern/muted groups could speak for themselves and be heard. For the idea of reclaiming a voice is not, in fact, about speaking, but about being heard: women have always spoken, but what and how they spoke was classified in the cultural hierarchy as less worthy than a normalized male voice and therefore unworthy of attention. Tibetan refugees and displaced Hampi residents certainly belong to muted groups. The language with which they formulate their political demands—addressed, among others, to tourists—is unquestionably the language from a borrowed dictionary: human rights, the nation-state, cultural heritage. However, as I show in this section, this language has been skillfully appropriated by them and used as a tool in their struggle. Tourism has become a new platform through which these claims can be heard.

Engaging tourists in the role of allies appeals to their sense of morality, and is an attempt to induce empathy. Of course, this is not an easy task, because—as Anna Wieczorkiewicz (2008, 207) emphasized—a tourist, while on holiday, does not want to take on "responsibility for poverty, hunger, and suffering that prevail in poor countries," so while trying to satisfy the need for "hot authenticity" (e.g., by fraternizing with the natives), such tourist tries to keep a safe distance. This involvement in political struggle or a direct action for the local community can be read as part of the project of "moralization of tourism" (Butcher 2003), since it is an attempt to give tourism a deeper meaning and make it useful in a broader sense than just as a source of material benefits for individual service providers. Freya Higgins-Desbiolles in her article "Justifying Tourism: Justice through Tourism" (2010) writes about the idea of a just tourism that aims to counteract global injustices. Various forms of tourism, which according to Higgins-Desbiolles (2010, 200–201) have such potential, range from responsible tourism (e.g., environmentally conscious tourism) to transnational solidarity–based activism, through PPT, fair trade tourism, volunteer tourism, and reality tours. The most radical form of just tourism is to actively support local communities whose rights (human, political) are being violated, in their efforts to overcome the unequal power relations that limit them. Higgins-Desbiolles calls it "justice through tourism" and the people committed to it "active tourists for justice," using the moral argument that "privileged tourists should justify their enjoyment of tourism by using their opportunities to work for justice and better living conditions for those who live in the places they visit" (2010, 200).

One of the first researchers to introduce the concept of justice tourism was Regina Scheyvens (2002b). She pointed out the following attributes of such tourism: (a) building solidarity between visitors and those visited; (b) promoting mutual understanding and relationships based on equity, sharing, and respect; (c) supporting self-sufficiency and self-determination of local communities; or (d) maximizing local economic, cultural, and social benefits (Scheyvens 2002b, 104). This is how the idea of justice tourism coincides with the hopeful tourism mentioned in the introduction (Pritchard et al. 2011) and its study with the "anthropology of the good" (Robbins 2013). A good example is provided by the research by Aleksandra Gutowska (2015), focusing on slum tourism in the Kibera district of Nairobi. Gutowska goes beyond the victimization narrative, that is, interpreting slum tourism as a subordinating practice of poverty consumption by rich residents of the Global North. She portrays the people of Kibera as active social actors who have internalized this form of tourism and are using it, first of all, to change their image in the eyes of visitors, creating their own self-representations as entrepreneurial, active, agentive people. In doing so, they skillfully deconstruct the image of the slum dweller built with the categories of apathy and despair, in favor of development and hope. Secondly, they treat the guided tours they give to tourists from around the world as an opportunity to establish direct contacts, thus building social capital and raising economic capital to get them out of poverty. Gutowska demonstrated how the money raised from this type of tourism—provided that it is organized by the slum dwellers themselves—are used to support local social initiatives such as the implementation of vocational courses.

The idea of justice through tourism obviously brings with it certain risks highlighted by Higgins-Desbiolles: short-term contact inherent in tourist encounters, the decrease in tourist involvement following their return home, and above all, arousing pity instead of compassion, bringing forth a paternalistic attitude and strengthening tourists' sense of privilege (2010, 207; see Mahrouse 2008). In this section I look at how the idea of just tourism is implemented in Dharamshala and Hampi by attempts to engage tourists in the political struggles for the right to self-determination, community's own vision of heritage, and the right to the benefits generated by tourism.

"How Can I Support Tibet?": *Momo* with Nutella and the Performance of Torture Victim

Although Dharamshala has become such an important point on the tourist map of India, the refugee authorities did not seem to see tourism as a significant source of income for members of the diaspora. I was surprised that in statistics on the economic situation of refugees in India, the tourism sector was not taken into account and rather hidden in-between different categories, such as trade and services. At the same time, however, tourism in Dharamshala was very much supported by the Tibetan administration and diaspora leaders. It seems that according to the refugee elite, tourism fulfilled primarily the image-making function, as an element of identity politics and the mission of promoting awareness of the Tibetan cause, that is, the fight for free—independent or autonomous—Tibet. Dharamshala as a tourist attraction was therefore intended to be a stage for political self-representations of Tibetans as pacifist Buddhists, victims of the communist regime, with a clearly defined, distinct from the Chinese, identity (see Bloch 2014a). Thus, it was to be a place of winning international moral and political support. This is why the shift toward a new kind of clientele, that is, Indian tourists focused on recreational tourism, was not only problematic for Tibetan vendors and service providers, whose hitherto customers were fleeing to more secluded places, but also worried the local Tibetan authorities, as it threatened their image-making project.

Creating self-representations and engaging tourists in political activities for the Tibetan cause happened on several levels. The first, most visible one, referred to the cultural production for the tourism market in the form of "political souvenirs." They all referred in many different ways to the "Free Tibet" slogan, which has actually become a brand of the Tibetan diaspora. As a result, tourists paraded through the narrow streets of McLeod Ganj dressed in T-shirts calling to "Stand for Tibet" (with the image of a row of people holding hands or with hands folded in a request/thanksgiving) or announcing "Proud to be a Tibetan." On cold evenings they could put on a tracksuit sweatshirt with a patch with the Tibetan flag and the inscription "Tibet is not part of China," and during the monsoon season hide under an umbrella with the Tibetan flag and the "Tibet will be free" slogan. They could go shopping with a cotton bag decorated with the motif of yaks urging: "Let's come back to Tibet." Tibetan national symbols were placed on hats (a baseball cap in summer and woolen for winter, some of which were knitted by elder Tibetan women[2]), scarves (as in sports clubs), jackets, mugs, key rings, and accessories for smartphones. Creativity in this respect seemed to have no limits and every time I visited Dharamshala I discovered new motives and reinterpretations. One of the manufacturers, for example, Tibetanized global trademarks, making the North Face or Coca Cola call for "Free Tibet." There was also an image of an eye in which the iris was set against Tibetan flag with a tear rolling down over a call for the United Nations to "Make change for Tibet." Another local manufacturer used the Tibetan national flag as their logo, sewn on sweatshirts and T-shirts as a tag, adding to it the initials JC (first from his name, second from his wife's name) under it (figures 4.1–4.4).

It seems to me—on the basis of observations made in Dharamshala since 2002—that initially, the production of gadgets with national symbolism was meant for the Tibetans themselves and was implemented mainly by NGOs and individual activists.[3] Gradually, however, it became tourist-related, which did not mean that the symbolism had not been commodified previously. Already a long time ago, Tibetans produced and sold plastic headbands worn on their foreheads during demonstrations, which earlier had been homemade from a piece of cloth torn of a bedsheet. Only later on, such objects gained the status of souvenirs and became vehicles of the Tibetan cause—tourists, by purchasing, wearing, and offering them as gifts to their relatives or friends, made the political goals of the Tibetan diaspora visible (and audible), all the more so because these souvenirs were mostly very handy. Therefore, as some of my interlocutors pointed out, even if these were bought by Indian tourists not out of the need to show political support, but as an exotic souvenir, they still fulfilled their role.

---

[2] During one of my field trips to Dharamshala, I saw a very elderly Tibetan woman who was knitting the words "Human Rights in Tibet" on a cap.

[3] As was the case of a tracksuit sweatshirt with the inscription "Team Tibet" and number 08, referring to the lack of Tibetan representation at the Olympic Games in Beijing in 2008, which was produced by the local branch of the international organization Students for a Free Tibet, or T-shirts that were entirely made out of the Tibetan flag created by an activist of the Tibetan organization fighting for the animal rights' protection.

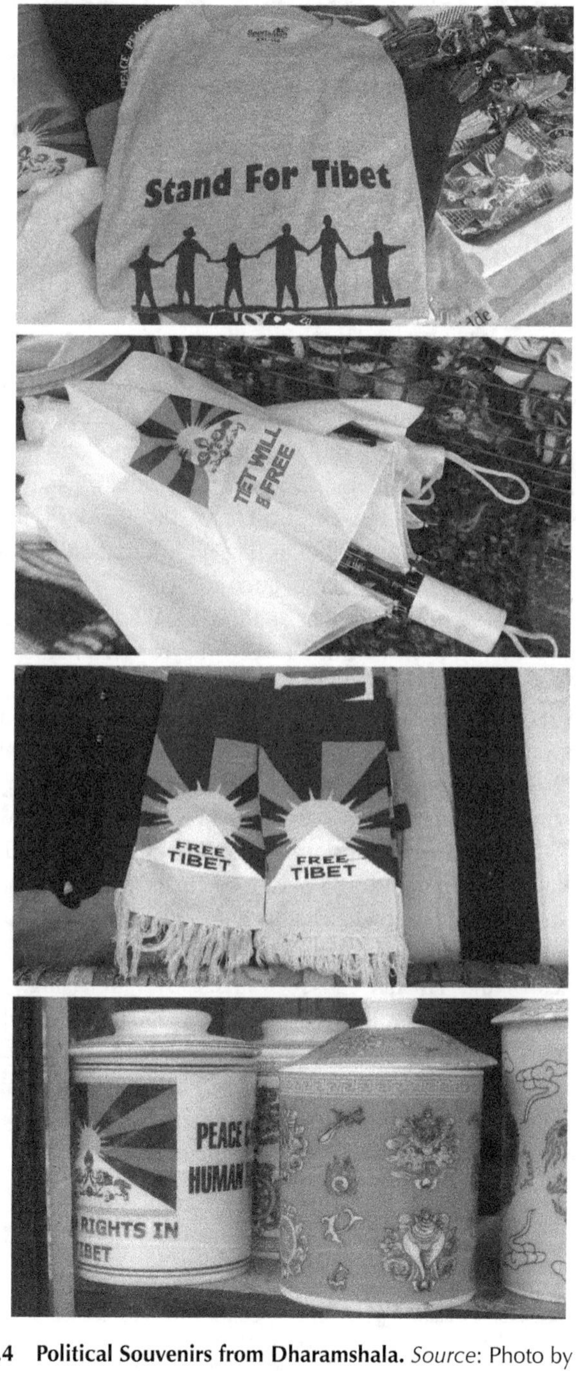

Figures 4.1–4.4 Political Souvenirs from Dharamshala. *Source*: Photo by Natalia Bloch (2018).

Figure 4.5 Political Street Art in McLeod Ganj by Coletivo RAGA and a Tibetan Painter, Karma Sichoe. *Source*: Photo by Natalia Bloch (2013).

Moreover, even if the tourists did not buy such political souvenirs, they still remained an audience of their messages as the tourist space of Dharamshala was saturated with political declarations. Even the Wi-Fi passwords, the prime need of tourists in restaurants or guesthouses, carried a political message: "gototibet," "savetibet," or "tibetwillbefree." Young Tibetans, most often already "born refugees," placed national symbols on their clothes, laptops, and motorcycles as part of their daily patriotism practice. The same symbols were used to decorate shops, restaurants, and reception areas in guesthouses. They took the form of murals or even tattoos, which have become very popular among Tibetan youth.[4] Therefore, political souvenirs also referred to the tourists' need for authenticity, that is, searching for objects "not meant for tourists," purchased and used by the locals (Wieczorkiewicz 2003) (figure 4.5).

The above examples indicate that production for the tourism market can have two recipients: tourists themselves, but also members of the host community, as I have already mentioned in chapter 3, discussing the phenomenon of ethnic entrepreneurship (Comaroff and Comaroff 2009). The political self-representations created for tourists fit perfectly into the refugee identity politics and the making of the Tibetan nation within the diaspora. A similar phenomenon was noted by researchers of Singapore, where a

---

[4] Susan T. Chen points out that Tibetan refugees in Dharamshala are very aware of how they are "seen" (2012, 279–280), hence consistently inscribing the politicized Tibetanness into one's body. On the other hand, many of the more conservative members of the community consider tattooing to be too far-reaching interference that, as a form of self-harm, is essentially non-Buddhist.

governmental campaign promoting tourism created images of multicultural harmony, with Chinese, Indians, Malays, and other Others, all symbolizing the new meaning of being Singaporean (Ang and Stratton 1995). Tourism in this case became an important instrument for building national identity and unity: "Tourists' fascination with the country's ethnic composition would foster a sense of civic pride, which in turn would help knit the ethnically diverse people together" (Chang 1997, 552, following Hall and Williams 2002, 21).

The second area in which Tibetan refugees tried to engage tourists politically was educational events, so numerous that the visitors could easily spend their entire holiday in Dharamshala attending them. Every Monday and Thursday, a local NGO organized screenings of Tibet-related films—mainly documentaries about human rights violations in Chinese Tibet. An additional attraction was the meals served during the screenings, which contributed to the organization's budget (participation in the screenings themselves was free of charge and everyone was offered free tea; meals were optional).[5] On Tuesdays, there were meetings with a former political prisoner sharing his experience of torture in a Chinese prison—on the occasion, participants were treated to a Tibetan noodle soup called *thukpa*.[6] In the meeting room, there was a large portrait of the Dalai Lama, accompanied by the leaders of two other sects of Tibetan Buddhism, a scroll explaining the meaning of the Tibetan national flag, another one depicting eight Buddhist fortune symbols *tashi tar gye*[7] and posters with English poetry by the Dalai Lama ("Never Give Up," "The Paradox of Our Age," and "Meditation on Dying"). There was also a bookcase with a small collection of Tibet-related books in the corner. After the screenings tourists were offered to buy a copy of the movie so that they could show it to their friends once they return home. The copies were made by the Tibetans themselves and were available everywhere in Dharamshala—in violation of copyrights but "for a legitimate purpose."

The effectiveness of this tactics of promoting the Tibetan cause among tourists can be demonstrated by the story of a volunteer in the abovementioned NGO, an Australian of Scottish origin, who at one of the screenings explained to the audience how he came to Dharamshala: "I knew nothing about these shocking things happening in Tibet until a friend who came

---

[5] I conducted participant observation during these screenings for two years in a row and the attendance never dropped below twenty person, which—given Dharamshala's size, its rich cultural offer, and the frequency of screenings—was a very good result. The majority of attendants were young people, mainly traveling within the gap year. Women participated in such events more often, many of whom were single travelers in their thirties.

[6] Wylie: *thug pa.*

[7] Wylie: *bkra shis rtags brgyad.*

back from holiday brought the *Tibet: Murder in the Snow*[8] movie with him." He then added, referring to meetings with refugees and listening to their stories, "What we receive from these people here, we should return when we get back home." Thus, he referred directly to the principle of reciprocity and the tourists' sense of morality—this experience created a desire to give something of himself in return (Russell and Russell 2010).[9] Scheyvens mentions "the 'hosts' telling their stories of past oppression" as one of the five forms of justice tourism (next to tourists learning about poverty issues; tourists undertaking voluntary conservation work [regarding traditional ways of living, natural environment, etc.] or development work, which will be discussed in the next section; and revolutionary tourism) (2002b, 105–119).

On Wednesdays, another NGO, located just next door to the one mentioned above, held "awareness talks" about the history of Tibet's independence and Chinese occupation. These lectures were held in a café run by the organization and the audience could follow a professional Power Point presentation accompanied by tea and snacks. The speaker was well prepared (after all, he graduated from political science): he peppered his talk with jokes in order to make the audience more comfortable after he had shared some drastic bits of information, he constantly made a connection between the Tibetan-Indian reality and the setting familiar to tourists, and referred to his own life experience. He awakened empathy through skillful cultural translation. When he described the hardships and dangers of escaping from Tibet—referring to his own story when, as a six-year-old, he crossed the Himalayas—he explained as follows (figure 4.6):

> When you travel by airplane and you need to catch another flight, you are a bit nervous. You also get nervous or irritated when you are required to ask for a visa. We have no chance of getting a visa. We need to travel illegally, on foot, crossing the mountains. The Himalayan climbers conquer peaks using professional gear and clothes, while we only have light jackets and cheap sneakers.

When the speaker discussed the installation of rocket launchers by the People's Republic of China in the Tibetan Highlands, he joked, "So if we

[8] It is a documentary (produced by 360 Degree Films in cooperation with the BBC in 2008) telling the story of international mountaineers who witnessed a Chinese border patrol shooting by to a group of refugees fleeing Tibet through the Nangpa Pass. The climbers managed to film the killing of a seventeen-year-old nun, which shocked the public. However, because this tragedy took place in 2006, many young tourists visiting Dharamshala no longer knew it from media coverage. The movie was often screened during the analyzed events.

[9] One of the popular slogans which appeared on the political souvenirs available in Dharamshala read "Tibet needs you."

Figure 4.6 Tourist Attending the "Awareness Talk" about Tibet in McLeod Ganj. *Source*: Photo by Natalia Bloch (2013).

regain independence, you should not mess with us! Of course, some of them may not work because they're made in China." Throughout the presentation he consistently demonized the Chinese Other as culturally different from the Tibetan Self: "The Chinese eat EVERYTHING"—he stated, suggesting that the decreasing yak population (the national symbol of Tibet) is the result of carnivorous tendencies of the Chinese. The last slide of the presentation made the audience ask themselves, "How can I support Tibet?" And it immediately provided answers:

- Be a member of different NGOs to know more about Tibet [pointing especially to the Students for a Free Tibet[10]].
- Give gifts that are Tibet related such as scarves, prayer flags etc. [the speaker's comment: "If you wish to buy a gift for your relatives or friends, go for something like this. It is that simple"; once the talk was over, one could buy stickers of several kinds, for twenty rupees per piece].
- Wear T-shirts, scarves, and something which is Tibetan [the speaker's comment: "There are many people already tired of helping Tibetan people. When they see other guys wearing a 'Free Tibet' T-shirt, they will be

10 SFT was founded in 1994 as an organization bringing together young people—mainly students—and aiming to support Tibetans in their political struggle. Today it has over 650 chapters around the world and is the most dynamic supranational pro-Tibet organization.

encouraged, you will give them new energy. It is like a handshake with those who support Tibetans"[11]].

- Write articles on your blog, Facebook, local newspaper, school magazine, etc.
- Organize events such as screening documentary, talks etc. [the speaker's comment: "Just like this one"; once the talk was over, everyone could get a copy of the Power Point presentation; always a few people were interested].
- Put Tibet stickers wherever you can [the speaker's comment: "If you place such a sticker on your laptop, on your car or in the bus, then people who will see it may one day sit down in front of their computers and google 'Free Tibet'"].
- Donate time, money and ideas [here the speaker commented on the ability of individuals to bring change].
- Become a volunteer.[12]

On the same day, in a nearby café run by Kelsang, whose story I told in chapter 1, and his newly arrived refugee colleagues, jam sessions were held on the occasion of *Lhakar*[13] (i.e., "White Wednesday"). During these events tourists had the opportunity to listen to Tibetan music and songs performed by local Tibetans. *Lhakar* on that time was a new form of resistance that emerged in Tibet after the suppression of the protests in 2008, when restrictions on any political activity of Tibetans became even stricter. The whole idea was very simple and could be implemented by everyone individually: it was all about doing "something Tibetan" every Wednesday—wearing a traditional dress, speaking Tibetan, listening to Tibetan singers, and so on. But also, if you were a tourist in Dharamshala, shopping in a Tibetan store, putting on a Tibetan-flag T-shirt, having dinner in a Tibetan restaurant,[14] or going to a café to listen to Tibetan music.[15]

---

[11] Tibetan refugees are aware that time is working against their movement. Although its structures are growing stronger, many activists may feel discouraged by the lack of progress in the ultimate goal of liberating Tibet from the Chinese domination (either in the form of independence or "genuine" autonomy). The more emphasis is put on the constant acquiring of new supporters, including tourists. Nevertheless, the global recognition of the "Free Tibet" slogan and the widespread moral support for the Tibetan cause is a huge success and an evidence of the Tibetan diaspora's agency.

[12] I write more on volunteering in the next section.

[13] Wylie: *lhag dkar*. The term was coined from the words *lhag pa*, meaning "Wednesday," and *dkar po*, meaning "white." Wednesday is a day, according to Tibetan astrology, considered auspicious for the 14th Dalai Lama (on this day he was born in the present incarnation).

[14] In the wake of the *Lhakar* movement, Tibetan culture became even more politicized, since "through de-collectivization of activism, *Lhakar* sustains the resistance by empowering the individual" (Dorjee 2013). Of course, T-shirts with the *Lhakar* slogan immediately appeared on the Dharamshala's market.

[15] These jam sessions were also based on the idea of cultural exchange as tourists were encouraged to perform songs from their own countries (once I even had the opportunity to listen to a Polish song *A Dream about Warsaw* sung by a backpacker from Poland). Similar events were organized

Diasporic reinterpretations of Tibetanness were interesting in this context. For example, food served during events promoting knowledge about Tibet was an extremely creative interpretation of the so-called traditional Tibetan cuisine. Tourists were delighted with *momo* stuffed with nutella[16] (called "chocolate *momo*"), "Tibetan pizza," mashed potatoes served with *thenthuk*,[17] or popcorn with *khabse*.[18] When I asked the organizer of one of the events where the idea of serving popcorn with Tibetan pastry came from, he replied that such meetings were to be an opportunity for "exchange." Cultural appropriation of pizza,[19] on the other hand, had more to do with India, since pizza has become one of the most popular dishes among Indian middle class (at the Tibetan events discussed here it was mostly served with spinach, paneer—an Indian cottage cheese—green peppers, and mushrooms). Moreover, the very fact that all other types of *momo* that appeared in the offer for politically engaged tourists were vegetarian (vegetable *momo*, potato *momo*, or potato-cheese *momo*) suggests that the new tradition of vegetarianism, which I have already mentioned in chapter 2, plays an important role in the self-representations created by Tibetan refugees, especially in opposition to the Chinese carnivore Other.[20] Young men who sang songs during the jam session were not dressed in *chuba* ("it's too hot for this in India")—their new "traditional" outfit was a "Free Tibet" T-shirt or a hoodie with the Tibetan flag. These examples show—in relation to bargaining Tibetanness discussed in the previous chapter—that the competition over the right to cultural heritage is not about cultural conservatism. However, it is the Tibetans—seeing themselves as legitimate heirs—who reserved the exclusive right to

---

in Dharamshala already in the 1990s and included screenings of documentaries about the Chinese invasion of Tibet, talks by former political prisoners, and multiethnic jam sessions. The only difference is that they were then centralized in the so-called community center run by the settlement's authorities (Tsering 1998).

16 Nutella is produced, among others, in India, where Ferrero—an Italian owner of the brand—employs the third largest number of employees after Italy and Germany (see http://www.ferrero-india.com/fc-2517).

17 Wylie: *then thug*. A thick vegetable or vegetable with meat soup similar to *thukpa*, but with hand-pulled noodles.

18 Deep-fried pastry served usually at Tibetan New Year or other special occasions, like wedding ceremonies.

19 In front of the Tibetan guesthouse where I lived during one of my field trips, there was a big sidewalk sign with an advertisement which read, "The best Tibetan pizza."

20 Tibetans in exile—regardless of the fact that Tibetan cuisine was shaped by harsh climate and largely pastoral culture—are expected to be vegetarians, which in part is the result of the identity politics implemented by the diaspora's elite and the image of Tibetans created within it. The emphasis made by the 14th Dalai Lama and other Buddhist teachers on the category of "compassion for all sentient beings" in the practice of Tibetan Buddhism, as well as essential for Buddhist doctrine concept of reincarnation, in an interpretation based on Western logic were synonymous with non-eating meat (more on the invented tradition of vegetarianism and animal rights' protection in the Tibetan diaspora see Bloch 2008).

reinterpretation, which they denied to the Nepalese, accusing them of commodifying Tibetan culture.

Educational events which took place in Dharamshala also included various photographic exhibitions, organized by the NGO sector, the government-run Tibet Museum, and the owners of restaurants and cafés. On top of that, there was extensive literature—concerning not only Buddhism but also the political situation in Tibet—available both in commercial bookstores and free of charge at the stand run by the Tibetan Department of Information and International Relations in the heart of McLeod Ganj. Publishing initiatives were also undertaken by NGOs. One of them initiated the *S.O.S. Stories of Our Students* project, publishing stories of escape from Tibet written down by the participants of language classes offered by this organization with the participation of foreign volunteers. This publication certainly had an educational function, but at the same time it attracted new volunteers with the promise of satisfying their need for "hot authenticity," suggesting them to come, meet the people, listen to their stories, and have personal, direct experience. This kind of exchange and inspiration could also work the other way round. In McLeod Ganj, there were sometimes meetings organized, during which tourists, who had the opportunity to visit Tibet, talked about this experience to refugees born in exile, contributing to their imagined homeland picture.

Included in the Tour Itinerary: Visiting the Prime Minister–In–Exile

Moreover, both Tibetan NGOs and the government-in-exile organized political excursions around Dharamshala, both for foreign and domestic tourists. These tours, like all the events and cultural production discussed in the previous section, aimed to involve participants in the Tibetan cause. Such initiatives can thus be read as attempts to reinterpret the tourist sightseeing practice and give it a deeper meaning, which is part of the already mentioned process of making tourism more moral (Butcher 2003). The difference is that here it was initiated from the bottom up, that is, by the host community, and not—as is the case of various forms of just tourism—by the organizers and participants of such tourism. If there were no such activities, the opinion of one of the local activists would probably ring true: "Tourists come here, and when they leave, the only thing they know about Tibet is *momo*." But this offer also involved skillful use of the more standard Dharamshala attractions, both nature and culture-related. One of such trips was advertised as follows:

> If you want to explore Tibet's current political situation and experience its socio-cultural heritage, the Little Lhasa program will be a great opportunity for you. We will spend one week in beautiful McLeod Ganj, learning about the different organizations within the Tibet movement, meeting activists, Tibetan government in exile and monks and nuns. You will also have time to explore the amazing natural surroundings of the Himalayan mountains, visit Buddhist monasteries and eat traditional Tibetan food.

I had the opportunity to conduct participatory observation during such a trip organized by one of the NGOs in summer 2014. Two groups were conducted in parallel: one for college students from Edinburgh and the other for young Indians. I decided to accompany the second group, because I was interested in the turn toward the Indian society, especially in view of the change in the profile of tourists visiting Dharamshala. During my doctoral research, conducted almost a decade earlier, Tibetan refugees sought support for their cause almost exclusively in the so-called West, consistently ignoring their country of residence and birth (which was already then criticized by some activists). Meanwhile, although Indian tourists who came to Dharamshala did not show as much sensitivity to the situation of Tibetan refugees as the activists would have desired (see chapter 2), Indian volunteers and participants of political tourist trips started appearing at McLeod Ganj as well.

Eight participants took part in the tour, mostly students in their first years of humanities departments from large Indian cities, who were required to apply and motivate their desire to participate. To a large extent, their ideas about Tibet and expectations of this trip were not significantly different from those of the tramping tourists from Poland with whom I worked doing mobile ethnography. In their views Tibetanness was confined to Buddhism and Tibetans to monks. Many students were attracted to Dharamshala by "Tibetan spirituality"—a young martial arts instructor from Delhi admitted that he supported political aspirations of the Tibetans because "he would very much like to see in this materialism-driven world at least one state that was based on spiritual values." Nonviolent struggle was also highly appreciated by the Indian participants of the tour, who explicitly referred to the Gandhian tradition, still very much alive in India[21] (for this reason, internal discussions within the diaspora about the need to radicalize methods of resistance were kept away from these tourists). The 14th Dalai Lama was seen as a kind of Tibetan Gandhi—a living guru,[22] a great holy man. Moreover,

---

[21] Tibetan refugees often refer to the authority that Mahatma Gandhi enjoys in India in order to acquire Indian public support for the Tibetan cause: Gandhi's portraits are present during most Tibetan political demonstrations.

[22] Sanskr. *gūru*: spiritual teacher, master.

many participants emphasized that such a form of tourism gave them the opportunity to "interact with local people, which is not provided by ordinary trips"; to "gain an inner perspective"; to "see how it really is." Their desire for authenticity, to enter a "backstage" of a tourist spectacle, was expressed in the same words as was the case with tramping tourists (see Bloch 2016a). Finally, some participants were guided by other motivations, such as accompanying their husband or friends, or—with regard to young women—simply getting out of family home under the pretext of an organized, that is, safe trip and "experiencing something new." At the same time, participants were fully aware of the expectations that the organizers set for them, they knew that they had a kind of moral obligation. One of the Indian women said straightforwardly: "They want us to increase other people's awareness of what happened to the Tibetans."

The first point on the tour's itinerary was the political introduction to the Tibetan cause, that is, the screening of *The Sun Beyond Clouds: Tibet's Struggle for Freedom*[23] in the organization's headquarters. It is a very emotional documentary shot in 2008, when the mass protests against the government of the People's Republic of China erupted in Tibet, including dozens of self-immolations. The movie contains scenes of brutal suppression of demonstrations by Chinese troops, as well as the mass political mobilization of Tibetan refugees. After the screening, the participants were invited to a discussion moderated by the group's mentor, an Indian activist, facilitating the meeting, who did a skillful cultural translation and, moreover, was herself an example of the commitment expected from the participants. On the first day, there was also a visit to the Tibet Museum and its exhibition—being itself a part of the government-in-exile's identity project, which clearly indicated the goals it attributed to tourism. The museum told the story of Tibet's political and cultural independence and the brutal Chinese occupation through a series of illustrated boards and documentaries. The only actual exhibits available on display were a bloody shirt of a tortured political prisoner and the torture tools used in Chinese prisons.

The importance of this type of tourism for the Tibetan political elites was evidenced by the fact that the itineraries of these tours included not only meetings with local activists of all major NGOs[24] but also sessions with

---

23 A documentary made by the most famous Indo-Tibetan directing duo: Ritu Sarin and Tenzing Sonam (White Crane Films, India and Great Britain, 2009).

24 The participants of the tour, which I took part in, visited the offices of the Tibetan Youth Congress, Tibetan Women's Association, Gu Chu Sum (association of former political prisoners), the Indian editorial office of a radio station Voice of Tibet, and the Indian branch of Students for a Free Tibet. They were hosted everywhere by important activists, introduced to the area covered by a given organization, and admitted to organizations' "backstage"; they could ask questions, discuss, interact. In addition, they had meetings with the icons of the Tibetan freedom movement such as the activist and writer Tenzin Tsundue and the poet Buchung D. Sonam.

respected former political prisoners and even with the head of the Tibetan government-in-exile himself, whose schedule was otherwise filled with events—as one could suppose—of much greater rank. Two such meetings were experienced by the participants as particularly remarkable. The first was a visit to a former political prisoner, Ama Adhe, a then eighty-two-year-old Tibetan woman, who spent twenty-seven years of her life in Chinese prisons (Tapontsang 1997).[25] Ama Adhe—a good-natured, cheerful woman and a committed, passionate activist, openly and emotionally recounting the torture and humiliation she suffered—deeply moved young Indians. The meeting was focused on mutual contact, and Ama Adhe herself had a very direct approach to the participants, whom she embraced, patted on their shoulders, held their hands. The second such event was a 1.5-hour visit to the seat of the then Sikyong,[26] Lobsang Sangay, the first democratically elected political leader of the Tibetans in exile.[27] Sikyong, an India-born doctor of law from Harvard, knew perfectly well how to talk to Indian middle-class youth. In particular, the way in which he referred to the threat from China, whose economic domination in India was becoming more and more visible (e.g., in the closure of Indian factories or the ubiquitous presence of "made in China" products); made references to Indian mass culture; and joked were very well received by young Indians (who were encouraged to make personal commitments, by direct questions asked by Sikyong whether they had already been learning Tibetan language, to which they answered, "No, but I will!") (figure 4.7).

Moreover, one of the main attractions of the tour was a visit to a Tibetan Children's Village in Upper Dharamshala—the oldest and largest Tibetan boarding school in India. Meeting hundreds of children who were "told by the Dalai Lama to learn English and science" was a moving experience. A visit in the nursery was particularly touching, where everyone melted at the "sweetness" of the Tibetan toddlers, who—as they were used to the visits paid by sponsors and other foreign guests—embraced the guests, which caused additional waves of admiration. On top of that, the guide taking participants around the village was a former pupil, then working as a volunteer at the organizer's institution—with whom most of the participants had already developed a friendly relationship during the evening beer outings—which

[25] Ama Adhe passed away on August 3, 2020.

[26] Wylie: *srid skyong*, the term translated as secular political leader, in contrast to the 14th Dalai Lama's joint political-religious leadership. Sikyong is the official head of the Central Tibetan Administration, that is, the Tibetan government-in-exile.

[27] It was after the 14th Dalai Lama officially had retired from his political activity. Lobsang Sangay had previously hosted the post of *kalon tripa* (Wylie: *bka blon khri pa*), that is, the head of the council of ministers (a kind of prime minister), but in 2012 the more adequate term—according to the Tibetan diaspora's elites—was introduced. *Sikyong* is to be translated as the president of the Central Tibetan Administration.

Figure 4.7 Participants of the Political Reality Tour in Dharamshala during a Meeting with a Former Political Prisoner, Ama Adhe. *Source*: Photo by Natalia Bloch (2014).

gave the visit a very personal, human dimension, and opened many doors, for example, when participants visited the house where the boy grew up. The warm welcome he received from the house foster mother (called *ama-la*[28]) was captured in many photographs.

The itinerary of the tour included also points which required direct involvement of the participants, for example, in the promotion of the event which was just being prepared by the organization (figure 4.8). Divided into teams of two or three, the participants had to go out into the streets of Dharamshala armed with banners, flags, loudspeakers, and leaflets, and encourage other tourists to take part in patriotic Tibetan poetry session. Additionally, they were asked to speak to the camera about the impact of the tour and how they intended to use it in the future, which was a kind of enforced declaration based on the principle of reciprocity (these recordings were then used to further promote political tourism). Each meeting was meticulously documented with numerous photographs—mostly taken with smartphones—by both sides. For the participants, the joint photo with the former political prisoner was not only a souvenir but also a kind of a distinguishing feature in the world of mass tourism, through which they could construct their own unique self-representations, for example, in social media. For the organizers—and the partner associations that cocreated particular meetings—it was a tool to

[28] *Ama* means "mother" in Tibetan and the suffix *la* is added to emphasize a respect for the person. The homes in which foster children of the Tibetan Children's Villages live are run by women, sometimes together with their husbands.

Figure 4.8 Participant of the Political Reality Tour in Action, McLeod Ganj. *Source*: Photo by Natalia Bloch (2014).

"disseminate" their activities and attract new engaged tourists. At the end of the program, the organizers and participants—including myself—took part in the evaluation of the whole tour, which made me all the more aware that it was treated by Tibetan activists as a project. At the same time, the event ending both tours—English and Indian ones—was an opportunity to "fraternize with the natives" in a more relaxed atmosphere, with beer and open mic sessions, providing a great opportunity for cultural exchange.

The program of cultural attractions was equally rich and included a visit to the main temple (although participants were somewhat disappointed that the Dalai Lama was not in his residence), an excursion to the Norbulingka Institute in Sidhpur near Dharamshala (where traditional Tibetan crafts are taught and made), a visit to The Library of Tibetan Works and Archives (where participants could touch several 100-year-old manuscripts) and a visit to the Tibetan Institute of Performing Arts (where they took part in the traditional dance, music, and singing performance). There was, of course, time to buy cultural souvenirs (the political ones—like "Free Tibet" T-shirts, badges, or stickers—the participants received from the organizer). Two tourists decided to purchase more durable souvenirs and left Dharamshala with tattoos. They chose the Buddhist motif of the endless knot, *palbeu*,[29] which

[29] Wylie: *dpal be'u*. One of the eight symbols of fortune in Tibetan Buddhism, already mentioned in this chapter, with a graphic pattern symbolizing the interconnectedness of the universe, having neither beginning nor end, as well as the boundless wisdom and compassion of the Buddha. The symbol is also very popular in other traditions (like Hindu, Muslim, Celtic, etc.).

was tattooed on the neck of one of the Indian activists of the organization (however, the female tourists decided to hide their tattoos, so that their parents would not notice them). The participants of the tour became my friends on Facebook, so I could follow them. The martial arts instructor was strongly involved in the actions for Tibet organized in Delhi, a girl from Pune opened a local branch of the Students for a Free Tibet at her university, others were giving "likes" to various pro-Tibetan initiatives online.

Such politicization of tourism by subaltern groups can also be encountered in other places. Researchers write about solidarity tours (Mahrouse 2008) and about the already mentioned reality tours (Higgins-Desbiolles 2010). Sometimes these are organized by international organizations from the countries of the so-called Global North which focus on human rights issues, such as the Global Exchange from the United States, which advertises its offer in the following way:

> Reality Tours are meant to educate people about how we, individually and collectively, contribute to global problems, and, then, to suggest ways in which we can contribute to positive change locally and internationally. We offer participants an opportunity to journey to other countries to examine a situation first-hand, to see beyond what is communicated by the mass media.[30]

Tourists can take part in tours dedicated to the situation of women in Afghanistan, the militarization of the American Mexican border, the LGTBQ movement in Cuba, fair trade and alternative economies in Palestine, and so on. Global Exchange makes direct reference to the idea of global justice and the possibilities of achieving it through tourism.[31] Very similar to the Tibetan example of using tourism for the political purposes by groups seeking international justice are the initiatives undertaken by Palestinians. As in the case of Dharamshala, tours aiming to explore the reality of Palestinian life under Israeli control are offered by local nongovernmental organizations. Some, just like the Alternative Tourism Group discussed by Higgins-Desbiolles (2010, 203), focuses on making tourists aware of the lives of people deprived

30 https://globalexchange.org/realitytours/.

31 Higgins-Desbiolles in her analysis of reality tours recalls other interesting variants, such as tours organized by the Australian branch of the international organization Oxfam to India, during which the participants from Australia, envisioned as future local leaders, get to know different kinds of bottom-up initiatives for social justice or sustainable development. They are then to implement such solutions in their own communities, thus reversing the vector of the "civilizing mission." Another organization, International Society for Ecology and Culture, arranges trips for young people from Ladakh—the Himalayan region in India—to countries of the so-called Global North to show them how the life in the West "really" looks like, in order to deconstruct their idealized image of the West created on the basis of observations of foreign tourists. This project aims to restore the participants' pride in their own culture (Higgins-Desbiolles 2010, 202).

of their right to own land by creating direct contact opportunities with the inhabitants of the occupied territories. The organization presented its offer on the popular travel forum TripAdvisor: "Discover justice tourism in a vibrant culture with friendly people. Resilience under Israeli occupation. Come and see for yourself."[32] Other organizations, such as the International Solidarity Movement, which focuses exclusively on freedom activism, directly involves tourists in acts of passive resistance such as sit-ins at house demolition sites or during street protests. On the one hand, the presence of foreigners increases safety of Palestinian activists, because when observed through lens of tourist cameras, Israeli soldiers are more restrained in suppressing protests. On the other hand, such tourists voice Palestinian demands, making them heard and thus breaking the political isolation and dominant media coverage. Higgins-Desbiolles (2010, 204) quotes an excerpt from an interview with the cofounder of the organization, who directly evokes the category of hope that this type of tourism offers: "When they see internationals who have come, Palestinians feel hope, that others have come to share their hardship. Hope is very important for a people who feel their pain ignored, their voice unheard, their land taken away every day." As one can read on the organization's website, one of the three reasons why the presence of foreigners is important, it is "sending a message to the Palestinian community—'we see, we hear and we are with you.'"[33] The above quotes demonstrate both the assumption underpinning "hopeful tourism" and the hope that through tourism the subalterns are able to speak.[34] In this context, tourism is in a way appropriated by communities which are attractive to tourists as a means of communication, to showcase their existence and demand of recognition (Lanfant 1995).

However, political tourism in the Tibetan context has one more dimension. Some researchers suggest that the way Tibetan refugees perceive their own cultural heritage—reduced to Buddhism—in relations with the external world is a modern reinterpretation of the historical "patron-priest" relationship (*chö-yon*[35]). It was a specific political and cultural institution born in the middle of the thirteenth century and based on the principle of reciprocity (Shakabpa 1967, 64–70). On its basis, secular, external rulers—initially Mongolian, and

[32] https://www.tripadvisor.co.uk/Attraction_Review-g1902315-d6158795-Reviews-Alternative_Tourism_Group-Beit_Sahour_West_Bank.html.

[33] https://palsolidarity.org/about.

[34] At the same time, we should remember that tourism can also be appropriated and used by the dominant group in order to strengthen its discourse. Tourist agencies in the People's Republic of China not only try to present China to foreign tourists as a harmonious conglomerate of ethnic minorities (including Tibetans) but with a similar political tourism offer have been targeting the minorities themselves, whose leaders were in the past invited to attend "real, wonderful China" tours (see Bulag 2012).

[35] Wylie: *mchod yon*. Some scholars claim that the term should be rather translated as "donor-spiritual teacher."

then Manchurian—provided protection (including the military one) to the leaders of selected sects of Tibetan Buddhism, in exchange receiving spiritual services and, over time, also the moral legitimacy of their rules (Norbu 2001, 7). Tibetan refugees, having found themselves in an unprivileged political situation, seem to perceive their relationship with those they see as influential "patrons"—that is, "the West" and more and more recently India as well—in similar terms. They create attractive representations of themselves as those who offer to the world the ancient, priceless heritage of Tibetan Buddhism, which they have been protecting from the devastating effects of the materialistic occupant, expecting in return economic and political support for their cause (Klieger 1992, 84–120). This in turn allows them to position themselves as equal partners in this symbolic exchange, and not as poor refugees begging for help. According to Kesang Tseten, "It was because of the dharma element that this sympathy for the Tibetan cause went beyond sentiment," and has become compassion instead of pity, changing "the relationship between the sympathizer and victim to one in which they stood at par" (1993, 30). In this way Tibetans—in their own eyes and in the eyes of "the West"—have ceased to be "ordinary refugees," skillfully turning their cultural capital into a political one.[36] Tourists—as the representatives of the powerful political players (the European Union, the United States, or India)—have been involved in this relationship of reciprocity adapted to the new political context.[37]

"Save Hampi People": Struggle against Displacement

The same process, albeit on a much smaller scale, took place in Hampi since the first evictions and demolitions in 2011, when the villagers became subordinate to the dominant discourse of protecting material cultural heritage. In their work, postcolonial critics try to draw our attention to the internal forms of domination and subordination within those who are usually homogeneously perceived as colonized/postcolonial. They point out to the areas of exclusion created and maintained by the elite in relation to the "non-elite" (e.g., Loomba 2015). Hampi is an example of how the postcolonial state and its elites, in the name of the "higher" national good of tangible cultural heritage, have silenced the voices of certain groups of their citizens, defining

[36] Cf. the argument made by Victor Azarya (2004) that for some ethnic groups forging their marginality and cultural capital into a financial one made it possible to gain access to the global tourism industry (following Comaroff and Comaroff 2009, 155, footnote 12).

[37] Not all refugee groups that have become a tourist attraction succeed in this regard. For example, Kayan refugees from Myanmar who fled to Thailand, and whose brass coils worn around women's and girls' necks are a magnet attracting foreign tourists, have no control over tourism to their villages, the latter being managed by the Thai government and external tour agencies, who exploit Kayans' cultural capital. Nevertheless, they still try to "develop an encounter with tourists based on respect and dignity" (Cole and Eriksson 2010, 115–117).

Figure 4.9 The Bazaar in Hampi Before the Demolitions in 2011. *Source*: Photo by Surendra Kumar (2008).

Figure 4.10 The Bazaar in Hampi After the Demolitions in 2011. *Source*: Photo by Surendra Kumar (2011).

them as driven by egoistic and immoral greed. Paradoxically, it is the tourists, who often come from former colonial metropolises, who became allies of the displaced residents in their struggle for their own vision of heritage, their right to homes, and their right to benefits from tourism (figure 4.9, figure 4.10).[38]

However, while Dharamshala has well-developed structures of political tourism and it is the Tibetans who skillfully create them (having acquired the competence to do so in the course of more than half a century of exile), in the case of Hampi, transnational solidarity–based activism for justice was only being born, initiated by tourists themselves. As Higgins-Desbiolles (2010,

[38] This struggle, in its rhetorical dimension, took shape with the support of an Indian NGO Equations (Equitable Tourism Options), which operates in the capital of Karnataka state, Bengaluru. The organization conducts fact-finding research, compiling reports on places, where the development of tourism collides with the needs of local communities and the natural environment as well as organizes campaigns for sustainable tourism, thus supporting various grassroots initiatives. See http://www.equitabletourism.org/.

200) points out, such activism aims to support local communities experiencing human rights violations and injustice in their efforts to reverse the unequal power relations in which they have been imprisoned. In the absence of an institutional framework, in which this activism could take place on-site (as in the case of Dharamshala and its educational offer for tourists), these activities are undertaken mainly in virtual space, that is, through social media and online tools of political struggle (e.g., portals where signatures for various petitions are collected), on which I focus in this part of the book. I was often an intermediary in these activities: tourists, worried by reports from the village residents about another threat of eviction and demolition, found my email address by browsing for "Hampi evictions" and contacted me asking for updates and advice on what they could do for the community.

Already during the first wave of displacement and demolitions that affected the residents of the bazaar in July 2011, the tourists present in Hampi spoke to the cameras about the unique atmosphere of the place, created by its inhabitants.[39] A journalist from France, a friend of one of the guesthouse owners and a regular visitor to Hampi (gifts for children, shared meals at home), tried to flag the subject of the demolition in the French media. She even managed to reach the person who issued the eviction notices and "asked some uncomfortable questions." Another tourist, from Australia—a friend of one of Hampi's guesthouse owners—initiated a Facebook campaign "Save Hampi People." Together, they created a video that clearly showed the destructive nature of the actions taken by the authorities against the inhabitants and called on the international community to prevent further demolitions. In the description of the video they posted, "This short documentary film has been made as a direct appeal to UNESCO to step in now and stop the demolition of Hampi's long-established community." In the nine-minute clip, Hampi was shown not only as a collection of archaeological sites but as "a vibrant community who have lived and worked here for generations," which "preserves Hampi's intangible cultural heritage," as "the evidence from the Vijayanagara Empire shows Hampi as a living community with a busy bazaar . . . full of people selling all kinds of goods."[40] Also under the petition "Save the village of Hampi and the citizens life" addressed to UNESCO and posted on change.org by a young man from the United States, tourists expressed their support for the vision of Hampi as an intangible heritage represented by its inhabitants, while at the same time clearly condemning the actions of the authorities:

---

[39] These videos never appeared on local television, which did not provide the coverage of the "spatial cleansing" of Hampi. According to the evicted residents, this is an evidence of political involvement and powerlessness of the local media.

[40] All quotes come from the "Save Hampi People" video: https://www.youtube.com/watch?v=N9docvUG0mM.

> "Save the people of Hampi like the Architecture. They are living treasures" (tourist from Australia); "i visited hampi just recently, and loved it. the people the atmosphere, the sights the markets is what makes . hampi . . . people lived there 600 years ago. why not now. preservation is one thing, but to make it a deserted coffin is not the feel anyone wants" (Australian tourist); "Removing the people will turn Hampi into a dead museum Please don't let this happen" (British tourist); "I think it is criminal to move people from their home. Who is representing the people in this tragic event. We as civilised human beings should support these poor people and not let the money win. This world is tough enough without wanted destruction of innocent peoples way of life. STOP THIS NOW!!!" (tourist from the United Kingdom).[41]

Although this petition had not brought the desired effect, tourists did not seem to give up. Four years after the first demolitions, Jake from the UK decided to create another appeal through change.org, this time addressed not only to UNESCO but also to the Government of India and the Archaeological Survey of India,[42] entitled "Stop the Destruction of Hampi." This time, the initiative came from rock-climbing tourists, reacting to the news that a settlement across the river is also to be demolished. The appeal stressed that climbing in this region "directly supports the local community—who's other main source of income is rice-farming and are often living below the poverty line." In this way, tourism—assuming the elimination of leakage—was shown to be a highly ethical activity, reducing global economic inequalities. The petition recalled the 2011 bazaar displacement and demolitions and pointed out that they caused public outrage, which was ignored: "So we must act! We must do everything we can to protect the fragile livelihoods and culture of the local people, and preserve this precious place for future generations. Please sign and spread this petition to show UNESCO, the ASI and the Government of India just how many people across the world support the existence of this precious place" (figure 4.11).[43]

There were three motives that underpinned the efforts initiated by tourists for the benefit of the Hampi community: the "living heritage" as opposed to "dead monuments," the human right to housing, and the right to livelihood through work in tourism. In the comments supporting the abovementioned petition, the residents were portrayed as those who continued the traditions of the

---

[41] All the quotes come from the comments posted by people who signed the petition "Save the village of Hampi and the citizens life": https://www.change.org/p/unesco-save-the-village-of-hampi-and-the-citizens-life.

[42] The Archaeological Survey of India (ASI) is the main institution in the country that supervises "the monuments of national importance," including the main bazaar and Virupaksha temple in Hampi.

[43] All the quotes come from the petition "Stop the Destruction of Hampi": https://www.change.org/p/unesco-government-of-india-archaeological-survey-of-india-stop-the-destruction-of-hampi?recruiter= -286664501&utm_source=share_petition&utm_medium=facebook&utm_campaign=share_facebook_responsive&utm_term=des-lg-new_account-custom_msg&fb_ref=Default.

Figure 4.11 Demolished Houses at the Bazaar in Hampi. *Source*: Photo by Natalia Bloch (2013).

Vijayanagara period, offering goods and services to visitors. Tourists expressed their preference for contact with people over admiring the ruins. In their opinions, there was a recurring theme of residents cocreating the site's heritage and the assumption that a balanced coexistence of the material past and the living present is possible, as opposed to the Archaeological Survey of India positioning people as a threat to monuments (see Bloch 2016c): "These villages which will be raised to the ground are an integral part of the living, breathing heritage of Hampi! I don't believe that the surrounding ruins are at risk from the tourist, climbing or local communities. . . . I found a great respect and reverence for the heritage of the place" (tourist from South Africa). Mainly, though, the tourists expressed their support for the right of inhabitants to Hampi and its heritage:

> "I'm signing [the petition] because apart from the beauty the residents are the trustees of Hampi and care for it and should be the ones to manage it" (tourist from Australia); ". . . The local businesses and people respect the temples and keep the area well maintained. I am against this idea of clearing people from their native land. What is going to make Hampi a good place after all? Only its people" (tourist from Hungary); "The people here have created an area that is incredible. Not only should this area not be destroyed because people from around the world won't get to see it, but because of the families living here who have tried so hard to make it into the area it is, so they deserve to keep it that way" (tourist from India).

There was also a fundamental question asked about the very idea of heritage and to whom it belonged. There were considerations about the possibility of reconciling the protection of monuments with the development of the sites where they are located (see Fritz and Michell 2012; cf. Shepherd 2012; Araoz 2011): "What is heritage worth if it cannot be enjoyed by locals and the world at large? There has to be a better way to both conserve the heritage and support local livelihoods without compromising access to this incredible place. Please don't tear these villages down!!!" (South African tourist). Tourists often linked the right to heritage with the right to housing and expressed outrage at the act of forced displacement:

> "Destroying this marvelous heritage and expelling its peaceful citizens would be incredibly atrocious, extremely cruel and just downright stupid!" (tourist from Sweden); "It will be criminal to remove the local population out of their livelihoods & homes" (tourist from India); "I got to know the locals quite well and they are suffering from the government's act of destruction" (tourist from Hungary).

Displacement often results from governmental projects aimed to protect material and natural heritage or develop tourism. Stroma Cole and Jenny Eriksson (2010) cite numerous examples of such actions in their analysis of the relationship between tourism and human rights. The "beautification" project of the historic center of Cairo implemented by creating open spaces, luxury hotels, and shopping areas resulted in the displacement of residents "illegally" settling in the old cemetery (Fahmi 2008). Michael Herzfeld (2006), in his study of the analogous case of the Pom Mahakan district in Bangkok, from where residents were removed after the eighteenth-century citadel was included on the UNESCO World Heritage List, described the process of discursive demonization of residents conducted with the categories of illegality and immorality. In the case of Hampi, this stigmatization was followed by an argument about the inhabitants being "non-natives" and their community being "artificial" (Bloch 2016c, 2017a, 2017b). In addition, state agencies responsible for shaping tourism began to push for plans to attract more affluent tourists, pointing to the lack of professionalism of local tourism workers and the poor quality of their services.

Meanwhile, one of the arguments in favor of the inhabitants' right to Hampi and its heritage, which appeared in the abovementioned video, was the fact that they made Hampi "a safe and welcoming place for tourists." The Hampi residents themselves—a motif that often appeared in my conversations with them—presented themselves as those who had developed tourism in Hampi. Not only did they build the necessary infrastructure (homestays and guesthouses, restaurants and cafés, shops and stalls), doing the work of the authorities, which neglected the site for many years, but above all, they created a friendly atmosphere, thanks to which tourists began to visit the place in such big numbers. In the opinion of the Hampi inhabitants, without them Hampi would lose its raison d'être as a tourist destination.[44] The owner of a family-run guesthouse justified his right to Hampi and the benefits from tourism in such a way:

> If there are no locals, there are no tourists. Where were the tourists 20–25 years ago? These bastards [generally, the government, that is, central, state, and district authorities] did not attract tourists, we did! And now they are saying: "It belongs to us."[45] They have no idea what tourists need—we do, because we offer them our services. It was the tourists who showed us how to do business.

The inhabitants of Hampi positioned themselves as experts in tourism: equipped with intercultural competences experts in tourist tastes and expectations, who responded to the demand at the right time, doing the job of the authorities. The person quoted above explained to me, "50-60 years ago there was no government here. They were just pilgrims, so the residents provided them with coconuts, flowers and incense for *puja* [at the temple] as well as food. Then, in 1992, tourists began visiting Hampi, and locals started selling them various things because the visitors needed them." Small business owners in the informal tourism sector in Hampi, what I have already discussed in chapter 2, were well aware of different financial resources available to different groups of tourists, and especially the fact that they handled low-budget tourism. They found it naive and unprofessional to assume that "their tourists"—mostly backpackers—could afford a multi-star hotel in Hospet. Above all, however, they stressed that no one in such a hotel would treat tourists "as friends, not just customers." This view resonated with the feelings of tourists who supported the struggle of residents to stay in Hampi: "They [the local villagers] are so nice a bunch & treat the tourists & back packers really well in a way I have not seen in any other tourist destination" (tourist from India); ". . . And the friendly, lively and warm atmosphere which is culturally rich and vibrant. There is no place in India which can be compared to this place" (another tourist from India).

---

44 One of the frequently recalled theories that villagers used to explain what had happened to them was the lobby of large, multi-star hotels' owners from Hospet and luxury resorts built along the road from Hampi to Hospet. After the collapse in the wake of the corruption scandal of the highly profitable mining industry in the region, the businessmen were to invest in the tourism industry, and then, together with local politicians, remove the competition.

45 One of the allegations made by the authorities against the Hampi villagers is that that they illegally resided on government land (among others, in the area supervised by the Archaeological Survey of India). They were also accused of having developed their businesses in residential premises without required commercial licenses, although the latter were issued by the local council (Gram Panchayat).

The right to satisfy one's livelihood through work in tourism also emerged in the statements made by tourists, which strongly contrasted with the Occidentalist rhetoric of religious nationalism portraying the work in tourism as a disgrace: the immoral exploitation of a holy place (see Bloch 2017a). In the abovementioned video, tourists expressed empathy with residents who had lost their businesses:

> "I've talked to some of the people in the town,"[46] says the British middle-age female tourist, "who are being forced to move. One guy spent his life from having a street stall to finally renting a tiny shop. He's so distressed. It distresses you hearing this story"; "Although we can understand it's supposed to be an archaeological site," says a family with a child, "we really hope the people of Hampi can keep their little businesses."

The comments supporting the second petition analyzed above also showed an awareness of the role that tourism plays in the economic survival of the Hampi inhabitants:

> "The local community, many who have lived here for generations, depend on tourism and the climbing community for their livelihoods" (South African tourist); "Hampi bizarre [should be: bazaar] brings in a lot of revue [revenue] from the European tourist trade that directly feeds into the local community" (British tourist); "It would be a massive blow to the local economy and to the world at large if this goes through. Many international climbers support the local community by coming back year after year to this special place" (tourist from the UK).

These arguments clearly contradict the theories about the tourists' blind need for "cool authenticity" and the resulting "invisibility" of tourism workers as those who remind tourists of the commodified nature of their journey (Lozanski 2010, 754–755). Tourists were fully aware that removing residents from Hampi would deprive the latter not only of their homes, but also of their livelihoods. In this sense, they joined the local community's struggle for a bottom-up PPT—another form of tourism against global injustices distinguished by Higgins-Desbiolles (2010). PPT is about increasing the share of poor people in the benefits of tourism, and especially enabling their active participation by creating employment opportunities and developing their own forms of entrepreneurship (Ashley et al. 2001). Proponents

---

[46] Hampi is officially a village, but its infrastructure looks town-like, therefore it is sometimes referred to as a town.

of such a potential for tourism[47] draw attention to a broad definition of poverty which also includes lack of freedom (as is the case with refugees), power (i.e., subalternity, vulnerability, marginality), opportunities, skills or education (Harrison 2008, 856). This type of tourism should be developed in cooperation between local communities, the private sector, and governmental tourism agencies, whose role is crucial in creating a policy framework that removes barriers to accessing the benefits of tourism for the poor (Ashley and Roe 2002, 62).

In the case of Hampi, we are dealing with a situation where the villagers themselves developed tourism on their own, without the support of the state, what allowed them to transform from poor farm laborers into small entrepreneurs. Meanwhile, the authorities stigmatized these activities as illegal and decided to stop them, thus closing the villagers access to the benefits from tourism and offering no real alternative to satisfy their livelihood needs.[48] Aiming to develop upmarket tourist enclaves in resorts and multi-star hotels, the government departed from the basic PPT recommendations, that is, avoiding enclaves (as they increase leakage), stimulating the development of small businesses, and making the local population into shareholders rather than passive recipients (Ashley and Roe 2002).

The petitions formulated by tourists did not receive any positive response from the Indian authorities nor UNESCO. The same is true of the interventions by the NGO Equations promoting equal access to the benefits from tourism or the appeals made by researchers (see Bloch 2016c for a full analysis of the "spatial cleansing" of Hampi and the various actors involved). Nevertheless, the fate of Hampi revealed the potential for the development of transnational solidarity–based activism with the participation of tourists who felt morally obliged to get engaged on the part of displaced residents. Perhaps their eyes turned to Hampi is one of the reasons why the last remaining part of

[47] Pro-poor tourism (PPT) is more a postulate, a practical proposition related to the tourism management rather than an attempt to conceptualize it, which is often an argument raised against the proponents of PPT.

[48] Displaced villagers, after a long battle in the court, received a resettlement site located a few kilometers away from Hampi. However, small compensations they were offered for their demolished property, were not sufficient to build new homes. Thus, they were forced to take mortgage loans from the state bank. Soon many were unable to pay the loan installments as this ad hoc rehabilitation program did not take into account the fact that displacement deprived villagers not only of homes, but also livelihoods (it was a similar case as with the displaced inhabitants of the historic center of Cairo, who were moved to the outskirts of the city, which cut their access to tourists off, while most of them were itinerant vendors offering their goods to tourists; Fahmi 2008). The authorities allowed the displaced Hampi inhabitants to buy a piece of land next to their resettlement site where they could build new guesthouses, restaurants, and shops, but due to the loss of jobs and businesses in Hampi, as well as bank loan obligations, they did not have money to invest in such development. This meant-to-be commercial site, located in the middle of corn fields, remained thus unused, while some fifty families received warnings from the bank about the possibility of their new homes being repossessed due to rising debts.

the village has not, so far, been demolished, although the business possibilities there have been widely limited.

VOLUNTEER TOURISM: TOURIST AS AN ACTIVIST

Another area where the residents of Dharamshala and Hampi had a good track record of enforcing reciprocity in their relations with tourists was engaging them in direct initiatives for the community benefits implemented in the form of volunteer tourism. In this formula, by doing voluntary, unpaid, temporary work for the benefit of the community, tourists were to contribute to poverty alleviation and empowerment of the hosts, thus becoming part of the PPT. For tourists themselves, volunteering was, on the one hand, an opportunity for self-development and satisfying their need for "hot authenticity," achieved by establishing non-commodified relations with the locals, while, on the other hand—as in the case of political tourism—it provided an opportunity to become a "new moral tourist" (Butcher 2003, 5–30, 77–94). The idea of PPT is justified here in that the concept is not only about the individual economic benefits that individuals can derive from the presence of tourists, but also about the collective, community-wide benefits, which are often nonmaterial (such as increasing educational opportunities, developing skills, strengthening social cohesion, etc.; Ashley and Roe 2002; see also Sin 2010).

Volunteer tourism developed in parallel with the growing popularity of backpacking in the 1990s, mainly in the countries of the so-called Global South, and within the institutional framework created by international or local NGOs (Callanan and Thomas 2005; Wearing 2001). Nowadays, volunteer tourism more often than not takes on a more informal, unstructured, spontaneous, and flexible form (Hustinx and Lammertyn 2003). In Dharamshala and Hampi it took an intermediate form. On the one hand, volunteering there was done within an institutional context, although the organizations that benefited from volunteer work were small, grassroots and tended not to have clearly marked hierarchy. On the other hand, many tourists who did this work did not come to Dharamshala and Hampi intentionally in this capacity—they were "recruited" on spot and often made spontaneous decisions. In this section I look at the implementation of volunteer tourism in both communities and reflect on its contribution to both parties, the tourists and their hosts.

"Think Globally, Act Locally"

For many years the NGO scene in Dharamshala had been dominated by a few clearly recognizable, "mass-scale" organizations with a primary focus on politics (such as the Tibetan Youth Congress or the Tibetan Women's Association) and some few organizations which focused on social work and benefited from the support of volunteers (such as the LHA Charitable Trust or Volunteer Tibet). However, after the 2008 Olympic Games in China, which again drew the attention of international community to the Tibetan cause, the diaspora reacted with a plethora of small, grassroots NGOs.[49] They shared similar objectives: increasing the potential of refugees by developing their competences (linguistic, technological, or professional), and achieving these objectives through voluntary work. Although Tibetan activists of these NGOs mostly agreed that long-term volunteers were most desirable, they addressed their volunteering offer to all tourists. Advertisements posted on the walls of buildings and leaflets distributed in the streets or restaurants read: "Needed Volunteers! Appreciate to join with us, no matter how long you can do volunteers" or "VOLUNTEERS NEEDED!! [The name of the organization] is looking for conversation partners for our students. Our conversations class runs from 11 to 12.30 pm, Monday to Friday. No commitment is required. Come to make a difference in somebody's life." As the motto adopted by one of the organizations read: "Think globally, act locally."

In this way, every tourist who came to Dharamshala could engage their time, energy, and/or skills to work for local organizations, even if only for a few days or hours. Tibetan refugees employed a number of tactics to attract tourists to volunteering. First of all, they used awareness-rising events discussed in the previous section, such as "a movie night with Tibetan pizza," at the end of which one of the organizers usually announced: "Tomorrow at 2 p.m. we have English conversation classes, where you are all welcome. You don't have to prepare in any way, just show up."[50] It was also popular to offer tourists cooking classes of "traditional Tibetan cuisine"—vegetarian, of course!—which had previously been the domain of rather individual entrepreneurship in the informal tourism sector (as I discussed in the context of Hampi in chapter 3[51]). This activity, apart from supplementing the organization's

[49] During my fieldwork, I had the opportunity to work with four such organizations.

[50] I often had the opportunity to observe how effective these activities were. For example, after one movie screening accompanied with a dinner, an enthusiastic Italian couple, who had just arrived in McLeod Ganj, came forward. They said they were just launching an online massage courses, but at the same time wanted to "do something charitable," for example, to donate every fifth course to some organization that could thus support the skills development of their proteges. Since the movie touched them a lot, they decided to ask whether this particular organization was interested in such donation. The same organization was so successful in acquiring volunteers for conversation classes, that they did not fit into the room, so that some of the classes had to be carried out outdoor.

[51] The offer itself was very similar here: one class per person costed 250 rupees (about 6 dollars) and included *momo*, *thenthuk* and Tibetan bread, though "original Tibetan pizza" also came into play if there was such a desire. The price included a "recipe book" and a photograph taken with the instructor. Usually, there was no specific schedule—whenever there were a few tourists interested, the class was organized for them.

budget, was intended to create a space for recruiting volunteers, using as a bite a tourist attraction related to the need for "hot authenticity."[52] At the same time, it was a kind of cultural exchange offer, thanks to which Tibetans did not only put themselves in the position of recipients, but also donors. Volunteers were also recruited among tourists attending folk performances organized by one of the NGOs, during which tourists were offered "the authentic Tibetan music" (performed by young Tibetans with tattoos and fashionably shaved and dyed brown hair, dressed in *chuba* and fancy sneakers, and wearing a "Make it easy" headband). At the end of each show, tourists were encouraged to perform a song or dance from their region, because—as the founder of the organization and the shows' MC emphasized during the performances—"we must share": "All is interdependent: if there are only artists and there is no audience, it can't work; it is the same if there is an audience, but there are no artists; and if there is an audience and there are artists, there must be [the name of the organization] to connect them." As a result, the participants learned the Romanian songs as well as a dance of the fishermen of Maharashtra,[53] and everyone seemed delighted. Finally, a common way to attract volunteers was to involve other volunteers: the leaflets they distributed, together with first-hand stories about their work in the organizations, were much more effective than leaflets simply left in different public places (figure 4.12).

Additionally, one organization published—also with a significant contribution of its volunteers—an English language magazine distributed free of charge throughout Dharamshala. Apart from articles on occupied Tibet and Tibetan diaspora, which contributed to promoting the Tibetan cause, the magazine included a tourist information section (bus timetables, schedule of upcoming events, list of places offering meditation courses, etc.[54]) as well as a list of Tibetan NGOs. Above all, however, it advertised its "Traveller Services & Volunteer Opportunities," and the list of the latter was long, as the NGO sought "long or short-term, skilled or unskilled" language teachers, tutors and conversation partners (mainly English, but also French and Chinese), computer (software) teachers, web designers and IT professionals, writers (for the magazine), environmental professionals, medical and healthcare professionals, fundraisers and grant writers, project professionals, chefs and cooks, human rights and legal professionals, personal assistants "and much more!"[55]

---

52 Another requirement tactic was offering Tibetan language classes, but these required more time and determination from tourists, which is why cooking classes were more effective in this regard.

53 One of the southern states of India.

54 About 20 percent of the magazine content were advertisements aimed at tourists: "Tibetan Kitchen Restaurant & Beans Café," "Stupa Guest House," "Crepe Pancake Hut," "Yoga class," "Need a dentist?," and so on.

55 Magazine *Contact* published by LHA Charitable Trust, Institute for Social Work & Education, June 2014, 3.

Figure 4.12 Volunteer ID, Dharamshala. Photo by Natalia Bloch (2018).

Although volunteers in Dharamshala were involved in various types of work, by far the greatest demand was for English language teachers, tutors, and conversation partners. Offering free foreign language classes to Tibetan refugees was a leading area of activity for most local organizations. In many of them, conversations with tourist-volunteers took place every day at different times, so really determined students could fill up their daily schedule by circulating between different NGOs. The scale of this activity was demonstrated by the fact that Dharamshala became known in India not only because of its new national cricket stadium, but also as a place to attend free conversations with foreigners. Some people who wished to learn a foreign language were even looking for teachers on their own, and I myself was a frequent object of their inquiries. All I had to do was drop in to a befriended bookseller, where a young Tibetan was about to buy an English grammar textbook for beginners; we joked with the bookshop owner that it was the most popular book in McLeod Ganj, at which I was immediately asked whether I was teaching (when I replied that I was doing research, the man—not discouraged at all—assured me that if I changed my mind, he was very keen to learn English with my help). Another time, when I was waiting for a meeting with the head of one of the NGOs, a young boy appeared before me out of the blue and started talking about "everything and nothing," apparently taking me as a volunteering conversation partner. Sangmo, the owner of a small bakery in McLeod Ganj—whose story I tell in chapter 5 and whom I often visited to listen to local gossips—having probably considered that I had a lot of free time, tried to set me up as a teacher for her friend, a monk. At the same time, she decided that *The Great Gatsby* left by one of the tourists at her café would be a great help in learning to read English. The monk was an absolute beginner—he had lived in India for eleven years, but in the state of Uttarakhand, where he had no contact with foreigners; he attended classes taught by a volunteer from the Netherlands in one of the organizations but needed more practice. Thus, we practiced on available reading and words such as "sexual" and "luxury." Sangmo almost died laughing, looking at this bizarre scene (which she created herself) from behind her counter, as it took us an hour to get through the first paragraph. At the end of this test-drive lesson "my student" insisted on paying for the coffee I had drunk, but I managed to stop him.

According to the statistics from just one organization, they worked with almost 9,000 volunteers from sixty-five countries in just six years. This demonstrates not only the scale of the phenomenon, but also that volunteering in Dharamshala is definitely holiday-related and the volunteers do not stay for long (most of the organizations actually suspended their activities for two or three off-season months, when cold winter kept tourist-volunteers away). Dharamshala's popularity as a tourist destination and Tibetans' effectiveness in engaging tourists in their community's activities have caused supply to exceed demand, resulting in the phenomenon of unemployed volunteers, who circulated from one NGO to another, desperately looking for any job, even the smallest one. I witnessed a situation when a professional teacher of English from the United States lost his "job" because he had to make space for other volunteers (as a result he had to shorten his stay in Dharamshala by a month and a half—he used this time to "wander around India"). In another organization, four young people from France, who eventually came to volunteer for a month, were doing one hour of language classes every four days, for the rest of the time distributing leaflets and serving food at events aimed at attracting more volunteers (and lounging in a pub next to the organization office). A volunteer from Estonia did not get to work for the organization she planned to involve with because it turned out that they were "overbooked" (eventually she decided to take a yoga course "so as not to sit idly").

This situation seemed absurd, but it can be historically explained. Many of these organizations emerged during a wave of patriotic elation after the brutal suppression of the 2008 protests in Tibet. Those NGOs aimed to support newly arrived refugees in their adaptation to life in exile, mainly by improving their language skills, which in turn was to foster social and economic integration.[56] In this respect—the Tibetans were in agreement—learning Hindi did not make any sense, as English offered much greater opportunities. First of all, in terms of being able to communicate with most Indians, including workers in the tourism sector in Dharamshala, as well as with representatives of the so-called Western world. Secondly, even passive knowledge of English provided access to information via the Internet,[57] that is, the possibility to enjoy this dimension of freedom in exile (and, for example, to build social links through participation in social networking platforms). As expressed by one of my research partners, English was "very useful in everyday life in India." And although knowledge of this language seemed to be primarily a tool for possible emigration to "the West," many of my interlocutors pointed out that it could also be helpful when returning to Tibet and reintegrating into the local labor market, as people who speak Tibetan, Chinese, and English were said to have a good chance to find employment in the rapidly growing tourism sector in Chinese Tibet.

Soon, however, the number of newcomers fell dramatically as a result of the tightening border control policy enforced by the PRC authorities and a turn in political sympathies of Nepal—which earlier served as the main transit country for Tibetan refugees on their way to India—toward its Chinese neighbor. As a consequence, at the time of my fieldwork, conversation classes held with the volunteers served primarily refugees with longer seniority and especially monks, including those born in exile and coming from monasteries where there was no secular education. When asked about their motives, they usually replied that they "want to speak to the world about Buddhism," which was a direct reference to the reinterpretation of the *chö-yon* relationship already discussed in this chapter.[58] Students taking conversation classes also knew that the basic tool for communicating with "the world" were new communication technologies, which mainly operate in English (the Internet, especially Facebook and other social media, applications for mobile phones, iPods or notebooks, the latter usually received by monks from their foreign

[56] Voluntary tourism was born in Dharamshala in the 1990s in the community center mentioned in the previous section. Tourists were encouraged to teach English and computer skills to refugees. The difference, beyond the scale, was in the recipients: then these initiatives were aimed primarily at young Tibetan drug addicts. Drugs, at that time, posed a new challenge to the Tibetan community and the settlement authorities tried to counteract somehow (Tsering 1998).

[57] At that time there were not many Tibetan language websites.

[58] It is also visible in the approach of the monasteries (re)constructed in exile, which more often introduce elements of modern education, including English language classes, into their curriculum.

sponsors).[59] Students also included people from culturally Tibetan regions in Northern India, such as Ladakh or Sikkim, and even Buddhists from Mongolia, Myanmar or Thailand who tried to benefit from volunteer tourism built on the attractiveness of Tibetanness, it being their own form of agency.

New Moral Tourists

NGOs that recruited volunteers among tourists appealed directly to their sense of morality—asking them to "give hope to those in need"[60]—with which these tourists cut themselves off from their mass alter ego. Jim Butcher calls them "new moral tourists," interested in the ethical aspect of traveling, combined with personal development (2003, 5–30, 77–94).[61] Their commitment—even if temporary—to the local community is a way to add meaning to their tourist experience, to make their holidays more than just pleasure. Such tourists not only take, but also give something back; they take action, while mass tourists only merely "consume" places visually (Urry 1995). Volunteer tourism is therefore an opportunity to transform oneself from "see-er" into "doer"—the binary opposition proposed by Eric McGuckin to show how those who classified themselves in the second category differentiated themselves from tourists they classified in the first group (1997, 361–362).

Who were the tourists involved in the volunteer projects discussed in this section? Of course, their motivations were diverse, complex, and subject to change over time, and the experience itself varied according to the organization, the type and period of work, the so-called beneficiaries, and the biographies of the tourists themselves, although they had surprisingly much in common with other tourists I worked with. Some of the volunteers in Dharamshala were rather typical volunteer tourists who undertook this venture before joining the university or taking first job, either on short-term assignment (during the holiday period) or as part of a one-year break (during the so-called gap year). This was the case for four French youth in their

59 This was also the reason why computer classes were so popular among monks (who learned Office, Power Point, and Excel programs, but also the very basics of how to start a computer, how to use Internet browsers, or how to set up an email account). However, also more advanced computer courses were organized in Dharamshala. For example, the government-in-exile-run Tibet Career Center offered a course in website designing—a Tibetan young women, who did her diploma in computer science from an Indian college, taught the beginners, while a foreign volunteer was a teacher on the advanced level.

60 The motto of the NGO called Tibet Hope Center. The name of the organization itself made an interesting reference to the idea of hopeful tourism, discussed at the beginning of this chapter.

61 These changes in the tourist profile have been observed since the late 1980s, when tourists who were more (at that time mainly) ecologically aware and critical of their consumer needs appeared (Krippendorf 1986). Some scholars suggest that the beginning of the phenomenon can be even drawn back to the 1960s, when the "alternative" to mass tourism forms of traveling started to develop (Lanfant and Graburn 1992).

early twenties, who went directly to McLeod Ganj for a two-month volunteer service at one of the Tibetan NGOs, where they were supposed to teach French and help with educational events organized for tourists. They knew each other from school and traveled to India together, so as to support each other. They were eager to combine their work—which was not very tasking as there were loads of volunteers in town at that time—with holiday pleasure: they went trekking, stayed late in restaurants, paraglided,[62] and had tattoos made (just like the young Indian participants of political tourism discussed in the previous section). Like the tramping tourists with whom I worked while doing mobile ethnography, they fell prey to the overheating of authenticity in a multisensory experience of Otherness (see Bloch 2016a)—diarrhea greatly decreased their initial enthusiasm and dominated all other experiences for nearly two weeks. Other young volunteers coming to Dharamshala in groups from the countries of the Global North combined volunteering with spiritual attractions, attending courses in yoga, meditation, or reiki. Volunteering was for them a somehow natural part of their self-development, achieved by gaining experience of living abroad, skills, and intercultural competences (Butcher and Smith 2010, 28; Heath 2007; cf. Aitken and Hall 2000). Most of their colleagues did the same, so the question was not whether to volunteer, but where.

Kristin from Germany was about the same age as the French tourists introduced above, but she came to Dharamshala for nine months. She wanted to gain "life experience"[63] and was wondering about what she wanted to do next in life (that is, firstly, what studies to choose). She came alone, but through a German organization that sponsored the Tibetan organization where she was to volunteer. As a result, she was more "looked after" than other volunteers, in a sense that she got an accommodation in a monastery, which was also a beneficiary of the German sponsor. Kristin's motivation for choosing the "developing country" for volunteering was not much different from that confessed by tramping tourists when asked why they decided to visit India, that is, a fascination that has often been developed since childhood: "I always wanted to go to India. I've been fascinated by India for many years . . . it has always been so fabulous, fascinating country, and colorful, and spices, and exotic music and wonderful oriental people." Nineteen-year-old Kristin did not have enough money to make such a trip on her own, and as part of the volunteer work supported by the German organization she had transportation,

[62] Paragliding is very popular at tourist destinations in the foothills of the Himalayas—not far from Dharamshala, also in a Tibetan settlement called Bir, there is a world-famous paragliding center.

[63] Among other opportunities available in this regard to German high school graduates, Kristin mentioned: military service, volunteer work in Germany or another European country, work&travel, or volunteering in one of the developing countries (which she chose).

accommodation, and food provided. This was "the only opportunity to live here for a year," which also gave her a sense of security due to the institutional setting (cf. Laythorpe 2010).

David from Canada, a few years older than Kristin, came to volunteer for a month at another Tibetan organization. Initially, when we just met, he told me about his need to help people—I must admit that I felt like I was a jury member in a beauty contest. With time, however, he admitted that he got tired of "just travelling around the world": "It doesn't make sense, it's just touching the surface, it doesn't give you a chance to get to know places better, more in-depth." His impressions of being in Dharamshala—shaped by the contrast with Delhi that he had visited before—were strikingly similar to those of tramping tourists: Tibetans were friendly and open-minded, and the Indians in the capital city were only seemingly friends, as "they just wanted you to buy something from them." Dharamshala and his status as a volunteer gave him the opportunity to establish close relations with the local population, that is, to satisfy his need for "hot authenticity." Volunteer tourism is thus a way of being a long-term (non)tourist immersed in an authentic everyday life. At the same time, it allows one to fall into the already mentioned category of "doers," that is, those who act, and not only gaze, who bring benefits to others, and not only to themselves, as is for instance the case with tourists who seek spirituality—they also get involved, taking different *dharma* courses, but do that only for their personal development (McGuckin 1997, 361–362). This kind of engaged tourism allows one to position oneself on the highest level of this imagined hierarchy, being at the same time an identity project (cf. Hustinx and Lammertyn 2003).

The choice of a particular place to volunteer is rather secondary to the country itself and is often decided by chance. Kristin chose Dharamshala out of several options in India, although, as she admitted, she had no idea about the social and political situation of the Tibetans at the time she made the decision. However, working with refugees seemed more attractive to her than volunteering in an orphanage in Delhi, because "everyone now is helping children." However, she knew about the Dalai Lama and Tibetan Buddhism, so she welcomed the opportunity to live in a monastery. As was the case for young Indian participants in political tourism—and tramping tourists as well—Buddhism proved to be the main attraction of Dharamshala and the cultural resource that refugees so skillfully translated into political support and visitors' involvement in their community tasks. David ended up in McLeod Ganj accidentally. He wanted to work in an Indian orphanage (but was unable to give the name of the town or even the state where it was located), but the institution turned out to be a center for girls only, which did not accept male volunteers. As an alternative solution, he was offered (it was the case of a volunteer tourism brokerage already mentioned above) an opportunity to work for an organization focusing on Tibetan refugees' education. He only had a very general idea of Tibet: "the same as about the situation in Africa or North Korea."

However, Dharamshala was also visited by non-accidental tourist-volunteers, usually older than holiday-time volunteers and those who came within a gap year framework. This group included two categories of people. First of all, there were tourists who had visited Tibet before, and that experience gave rise to a need to learn more about the "real" Tibetan culture and to "do something" for the Tibetan cause. For James and Kathleen, an Australian couple of physiotherapists, their stay in Tibet was "a sad and disappointing experience": "it is actually China, there is hardly any Tibetan culture there, you can't move freely, and although our guide was a Tibetan, she couldn't tell us what she really wanted, and you could really see that the culture is extinct there." The refugee elites, as it was already discussed in the previous chapter, in their identity politics, are effectively promoting the diaspora as heirs and custodians of authentic Tibetan culture, which actually convinced James and Kathleen they should have visited Dharamshala. They had heard that there were opportunities to work for the local community, so they decided to come, look around, and "find something." Australia has long been granting asylum to former Tibetan political prisoners, so they knew of the existence of the Gu Chu Sum association in McLeod Ganj.[64] Bob from the United States, on the other hand, came to Dharamshala for six months' volunteer service because he had been a practicing Buddhist for years and wanted to be close to the Dalai Lama (he jokingly described himself as a "Dalai Lama's groupie"). Earlier in his life he had worked in the hospitality sector, which was a "very lucrative but not very self-fulfilling" job. Therefore, at some point he decided to do something with more developmental potential: "both for himself and for others." And since he completed a course in teaching English as a foreign language, he chose an organization that focused on educating refugees. Initially, most of his classes continued in one of the cafés, because "everyone wanted to have extra and free conversations," but over time Bob felt tired of being too friendly. Like the young French volunteers and Indian participants in a political tour, he too had tattoos made in the same studio (one was an image of the meditating Buddha Sakyamuni). In addition, he decided to take a *momo* cooking course run by a young refugee from Tibet.

The above examples show that on the practical level, tourist-volunteers often did not differ much from other tourists, and their presence, apart from being a form of work for the local community, brought very tangible benefits

[64] They, too, suffered from being "overheated with authenticity," and—as a result of diarrhea—were often absent from work.

to individual tourism workers. Due to relatively little institutionalization, volunteer tourism in Dharamshala was characterized by a high degree of openness. Volunteers were not bound to any particular organization, actually often they circulated between several of them, and were not very much "looked after." As a result, the money they left in McLeod Ganj was used by local guesthouse owners, restaurateurs, shopkeepers, cooking instructors, and tattoo artists. The fact that volunteers were an important group of customers was evidenced by advertisements about different kinds of tourist services that appeared in the NGO offices: paragliding in Bir or discounted bus tickets to Delhi.

Bob and Kristin were long-term volunteers, and unlike short-term and "drop-in" volunteers, spent a lot of time volunteering. They both worked as regular teachers (of English and German languages), so they taught classes every day, additionally spending two to three hours for preparation (searching for information, writing syllabuses, developing materials for students, inventing homework). They made all the photocopies needed for the classes out of their own pocket, as the organizations did not have such funds in their modest budgets. Moreover, they had tutoring sessions, for which they did not have to prepare, but which, added to classes accounted for an eight-hour working day. I often saw Bob literally dragging his feet out of fatigue while going back to his guesthouse at the end of the day. Kristin gave up on the conversation sessions at some point, but she was heavily involved in the organizational and office work, which also kept her busy for most of her day. They were both—when compared to short-term volunteers—more often disappointed with their collaboration with local organizations, as they put more time and energy into their work. They expected more "professionalism" and "organization culture," as well as a much greater appreciation of their work (Kristin, for instance, was irritated when the head of her organization asked: "But you speak English, so why can't you teach it?," or when her students were drawing doodles on the educational materials that she was copying for them at her own expense).

It was those most committed volunteers, more often than the others, who adopted the attitude of "friendly moral ethnocentrics": on the one hand, already familiar with the site and "open to contacts with the natives," on the other hand, "perceiving other culture in the Enlightenment's categories of 'conservatism' vs 'progress' and leaning towards rationality and modernization-driven projects in the spirit of civilizing mission" (Podemski 2005, 206). For example, Kristin was irritated by the Occidentalist image of the West that, in her opinion, her students had, which she attributed to their ignorance and unwillingness in obtaining information:

> They have no idea about Europe, and the West. They don't even talk about "Westerners," they talk about "foreigners," so basically everybody who doesn't basically look Indian, Nepali or Chinese, is a "foreigner" and they all [those foreigners] are from one big country [located] somewhere, that is rich and everybody drives a BMW there, and nobody has to work. . . . [They think] that we don't have morals, we don't have culture, we don't have religion . . . and they automatically assume that everybody is like that.

At the same time, however, Kristin did not blame only Tibetans for having such an image of the West. In her opinion, such an image emerged out of Western mass culture—especially Hollywood movies—so popular in India, and from observations of the behavior of some tourists in Dharamshala. Unlike her French peers from another organization, she tried to behave "decently"—she did not go out in the evenings (the monastery used to be locked shortly after dark), rarely drank alcohol, dressed modestly. She was annoyed by ego-motivated tourists, obsessed with "O yeah, it sounds so cool! I can add it to my blog." She felt strongly connected to her students, having worked with them a few months, and felt that the tourist urge to take a sneak peek at their lives as an indecent tanatotourism: "They [drop-in volunteers for conversation classes] come here, it's like sensational, it's only out of curiosity. Somehow, I got the impression that they want to hear horrible stories [about prisons, escapes]. . . . I really feel angry when they [my students] are so exposed to these tourists coming . . . it's like a zoo. And they [tourists] come with their cameras, they want to take picture of them. Damn!" Kristin seemed to be designing her own cultural boundaries and norms for the Tibetans, whom she treated as her responsibility and as victims of the tourist gaze. Victimization meant that she did not notice their own agency and the benefits they could derive from such encounters.

Despite some disappointment and "losing his job" before the end of his volunteer work arrangement, Bob planned to return to Dharamshala. He agreed to teach English at a nunnery in McLeod Ganj, but made it conditional on receiving free accommodation, as he was afraid that he would not be able to afford renting a room for a year. In his new identity project, there was no possibility of returning to the world of global capitalism—he saw volunteer work as a niche for himself among Buddhist followers. Such an extension of the volunteer work period can be seen as a safeguarding strategy against job insecurity in the postindustrial era (see Rojek 2010). After returning from India, Kristin began studying international relations at a British university. She also completed an internship in alternative economics in the United States. It can be argued that the cultural capital and self-confidence that she gained during her volunteer work in McLeod Ganj gave her the courage to develop her career in an international environment.

At the same time, volunteering was a difficult experience for Kristin, especially with regard to her self-perception. Negative emotions in relation to the generalized category of Indians contrasted with the perception of Tibetans. For a blond European teenage woman, traveling alone and for so long for the first time, the Tibetan diaspora became an oasis in which she could hide from what she perceived as the persistent pestering by the Other, in this case the Indian. Tibetans from Dharamshala—whose subaltern position required a certain flexibility—acquired cultural competence in contacts with the so-called West, as a result of which they appeared to be "more reserved" and "easygoing" in relations with tourists, when compared to "Indians." This is why Kristin, like other "Europeans," complained about McLeod Ganj being "Indianized" and was reluctant to appear in the town on weekend evenings ("too loud, too crowded, too annoying").[65] The earlier "Western" character of McLeod Ganj—which was being replaced by the "Indianness" of domestic tourists—was thus seen as an asset, as it gave the town a sense of familiarity and security in experiencing Otherness (see Bloch 2016a).[66]

Kristin, like many other tourists with whom I worked doing mobile ethnography, fell victim to the overheating with "hot authenticity" in the multisensory experience of India (see Bloch 2016a). In conversations with her, there was consistent talk of the difficulty to accept dirt (not only as an aesthetic category, but literally a sensual one),[67] invasive soundscape, and the sense of irritation or even danger caused by norms of physical and mental closeness other than those known to Kristin from her country of origin. She kept telling herself that she could not "think like a racist," but then there was ". . . this smell, and the cow shit . . . I have never expected so much garbage and rubbish . . . you are automatically getting annoyed by the horning and the cows, and people trying to sell you things, people trying to overcharge you, and

---

[65] It is interesting that an Indian journalist, writer and researcher, Prahland Singh Shekhawat, described his experience of McLeod Ganj in a similar way: "After the exhausting experience of dealing with all the heat and dust, touts, commission agents, rickshawalas, male-staring, and ogling in India, McLeod Ganj offer to the weary travelers an escape to a little heaven. The Tibetans in contrast are more relaxed and unobtrusive. There are no pushy peddlers and aggressive sales persons. . . . Tibetan boys are more easy mannered and bother the Western girls less. . . . Why is it that Indian shopkeepers, sales agents and rickshawalas hustle so much and are more off-puttingly persistent and less trustworthily?" (Shekhawat 1995, 14).

[66] Toward the end of her stay in Dharamshala, Kristin concluded that "it's actually sad that I went to India and I didn't get to know any Indians."

[67] Dharamshala, like most of Indian towns, was struggling with the problem of a lack of efficient waste management in the face of the market being flooded with low-quality goods in non-biodegradable packaging. However, in the diaspora's identity politics, there is a strong emphasis on the new Tibetan tradition of environmentalism (see Bloch 2008). Thus, as part of the image consistency, cleaning campaigns were undertaken by both the government-in-exile and McLeod Ganj's NGOs. As a result, tourists get the impression that, as Kristin put it, "Tibetans are cleaner [than Indians]."

people asking you annoying, personal questions and taking photos of you, and men touching you . . . It is hard to ignore it."

At the same time, in my conversations with tourists about the benefits of volunteering, they often mentioned the possibility to establish direct, close contact with another person. This is shown by excerpts from French volunteers' statements made during a focus group interview:

> "The people you meet during conversation classes have so much experience and are sharing it with you"; "I was a pretty ignorant [regarding the political situation of the Tibetans] before I came here, but the conversation classes are a great opportunity to hear all these stories"; "[these are] such sweet people, they went through such terrible things, many of them are monks, and they all share with you all these stories about their escape, and the knowledge about Buddhism they have."

At the end of her classes, a young volunteer from England was taken by the disciples-monks on a farewell trip to the monastery, where she was given blessed prayer flags while the whole event was recorded with an iPod. Other students gave her eco-friendly notebooks with the logo "Made by Tibetan refugees" and expressed their gratitude—she was sincerely moved. Mary Conran (2011), who conducted research among three different groups of volunteers in Thailand, pointed out that the category of closeness and intimacy was the leading motive indicated by tourists as the greatest benefit and most memorable aspect of their volunteering experience. Below, I wish to reflect to what extent this closeness translates into a real change in the lives of the so-called beneficiaries of volunteer programs.

Inequalities, Exchange, Capitals

The phenomenon of volunteer tourism, especially in the context of postcolonial countries, has met with widespread criticism as a form of domination implemented by imposing on the societies of the Global South the ideas of modernization developed in the countries of the Global North, that is, de facto as a continuation of the colonial civilizing mission.[68] The same accusations have been voiced against development projects that volunteer tourism often supports (Simpson 2004). Volunteers in this critical perspective appear to be modern missionaries, motivated by a moral obligation to lead by the hand those whom they feel need salvation (Conran 2011, 1464). In this lens, a fundamental question is often asked: whom volunteer tourism really benefits?

[68] Although there have also been other takes on volunteer tourism as, for instance, "an expression of love in tourism" (see Singh 2019, ch. 4).

Moreover, as critics point out, this kind of tourism, in its mass-scale form, can sustain global economic inequalities, as it is the poverty, in its aestheticized and depoliticized form, that seems to be the main attraction for tourists (Mostafanezhad 2013). A volunteer named Kit, one of the protagonists of John Hutnyk's book about Kolkata, aptly captured this paradox: "An Irish nun working at Mother Teresa's talked about the 'beauty of the poor,' and even though she meant well, it's as if 'the poor' are forever there and will never change, will remain forever 'beautiful,' and always poor" (1996, 57). Hutnyk used the term "voyeuristic consumption of poverty" to refer to those to whom global economic differences allow to live in India as if they were "Rajas" (1996, 9–11).[69] Conran (2011), quoted above, points out that the focus on emotional closeness in relations with the so-called beneficiaries of volunteering programs veils structural inequalities on which such encounters are based. Moreover, it transforms the question of institutional causes of injustice into a question of individual morality, making volunteer tourism become an apolitical cultural practice, as "intimacy, goodwill and compassion are used to justify and depoliticize the volunteer experience" (Conran 2011, 1465; see also Woodward 2004).

The image of India shaped in tourism imaginaries—as a great ancient civilization, but today a backward country that is not keeping up with modernity and requires constant assistance from the West (Echtner and Prasad 2003, 669–672)—seems to legitimize volunteer tourism. This, in turn, leads to a patronizing construction of the "ideal victim": weak, passive, lacking awareness, unable to act independently (Domańska 2008, 30). This seemingly noble motive leading to victimization is discursively objectifying, as it allows one to decide for the victim, to guide them. Another argument leveled against volunteer tourism is leakage, a phenomenon I discussed in the Introduction. Often the money that people pay for the opportunity to do volunteer work does not go to the target organizations, but to institutions specializing in recruiting volunteers for local organizations which do not have resources to do that by themselves. Critics point out that if these funds were channeled directly to the communities where volunteers are to work, they could be used much more effectively, for example, to create local jobs (Butcher and Smith 2010, 33).

Such critique resonates with the criticism of PPT, which does not deconstruct the structural foundations of global inequalities, nor creates alternative

[69] He admits that during his research he met volunteers who were very critical about the naive sentimentality of their colleagues—a kind of Patrick Swayze of the *City of Joy* movie style—though they believed that they themselves could avoid it (Hutnyk 1996, 59). *City of Joy*, dir. Roland Joffé (1992), is a British French American drama which enjoyed great popularity in the 1990s; Swayze plays a role of a doctor looking for meaning in his life, who ends up in the slums of Kolkata, where missionaries together with volunteers help the Indian poor.

scenarios, but only focuses on the inclusion of the poor in mainstream tourism, thus perpetuating the neoliberal capitalist order on which the North-South division is founded:

> Rather than attending to the need for structural change, redistribution of wealth and resources, and addressing international and national power structures, they tacitly accept a neoliberal approach to development and tinker with the capitalistic international tourism system at the edges, eking out a few resources for small, selected groups of the poor (or relatively poor) in destination areas. (Harrison 2008, 858; see also Schilcher 2007)

The proponents of PPT, in turn, accuse its critics of utopian idealism and of assuming a superior moral position. They also point out that in many alternative—"anti-capitalist" and "anti-mass"—forms of tourism, such as, for example, ecotourism, the benefits for the poorest inhabitants of tourist regions often become secondary to other priorities, such as environment or culture conservation. PPT focuses more on the development of regions that can benefit from tourism, rather than on conservation projects that often become altars on which the needs of the local population are sacrificed, as is very well illustrated by the example of Hampi and its inhabitants being displaced in the name of "cultural conservation" (Ashley and Roe 2002, 67; see Goodwin 2008 and his response to Harrison 2008 criticism).

However, in case of Dharamshala, the above criticism is largely unjustified. Tibetan refugees—as I have already shown in this and previous chapter—do not construct their attractiveness on the basis of their difficult economic situation, that is, the category of poverty, but by creating representations of a unique, precious culture and political struggle for a just cause. The pride upon which the Tibetanness is built in diaspora, combined with autonomy in the way volunteering is organized (which I discuss below), show that Tibetans do not position themselves as victims. They are even perceived by part of the host society as "rich refugees" (Prost 2006) or refugees who "don't behave like refugees." Of course, this does not dismantle global inequality structures, but Tibetan refugees have skillfully appropriated conventional tourism to fight against the global power, which is the People's Republic of China and its imperialist policy along with rhetoric challenging the universality of human rights. Moreover, in Dharamshala—at least at the time when I conducted my research there—most of the organizations were local and they directly recruited tourists-volunteers without any extra charge for the opportunity to do volunteer work.[70] Although it was certainly a problem that the high supply

[70] Only the oldest of these organizations charged volunteers for additional services, such as transportation from Delhi airport to Dharamshala (thus satisfying the tourist need for security),

of unpaid work made it difficult for Tibetans themselves to find employment in these organizations and—contrary to many declarations—even classes for beginners were taught by volunteers and not by Tibetan-speaking teachers. What is important, however, is that Tibetans defined their goals by themselves, although there is hard to deny that they tried to adapt to the prevailing global standards, such as the knowledge of English. These were, however, bottom-up initiatives, which—as one organization declared—aimed to "provide free education to Tibetan refugees and empowerment opportunities to the local community." This short quote reflects how individual and group goals set by the Tibetan diaspora are connected. Moreover, these organizations created a space in which social problems that had been present in the community for a long time, but often kept silent in order to maintain the idealized image of Tibetan refugees produced within the framework of the diaspora elites' identity politics, could be addressed. "Many refugees," the quoted-above organization explicitly admited, "find it difficult to integrate into India which can lead to high unemployment rates, alcohol and drug abuse, depression, poverty and feelings of disempowerment and isolation."[71] As one of the founders of another organization stated, "New refugees feel discouraged and socially useless." What is significant in this regard is that most of the NGOs referred to in this section were founded by Tibetans who were born in Tibet and fled into exile either as children, teenagers or adults, thus it was much easier for them to identify with the newcomers than it was in the case of India-born refugees (see Bloch 2014a). Moreover, it reflects a progressive change of thinking in the diaspora about the individual economic self-sufficiency, once perceived as an aspiration unworthy of a refugee, who

---

accommodation in homestays, that is, homes of Tibetans (which, in turn, was a response to the tourist need for "hot authenticity") or for participation in the local tours. However, there is a risk that Tibetan NGOs will start using intermediaries in recruiting volunteers. During my research, I met a Canadian volunteer, who arrived in Dharamshala for a month-long stay through Volunteering for India organization. He was charged 1,200 dollars, for which he received a modest accommodation, meals in the organization's kitchen, transport from and to Delhi, on-site "care" and participation in local excursions. Meanwhile, the actual costs of living in Dharamshala were at least half lower for regular tourists (and of much higher standard). At the same time, nongovernmental organizations working for local communities in the surrounding Indian villages function in a more classic formula, offering full-paid volunteering through institutions such as Working Abroad, an organization registered in the Great Britain, which at the time of conducting fieldwork for this book offered volunteering opportunities in over 150 countries.

[71] Central Tibetan Administration created for adult refugees from Tibet the opportunity to receive elementary education in a boarding school (formerly called Transit School, now Sherab Gatsel Lobling), located a few kilometers away from Lower Dharamshala. However, many newcomers gave up upon studying there for a number of reasons, which included the disciplining character of the school (e.g., no smoking rule or restrictions on intersex relations), its isolation from the rest of the diaspora (as well as Western tourists, contacts with whom offered many opportunities), and—most of all—inability to combine studies with work. See Bloch (2014a).

should put the good of the homeland before one's own existence and material needs.

The effectiveness of such programs can be, of course, questioned, and so can be the level of professionalism of the implemented initiatives. Volunteers' lack of preparation, resulting from their mostly random appointment, can be an obstacle to the effective use of their potential. For years I have heard opinions that before a volunteer gets to know the local community and learns about its specific character, the volunteering time is over. Many volunteers who taught languages in Tibetan NGOs did not have any qualifications or teaching experience, many conversation partners were not even native speakers, and neither volunteers nor students were regular in their class attendance, which was a source of growing frustration especially among long-term professional volunteers (a view which I also initially shared). The fact that the gap year took place more and more often after graduation from a high school—and not between completing the university education and entering the job market—meant that volunteers were in their early twenties (or even younger) and just started to accumulate life experience. As a result, Tibetan students were often older than their teachers, which in some way violated the traditional pattern of knowledge transmission from an (older) master to a (younger) disciple. I noticed that it was an uncomfortable situation for some volunteers as well, as they felt that they could not require homework or not being late from people in their parents' age or from monks. Even if teachers and students were peers, the "shyness" of Tibetans, often discussed by the volunteers, was a barrier, seen as their "natural" cultural trait. However, it is largely due to different educational experiences of both parties, about which Western volunteers were not aware. Until recently, Tibetan schools—and the monastic education system—used the rote learning as the main teaching method. This is the reason why Tibetans did not feel comfortable when faced with the need to strike a conversation or speak freely on various topics suggested by volunteers during conversation classes, especially as these topics often went beyond what they knew. While India-born refugees, beneficiaries of modern education system established in exile, were well versed in, for example, environmental issues, for newcomers from Tibet the topic of renewable energy sources was rather enigmatic. An additional problem was the emphasis placed in Tibetan education in exile on written language, in which most of the homework and exams are conducted. Meanwhile, especially volunteers from Western European and North American countries preferred a more hands-on approach, that is, learning that involved active participation of students, a way they knew from their own schools. The matter was further complicated by a somewhat haphazard attendance of the students: most of them, for various reasons in life, could not attend classes regularly. Some students, on the other hand, attended all classes irrespective of their level of competence, including beginners, intermediate and advanced courses.

"For them, English is English, there is no difference," commented Kristin, visibly annoyed.

However, as pointed out by Conran (2011, 1465), while such critique is necessary, it is not enough to stop there: we should try to move beyond the overwhelming criticism of volunteer tourism as an exclusively neocolonial practice, because, according to Hutnyk, "something must be done" (1996, 222). Volunteer tourism has the potential to contribute to the global movement for justice (Conran 2011, 1466), and in this sense to be a starting point for structural change (see, for example, McGehee and Santos 2005). Sometimes it is enough to change one's perspective in order to see other kinds of benefits that such tourism can bring. In the context of Dharamshala, it is worth remembering that Tibetan refugees have multiple aims. Just as events aimed at gaining political support were also aimed at recruiting volunteers, the aim of the organizations which acquired tourists-volunteers was not only to work for the local community, but also to promote the Tibetan cause. In this respect, the initial ignorance of volunteers was not an obstacle, but a potential. This explains the alleged paradox of constantly seeking new volunteers, including short-term ones, despite their oversupply. Secondly, even if volunteers had selfish motivations—such as enriching their CVs—it did not matter, as long as the other side benefited from their work. Thirdly, language classes arranged by NGOs created a platform for direct, informal interpersonal contacts between Tibetans and foreigners. Especially conversation classes, as opposed to more structured lessons, provided an opportunity for chats which most often related to the biographies of the interlocutors, so they could get to know each other and create their own self-representations. Such classes often moved to other public spaces, especially restaurants and cafés. This practice was so prevalent that one café even placed a handwritten notice next to the "No Smoking" one: "Please Don't Occupy Seat For Long. Classes Not Allowed. Thank You."

These contacts were then developed on other levels and could take the form of acquaintance, friendship, or even love, which in turn had a very real impact on the lives of individuals. Robert Rydzewski (2020), who conducted research among volunteers working with refugees on the Serbian-Hungarian and Serbian-Macedonian border, pointed out that even short-term involvement of volunteers had the potential to develop into long-term friendships, mainly due to new communication technologies—as I have already mentioned in the previous chapter—which made the relationships last and even deepen despite the departure of the tourist-volunteer. Rydzewski stressed that these relationships were not easy, as they created mutual dependencies and expectations. Also Brooke Schedneck (2017), who worked with volunteers teaching English to young monks in Thailand, observed that while the need for self-transformation through the experience of the Otherness was the driving force behind the volunteers' commitments, upon prolonged, closer contact the monks stopped being seen as merely representatives of Thai culture. The relationships that tourist-volunteers developed with the monastic community made them see its "unexpected familiarity," which extended "beyond the golden, glittering Buddha statues."

Personal relationships with volunteers can serve those who are on the other side of tourism encounters to diversify their social networks,[72] as well as a source of cultural capital,[73] or material support. Usually the attention of researchers is focused on the latter aspect, while the accumulation of cultural capital is mainly discussed in the context of volunteers and their motives (Heath 2007),[74] bypassing the fact that members of the communities hosting tourist-volunteers may be interested in benefits stemming from intercultural encounters. In this respect, encounters that take place during volunteering can serve the purposes of cultural exchange (cf. Uriely and Reichel 2000). Take, for example, the "Western culture course" offered by one of the Dharamshala-based NGOs. This popular course (designed by Kristin) consisted of the following modules: geography (taught according to the nation-state logic); history (with emphasis on the ancient Greek philosophy and the history of Christianity, with various festive rituals); and democracy (where it came from, how it developed etc.). On the one hand, such an idea could be considered a manifestation of cultural patronizing attitude or even the continuation of the colonial civilizing mission. On the other hand, Kristin did not hide her motivations, namely the desire to equip refugees with cultural capital to help them go "to the West": "They all [students coming to language classes] want to go there, . . . so I wanted to give them not only the information, coz they have access to information, but to show them how to use this information." Bob was also aware that for many of his students the main motivation to learn a language was to leave India.[75]

---

[72] According to Kristin, "conversation classes make sense at least if you have some basic communicational skills, otherwise they just sit there and look at you and copy, copy, copy whatever they can." However, she would be surprised how many relationships between tourists and members of the local community were born with very little verbal communication.

[73] I use the concept of cultural capital here in the broad sense of Pierre Bourdieu's (1986). It is a resource consisting of various competences (e.g., linguistic) and knowledge, which can then be transformed into other forms of capital: social (manifested by support—both instrumental and symbolic—resulting from embedding in social networks) or economic capital denoting access to financial resources and material goods. Particularly the cultural capital was recognized by Bourdieu as a tool of social domination and exclusion used by upper classes.

[74] Anna Romanowicz (2017), for example, who conducted research in Indian NGOs aimed at women empowerment, argues that these organizations primarily served the accumulation of cultural, social, and economic capital by middle-class women who were the NGOs' employees, and that the programs they implemented resulted in reproducing the difference, that is, a class-different Other who requires support.

[75] I also noticed that many young Tibetans took advantage of the Dharamshala NGOs' rich offer simply out of boredom, to "just do something." Tibetan diaspora is increasingly affected by the

All this—encounters, social networks, cultural capital—can result in emigration to "the greener pastures," that is, to "the West." One NGO openly admitted that "every year around 300 students leave the center speaking fluent English, many of whom now live in Europe, the United States or Canada, while others work in cities, holding different professional positions." For the same reasons, German and French languages were increasingly popular: the countries whose asylum policy, at the time of my research, was most favored by Tibetans were Switzerland, Belgium and France. Vincanne Adams analyzed the strategies used by Sherpas to build mutual commitments with Western trekking tourists visiting Nepal, which resulted in regular financial support and official invitations to foreign countries (1996, 206–232). Volunteer tourism is different in that it makes foreign visitors accessible to those who do not work in the tourism sector on a daily basis, making this particular sector more open and democratic. In other words, the local NGOs in Dharamshala helped to create a space and atmosphere conducive to establishing reciprocal relationships, a role that mountain expeditions played in the case of the Sherpa (Adams 1996, 221). All this, in turn, created opportunities to improve the disadvantaged position of individuals who belong to subaltern communities, which I discuss in chapter 5.

At the same time, the very creation of these organizations was often possible thanks to tourists and their social and economic capitals. Behind most of the NGOs I know, and which worked with tourist-volunteers, there was a history of a close relationship—friendship or even love—established between a member of the local community and a tourist. A spectacular example, that I had the opportunity to observe, was setting up a new NGO by the then manager of one of one of the oldest organizations of this type in Dharamshala. It took place with the active support—linguistic, instrumental, and emotional—of a volunteer who worked for this organization at the time and who assisted the man in establishing the new NGO. Above all, however, the said volunteer managed to convince the sponsor from her home country to provide funds to this completely new organization. In the following year, it turned out that the head of the organization and the (former) volunteer were no longer in touch, but the organization was doing well, to the extent that it had the largest office space of all local NGOs.

These processes also applied to Hampi, which, although much smaller than Dharamshala and not a "specialized" volunteer tourism destination, had its own organizations, which activities were assisted by tourists. One of them, which supported children from disadvantaged families, and, following the bazaar demolition, also children from displaced families, was founded by a local man

problem of unemployment, hence many young people, especially men, spend their days just wandering around.

and a tourist from the UK. Nelli—whose story I told in chapter 1—was a volunteer there, and that is how she came to Hampi. This was also the case with two Australian women who were directly involved in the residents' resistance against "spatial cleansing": a cofounder of the "Save Hampi People" campaign and a tourist who was a friend of Phurbu and Padma and often stayed at their home in Manali. When, following the evictions, more than eighty of the poorest families moved to the shack encampment on the outskirts of their former village, the abovementioned tourist organized a fundraising for them. She then bought a dozen or so metal beds through Nelli acting as a go-between, so that the slum-dwellers could survive the nights when the monsoon rains flooded their makeshift shelters. Middle-class Indian tourists were also involved in volunteering work there. Young architects from Bengaluru, who specialized in restoring traditional building materials and techniques, created a prototype of a cheap, eco-friendly house that was to solve the problems of the Hampi displaced residents. These examples show that political engagement of tourists and direct actions for local communities often overlap.

The same goes for grassroots social initiatives and paid work in the tourism sector. This was quite common in Hampi, where the labor market was much less diversified than in Dharamshala.[76] In Hampi, contacts with tourists and opportunities they offered in terms of social, cultural and economic capital inspired inhabitants to undertake pro-social activities. In their conversations with me, they often expressed visions to launch an NGO for the benefit of women, the disabled or the environment. The head of the abovementioned trust supporting children, apparently on the basis of the success of that initiative, made plans to expand its activities to include the elderly and even stray dogs. He was an example of a smooth, skillful operating at the interstices of the charity and tourism sectors. Even before he set up an NGO, he worked as an English- and French-speaking guide in Hampi, which gave him access to tourists and provided opportunities to attract sponsors and volunteers (which he continued to do as the head of the trust).[77]

[76] Tibetans, for instance, had the opportunity to find employment in the diaspora's administration.

[77] When we first met and the man did not know yet that I was not a "common" tourist, he asked me what I thought about the building where the NGO was located. I replied that it was okay, but a little bit cramped, to which he reacted immediately: "Maybe when you return home, you will find some sponsors for a bigger building!" The issue of purchasing land for the NGO in order to ensure more space, was often mentioned in his conversations with tourists.

Chapter 5

Tourism as a Source of Individual Empowerment

Stories of Encounter

In the course of my field research, I was often surprised by how different the perspective of hosts on the presence of tourists in their world was from the demonizing view of critics portraying tourism as a form of neocolonialism. Tourists were seen not only as allies in an unequal political struggle—either against a powerful occupant, their own government, or international regimes (as I demonstrated in the previous chapter)—but also on a more individual level, as an opportunity to improve one's economic or social status. One of my research partners, Mahendra, a guide to the scenic sunrise viewpoint and an owner of a small mobile eatery, used to say—while looking forward to the buses that were coming daily from the nearby city—"Every day I pray for tourists. Without them there is nothing to do in Hampi." With the end of the tourist season, life in the village was coming to a standstill—people were wandering aimlessly, idly planted in their chairs in front of their shops, staring at the road, and fiddling with their mobile phones. Outside Hampi, as another informal guide used to say, "you don't live at full throttle." For small entrepreneurs from Hampi, tourists were something "good," something which enabled them to earn a decent living, while UNESCO, the Archaeological Survey of India, and the "government" were "evil," as they "only cared about dead monuments, not people." Given the uncertainty and mutual accusations that emerged when people faced displacement and demolition, some voiced the opinion that "tourists were more to be trusted than locals." Many people who grew up among tourists were nostalgic about their childhood. Raj told me once: "I really liked white people. One used to buy me chocolate, another taught me English, yet another was helping me with my homework."

Such a perception can explain the great flexibility of the tourism workers, which I discussed in chapter 3, giving the examples of Geeta and Ravi. Running several businesses (even if some seem to be completely unprofitable) or frequent changes in business line serves to stay part of this sector, because it is valued and desired and not perceived as resulting from lack of other opportunities (cf. Ghodsee 2005, 58–59). Contact with foreign visitors and the opportunities it creates can be used to strengthen one's own economic, social, and even political position, thus resulting in empowerment. Recognition of this potential and its actualization can be interpreted as a manifestation of disadvantaged persons' agency (see Brettell 2002). I argue that such agency should not be equated with instrumental calculation, as it is the case of allegedly calculated non-commodified close relationships manufactured by the hosts (e.g., Lozanski 2010, 756–757), as these relationships are generally no different than those which people construct in different social contexts, always taking into account the available opportunities and benefits. In this chapter, I discuss the bottom-up forms of empowerment, which happen outside the official discourse of development and modernity.

At the same time, however, scholars point out that the hosts' agency is often not what tourists expect. This is particularly true of those members of the host communities who are very active in trying to persuade tourists to buy goods and services they offer, thus being "openly agentic and overtly serve[ing] their own—instead of travellers'—interests" (Lozanski 2010, 755). Kristin Lozanski calls them—touts, beggars, rickshaw drivers, or shopkeepers—"hypervisible Indians." This "visibility" does not allow tourists to forget the commercial nature of their journey and thus does not satisfy their need for "hot authenticity" (discussed in chapter 3), which requires contact with a "genuine" Other. The same applies to workers in the tourism industry who do unskilled work, which is necessary to support tourist infrastructure (such as cleaning, cooking, doing laundry, etc.). Lozanski describes them as "invisible" to tourists, because "their invisibility is . . . critical to maintaining the semblance of non-commodification" (2010, 754–755). However, the members of the middle class also seem "invisible" to tourists, because they do not satisfy their need for exoticism (i.e., "cool authenticity")—they are just as modern, mobile, and middle-class as tourists themselves.

Lozanski coined the "hypervisible" and "invisible" categories based on her research on how Indians were perceived by backpackers traveling in India. However, my field experience shows that tourists, although clearly motivated by the need for "hot authenticity," often involve themselves in close and complex relationships with the hosts who do not seem to meet this need, but who belong to one of the three groups identified by Lozanski in her analysis. These relations often go beyond a single tourist trip, as I already described in chapter 3 in the context of the principle of reciprocity. This chapter depicts such encounters and how they are used by those who are on the other side of the tourist contact to strengthen their often disadvantaged position. Apart from the unequal power relations inherent in tourism in the

Global North–Global South context, relationships with foreign tourists provide an opportunity to improve one's individual social and economic standing. This applies especially to people who belong to the most vulnerable groups: women (especially widows, divorcees, single mothers), people from the lowest social strata (from indigenous communities or groups outside the caste system), the newcomers from Tibet whom Tibetans born in exile push out of the labor market or people who function in any other way on the fringes of their communities.

Building close, interpersonal relationships in tourism is obviously not easy. Researchers point out that even though tourists declare their wish to "fraternize with the natives," in practice they are often afraid of these relationships and lock themselves in their own social bubbles, such as the backpackers' ones (see Tucker 2003a, 43–68). This is why they are often accused of forging only superficial illusory bonds: "Contact with the natives is temporary, purely external, based on mutual consent to a certain convention of 'encounter'" (Wieczorkiewicz 2008, 206). Excessive closeness, as Anna Wieczorkiewicz argues, would involve taking responsibility for poverty and injustice, which are often part of everyday life in the countries visited by tourists (2008, 207). Meanwhile, one of the main mechanisms of shaping the tourist's imagination is mystification—a trick consisting in omitting the reality of contemporary life while suggesting that there are other, higher values (Albers and James 1983). In such representations, the natives are shown as those whose only concern is to satisfy the needs of tourists, as those who happily offer their culture to their guests. This "framing" of the tourist reality makes everything that is outside—the whole complex political, economic, and social situation—nonexistent (Bruner 1989, 441). It thus also obliterates unequal relationships between the local population and tourists (Wieczorkiewicz 2008, 182).

As a result, tourists, confronted with economic inequalities either succumb to the mechanism of idealization and construct romanticized images of "happy poverty" (i.e., simple and poor people, who live in harmony with nature and are able to see the beauty of ordinary things, a skill that we, stressed-out, busy inhabitants of the (post)modern world, have lost), or turn a blind eye, often blaming the victims themselves for their deplorable position. Poverty is not only an aesthetic disruption of a touristic "imaginary paradise" (Dann 1996, 68–71), but is also experienced as multisensory (through smell or touch) and as such poses a difficulty for tourists originating from worlds where, although poverty and economic and social inequalities are still present, they are hidden within different institutions, creating the impression that they do not exist.[1]

[1] In India, as in the entire postcolonial world, the practice of "spatial cleansing" (Herzfeld 2006) is also employed in order to remove images of poverty from public view, especially in the context

All of this requires emotional distancing, and "the easiest way to achieve it is by taking on an observer's attitude—a typical tourist behavior, which is formed outside the area of moral obligations" (Wieczorkiewicz 2008, 207). Thus, according to Wieczorkiewicz, "traveler-observers" become "imposter-participants": "Their attention is attracted by characteristic dress details, gestures, and regional catchphrases; they learn local ways to respectfully address others, sometimes devote an hour or two to learning local dances or trying on regional costumes. They enjoy an orchestrated situation of 'fraternizing with the natives' or 'befriending the villagers.'" But they do not go any deeper. It is a "distanced friendship," which allows tourists to feel safe while at the same time "egocentrically" making their own traveling biography more attractive (Wieczorkiewicz 2008, 207).

I, however, would not be so unequivocal in blaming tourists for the superficiality of such contacts. On the one hand, it results from structural factors, that is, mobility inherent in tourism, which often makes it impossible to build meaningful relations, regardless of the intentions of both parties, as I had the opportunity to observe among tramping tourists. Despite the declared desire to "fraternize with the natives" and indicating the Other as the main reason of the journey, in practice, their focus was clearly and inevitably shifting toward fellow travelers—companions on an often intense, physically and mentally challenging monthly tour (Bloch 2016a). On the other hand, as Wieczorkiewicz also noted, tourists often feel lost in the new, alien reality, "they cannot let themselves be fooled, robbed, killed . . . , they must be both vigilant and friendly; open to contacts, but also financially independent" (2008, 207). Often, such anxiety leads tourists to retreat and seal off within the enclave of people "like me," that is, other tourists, as was the case for Kirstin, a volunteer, at the beginning of her stay in India. Kristin became unpleasant and inaccessible, did not leave the monastery unless she had to, and began to react aggressively when people attempted to talk her up or smile at her. By the other party, it was interpreted as an expression of colonial superiority or vain Western individualism. Teenage Kristin, craving safety, locked herself in a societal bubble created by other volunteers. After some time, however, she managed to open up again, although mainly to Tibetans. She began to learn the language (with varying degree of success, but the mere fact always triggered a positive reception), wore *chuba* on "White Wednesdays,"

of important international events. In the course of the preparations for the 2010 Commonwealth Games, the main streets of Delhi, including roads leading from the airport to the city, were cleansed of not only homeless people, but also stray dogs and holy cows. The same applies to the World Bank meetings or the Olympic Games. Herzfeld calls it "practical essentialism," and argues that it results from "crypto-colonialism," as it refers not only to those countries which actually fell victim to European colonialism. "Crypto-colonialism" is about striving to adapt to the Western imaginaries, that is, "making a good impression on foreigners" (2006, 135, 144).

built close relationships with several people, and when her nine-month volunteering work period ended, she extended her stay for another two months.

Hazel Tucker argues that, if we want to understand tourist encounters more fully, it is necessary to take into account their emotional and physical dimension, as it has the potential to positively disturb the otherwise inherently colonial relationship between a tourist and the Other (2009, 444). Tucker focuses particularly on the category of shame, which makes the visiting and the visited meet within the framework of an equal relationship, equally vulnerable (2009, 456).[2] Shame causes tourists to feel embarrassed, but also makes them subjects to productive discomfort. They feel moral obligation based on reciprocity and many overcome their fears by giving up the safe, emotionally uninvolved role of a distanced observer, instead opting for "intimate closeness" (Adams 1996, 220) and often risking they will be hurt. Conclusions similar to Tucker's, based on her research on backpacking in Turkey and to Adams's, based on her research on the relationship between trekking tourists and Sherpas in Nepal, have been drawn by Valerio Simoni (2017) in a book focusing on informal encounters in Cuban tourism (involving economic transactions, friendships, sexual contacts, etc.). Simoni sees the potential of such encounters—which happen across differences and inequalities—in redefining their moral basis and power relations inherent in them, despite both sides having concerns about the other party's intentions.

In this chapter, I present stories of encounter that contradict the thesis of "distanced friendship" between tourists and workers in the tourism sector, and indicate the benefits which such encounters can bring, especially to the latter. I discuss individual sponsorship, various forms of support offered by tourists to stimulate the development of local tourist-oriented businesses, the ways in which relationships with tourists can strengthen women empowerment and the role of love in overcoming structural constraints.

SPONSORING: ATTEMPTS TO ESCAPE ONE'S SOCIAL DESTINY

In the previous chapter, while pointing out the benefits of contacts with tourists facilitated by volunteer tourism, I mentioned the acquaintances, friendships, and love relationships resulting from such encounters, which can be a source of cultural, social or economic capital. Many tourism workers I got to know had managed to build close enough ties with foreign tourists and trigger

[2] Of course, shame in other contexts can work as a disciplinary category, as Tucker herself showed on the example of Turkish women from Cappadocia, whose opportunities for operating in the tourism sector were limited by the fear of shame (2009, 458).

such a sense of responsibility for global inequalities that the latter became their "private sponsors" (see Adams 1996, 219), more or less regularly supporting them financially. According to Vincanne Adams, whose research I discussed in chapter 3, in the late 1980s more than a third of the Sherpa families from Khum Jung village in Nepal, who worked with trekking tourists, benefited from foreign sponsorship, some of them having one main sponsor and a few minor ones; mostly the sponsors were unaware of each other's existence (1996, 221).[3] Adams suggests that these relationships were shaped along the lines of the *jindak* relationship that Buddhist monasteries had with their donors (1996, 219–220).

In the case of Tibetan refugees, such ties are often fashioned after a reinterpreted "patron-priest" relationship, which I discussed in the previous chapter, but here they are acted out on an individual level. It is important to emphasize that such a form of support is socially sanctioned in the diaspora, which has developed an extensive system of individual sponsorship, usually in the form of "distant adoption." Such individual sponsoring addresses students of Tibetan schools, monks and nuns in monasteries, and the elderly in nursing homes. Some Tibetan NGOs are also trying to develop similar programs.[4] Most Tibetans know someone who is or was sponsored by a "patron from the West" within a framework of institutional projects, so they try, having access to tourists, to seek this kind of support on their own. This is especially true for newcomers from Tibet who fled to India as teenagers or adults and whose position on the labor market is very weak in comparison to the Tibetans already born in India. The latter are (usually well) educated, know languages (English and Hindi), and are familiar with Indian reality. This is the reason of why newcomers are the most frequent clients of Dharamshala-based NGOs.

[3] Adams mentions that she herself was often asked by Sherpas for a financial support, such as paying tuition fees for a child's education (1996, 220).

[4] One of the organizations wrote about its sponsorship program as follows: "Currently, there are many Tibetan individuals, single parents and families in the town [Dharamshala] where they need financial support, thus we will talk responsible for finding a sponsor for them. We will help them get in contact with the respected sponsor" (http://tibethopecenterindia.blogspot.com/p/our-project.html). Another NGO, which runs a nursery for babies brought up by single mothers, offers an opportunity to sponsor a workplace for such a woman for twelve dollars per month, "which will naturally increase her confidence and potential for self-sufficiency" (http://www.tibetrogpa.org/Project-01.php). Here, again, we are dealing with the skillful combination of individual and collective goals of the diaspora, something that I discussed in the previous chapter in regard to voluntary tourism. These women, often having no professional qualifications, are employed in simple craft jobs which do not require large investments, such as making two-color braided bracelets. The organization calls these braids—referring to the tourist notions of Tibetanness—"peace bracelets," although among Tibetans they are known as "Rangzen bracelets" (Wylie: *rang btsan*), that is, "independence bracelets." Their weaving was to be initiated by Tibetan political prisoners, hence the trade in them is not only used to fight unemployment in the diaspora and contribute to the nursery's budget, but is also an expression of "solidarity with the Tibetan cause," as the NGO declares (http://www.tibetrogpa.org/Project-01.php).

Sonam used to be one of them. When we met, he already was the owner of a small but popular restaurant in McLeod Ganj. He left Tibet as a twelve-year-old, in the second half of the 1990s, under the care of his uncle. Initially they settled in Nepal, where his uncle worked in a carpet factory and Sonam took care of the housekeeping. He came to India at the age of eighteen and initially tried to study at a secular school, but eventually ended up at a monastic college, where he studied Buddhist philosophy. It seems that, like many other young Tibetan men who found themselves in India without any formal education, knowledge of the language or support from local social networks, he treated the stay in the monastery as a safety cushion and a way to adapt in a new place. He came to Dharamshala in 2007, joining his sister, who together with an exile-born Tibetan girl from a fairly wealthy family set up a small restaurant at McLeod Ganj (Sonam's sister contributed a know-how, as she had worked in Delhi for several years in an restaurant, while the other young woman invested money; over time, the girls' each went their own way, and set up their own restaurants). Sonam, trying to find some footing, took a yoga course, which he then taught himself—initially in Tibetan monasteries and nunneries ("monks and nuns have few physical activities, so it served them very well"), and over time also for tourists. He boasted that he never took money for it. How did he earn his living, then? Once, as he was hanging out in the main square in McLeod Ganj, an older American man approached him. He felt unwell and asked Sonam to take him to a hospital. The illness proved mild (some gastric problems), but the two started to spend a lot of time together. When the American was about to leave Dharamshala to visit the south of India, he suggested that Sonam should go with him (he said that he "was very old, in poor health and needed help"). The trip lasted three months, and the American tourist was so grateful for company and assistance, that for the next few years he covered all Sonam's expenses, which, as he admits, allowed him to have a relatively carefree life. It is worth mentioning here that life in McLeod Ganj is quite expensive for Tibetans, because the presence of tourists causes the increase of prices (rents, shopping, etc.). For people from the Global North, however, due to differences in currency values and purchasing power of the dollar and euro, these were small expenses. This private sponsoring lasted until 2012, when Sonam's sister married a Tibetan residing in a Western country, and Sonam took over her restaurant. To conclude, financial support from a Western sponsor allowed Sonam to survive the adaptation stage in Dharamshala, until he was able to get on his own feet and—due to his sister's entrepreneurial spirit—become a small entrepreneur in the tourism sector. The stories like Sonam's were a source of inspiration for other newly arrived refugees, who lacked other opportunities. They demonstrated that close relationships with tourists could prove really fruitful and were, therefore, worthwhile.

Pasang, the same age as Sonam, who has lived in India for ten years, knows this. Ambitious and determined, he managed to complete a four-year education cycle at a school for adult refugees from Tibet. Then he became a regular participant of all possible English language courses and conversation classes offered by NGOs in Dharamshala (like many newcomers who had already been in India for some time and had access to tourists who were native English speaker, Pasang now speaks very good English). In the meantime, he completed several IT courses at the Tibetan Career Center and a training for call center workers at the Tibetan Children's Village in Upper Dharamshala (he even had a job interview on the phone, but the company was looking for someone who could "speak with a Mexican accent").[5] He took various jobs at Dharamshala's NGOs, but the earnings there were just "pocket money" (from 50 to 100 dollars a month). He decided to continue his education and graduated from Kunpan Cultural School,[6] with an English certificate from Cambridge University.[7] However, he decided that English alone was not enough and decided to learn French. He had trouble finding good teachers, as his level of proficiency was more advanced than the classes offered by four French volunteers I wrote about in the previous chapter. When we met, I wondered (as I often did) how he was earning his living in Dharamshala, without the support of a family who stayed in Tibet. After all, I knew that renting a room (with attached bathroom, so thus a luxury one) cost him 4,000 rupees a month (about 90 dollars), and one day he complained that his basketball shoes—worth as much as the monthly rent—which he left at his room's door, were stolen. I laughed at him that he had other "bourgeois" habits, such as using toilet paper; I joked that he could have saved up if he had chosen the Indian method, but he said that it was "disgusting," thus showing

[5] In the 1990s, work in the so-called call centers—set up by companies from the Global North who outsourced their customer services to lower-cost countries—was an opportunity for young Tibetans to find employment outside the diaspora. Few people, however, stayed in this sector for longer, due to the gradual reduction of initially relatively high earnings, night shifts (caused by the time differences between India and the customers' countries) and the gradual transfer of call centers to the countries with even cheaper labor.

[6] The school was founded and financed by the Swiss-Tibetan Educational Support Tibet foundation. Its students were the best graduates from the school for adult refugees, mentioned in the previous chapter, so that they could obtain formal education to a level which corresponds with the ninth and tenth grade in Indian public schools, and learn English, Chinese, and computer skills, obtaining certificates useful in seeking employment. There, too, the work of Tibetan teachers was supported by foreign volunteers—teachers of English and computer programs. Students who got admission to the school were provided with accommodation, food, and small pocket money. It was assumed that most of them would return to Tibet as "future cadres for a free Tibet," working as English teachers or translators in NGOs, or finding employment in tourism sector as guides or travel agency owners.

[7] The students at Kunpan Cultural School can take several exams: KET (Key English Test), PET (Preliminary English Test), FCE (First Certificate in English), and CAE (Certificate in Advanced English).

a cultural superiority toward Indians.[8] In the end, it turned out that he made a living with the financial support of several former teachers whom he met through NGOs and who became his friends. He often hosted them in his previous, rented room, a tiny space without a bathroom and kitchen ("they were in shock, how can you live like this!"), thus evoking in them a sense of moral responsibility. He said he did not want to go to "meaningless work for which you are paid an embarrassing amount of money," he preferred to study as long as possible. This active self-development allowed him not to see himself as an unemployed person who just consumes the sponsors' money. He had a strongly internalized imperative to pursue education, which he attributed to the Tibetan national project: "Only with well-educated people we will be strong as a nation," he used to repeat the words he had surely heard many times at both the adult refugee school and the Kunpan Cultural School.

However, the prevalence of individual sponsoring in the Tibetan diaspora does not mean a simple, one-sided relationship. More and more Tibetans who have been economically successful become sponsors themselves. Tenzin's brother, who runs a travel agency in McLeod Ganj with him, sponsored a child of HIV-positive parents with 1000 rupees a month (about $22). I know of many cases where Tibetans, who became financially independent, wrote to their sponsors to thank them and ask them to discontinue their support as it was no longer needed. Various types of sponsorship programs implemented by NGOs are accompanied by initiatives targeting their own community. One of them is the "Adopt a granny" project, which seeks young Tibetans who could devote an hour a day to helping the lonely elderly person with small tasks (like shopping, cleaning) and keep them company.[9] Moreover, Tibetans provide support to the citizens of their host country. The NGO Tong-Len (Compassion in Action), backed by funds from European and Australian donors, supports Rajasthani and Maharashtrian slum dwellers from the areas around Lower Dharamshala. These are mainly climate IDPs[10] from extremely poor, drought-stricken regions of India. The NGO runs a boarding school their children, arranges scholarships, organizes hygiene trainings and tries to improve sanitary conditions on site (see Bloch 2019).

However, not all these stories lead directly to people overcoming their unprivileged situation, especially if other forms of internal subordination are involved. Parmesh is a poor relative of one of the guesthouse owners in Hampi. He completed two, maybe three years of elementary school—he does not remember exactly—in his hometown, after which he was taken to

---

[8] Toilet paper is quite a luxury item in India; in small towns like Dharamshala it is mainly bought by tourists. Water is mostly used for toilet purposes.

[9] See, for example, http://tibethopecenterindia.blogspot.com/p/our-project.html.

[10] Internally Displaced Persons.

Hampi as a seven-year-old boy, to help in a restaurant at the bazaar. That was over twenty years ago. When we met, he cleaned rooms in a guesthouse and lived in his workplace, sleeping on the rooftop of the building. He did not receive any regular salary for his work—the owner's mother, who actually run the business, paid his parents. Parmesh's situation was therefore a classic example of work arrangement in an extended family, based on a system of mutual dependencies. As a son, he had a duty to earn his parents' livelihood and his earnings were transferred directly to them, because, as the guesthouse owner said, "he is fully provided for here." Parmesh himself reproduced this logic: "When I need something, I ask for money and I get it." The owner was aware—being interculturally competent—that this situation did not look best in "Western" eyes (child labor, modern slavery), so he claimed that he paid Parmesh 10,000 rupees (which was an exaggeration even on a declarative level as such salaries were paid to qualified cooks or bakers, but never to cleaners[11]).

One day during my second field research, Parmesh told me excitedly that "his friends from Netherlands" were coming. This middle-aged couple visited Hampi regularly for nearly two decades, each year for three to four weeks. They knew all the waiters, cleaners, chefs, and auto-rickshaw drivers since they were adolescents, and everyone knew them. They asked questions: "How's mom? Is the house already rebuilt?" They brought Parmesh a green T-shirt as a gift (they were worried whether it would fit) and talked to him for a long time. They knew him and his situation well. Their political stand was socialist and they divided India into the rich and the poor, and argued that the former exploited the latter. They were strongly committed to fighting economic inequalities, but also very critical of the aid industry. Thus they avoided the intermediaries, that is, charity and development organizations and foundations, and preferred "direct assistance," which consisted in supporting those individuals and families they had known personally for years.[12] However, they were afraid of the dependency effect and already a little tired

[11] During my fieldwork, cleaners in Hampi earned about 3,500 to 4,000 rupees (approximately 75 to 90 dollars) a month. It is difficult to expect loyalty for this amount, so employee turnover was high (guesthouse owners usually referred to employees who quitted their jobs overnight that they "ran away"). In just two weeks, two cleaners "run away" from this single guesthouse, as a result of which the owner's mother had to clean the rooms and he himself was responsible for cleaning the toilets. In this system, "employing" a person who is bound by kinship ties is a kind of the guarantee of manpower.

[12] Among them a woman who used to run a small stall at the bazaar, but was evicted in 2011. She had had a drinking, violent husband, and finally decided to separate from him, but was left alone with the children (the Dutch couple met her when she was pregnant for the first time, today she is already a grandmother). Among those supported there was also a family who had ran a small *chai* stall at the bazaar. After the demolition, they tried to do their petty business from the roadside, but initially were not very successful, so the Dutch couple supported them with 10,000 rupees (about 200 dollars).

of people attempting to exert moral pressure on them. That is why they wanted to help Parmesh, because "he never wanted anything from us." The plan was simple: the first thing to do was to get the boy out of the network of family dependencies. The opportunity arose when Parmesh inherited a small piece of land in his hometown. The couple thought that if Parmesh sold or leased half the plot, he would get the money not only to build a house for himself and his parents on the remaining half, but also to get married. But this money turned out to be definitely not enough, hence the idea to support him financially. The Dutch couple decided to contact an English friend, who had also been visiting Hampi for years, to invite him to participate in this initiative. But then it turned out—although various rumors about it had been circulating in the village for a long time—that this person had already donated money to Parmesh a few years earlier. The substantial amount was actually an unpaid loan which the Englishman granted to the owner of the guesthouse where Parmesh worked. Since the Englishman was having trouble getting his money back in previously agreed installments, he decided that all the remaining debt was to be transferred to Parmesh, and he informed the interested parties about it. However, he had no way of ensuring that this actually happened, and in fact the money never reached the recipient. The guesthouse owner explained that it was impossible due to the financial difficulties caused by the demolitions in Hampi and suggested that Parmesh had "mental problems" and thus was unable to manage such funds. Parmesh himself did not have any tools to retrieve the debt, which would have been a chance for him to end his subordination. However, he opposed it in the only way he felt was possible, that is, through passive resistance. The owner often complained that "this boy has been spoilt, and he used to be such a good worker ten years ago."

"TOURISTS TAUGHT MY DAD HOW TO MAKE AN OMELET": SUPPORTING LOCAL TOURISM ENTREPRENEURSHIP

A very interesting phenomenon that I encountered at every step of my fieldwork was the involvement of tourists in the development of local small tourism businesses. According to those promoting the idea of pro-poor tourism, which I described in the previous chapter, lack of skills in running one's own business as well as lack of understanding of the needs of tourists and how the whole sector works are the main obstacles to the development of tourism that would include the poor among its beneficiaries (Ashley and Roe 2002, 77). However, it was due to tourists that many residents and migrants in Hampi and Dharamshala developed their management know-how and intercultural competences which helped them operate in the tourism sector, and gain—both directly and indirectly—investment capital. Tourists often suggested business ideas to their hosts, inspired and encouraged them, thus contributing to the development of local entrepreneurship. At the same time, the involvement of tourists in capacity building and support for small businesses stands in opposition to their apparent desire to satisfy the need for authenticity based on non-commodified relationships with the Other. However, the tourists I am referring to did not seem to perceive all those who were openly representing the tourism sector as unauthentic, as Lozanski (2010) suggested. They were well aware of the broader context of tourism, in which both them and the inhabitants of tourist destinations were situated, and understood that tourism served not only to satisfy their cultural needs, but also the livelihoods of the hosts.

In fact, the whole history of Hampi's transformation from an agricultural, peripheral village into an important destination on the tourist map of India resulted from this kind of support, which started with the very first endeavors in the area. A guesthouse owner, who originates from one of the oldest families in Hampi, once told me: "All of this is thanks to tourists. Without them there would be nothing here." Another, late owner of the first guesthouse in Hampi which hosted foreign tourists, started his business by launching a small stall with "Western food" in 1987. His daughter remembered it as follows: "My father was very open to visitors, very friendly. He absorbed all the ideas they shared and they enjoyed it, so they were happy to help him." The tourists taught her father how to make an omelet, oatmeal, and toast with jam. A befriended Dutchman gifted him a brand new fridge as a contribution to the development of the gastronomic business in the hot climate of Karnataka—the appliance which was then very rare (possibly, it was the very first fridge in Hampi). The Dutchman then kept visiting Hampi regularly for several years, and witnessed the business flourishing. At the same time, in the 1980s, a young Austrian man came to Hampi, and—like many drifting tourists from the hippie era—he wished to experience "something different." Like Buddha Sakyamuni, he abandoned everything (or rather deposited it with one of Hampi residents, Shankar) and as a beggar[13] set off toward the Himalayan Rishikesh[14] in search of the meaning of life. Upon his return, he decided to get involved in helping those "really in need" and Shankar became his translator. One of "the needy" were Lambani people living in the neighboring villages, who fascinated the Austrian to such an extent that

[13] Having experienced a prosperous and careless life in the family palace in Lumbini (present-day Nepal), Prince Siddhartha Gautama abandoned wealth, position and family in order to seek an answer to the question about the cause of suffering in ascetism.

[14] The holy town for the followers of Hinduism—already mentioned in the introductory chapter—and at the same time a very popular destination for spiritual tourists since the time when The Beatles themselves meditated in the ashram there in the late 1960s.

next time he came to India in the capacity of a cultural anthropologist and studied Lambani customs, with Shankar as his assistant. In the meantime, to earn money, the Austrian man began working as a tour guide. Shortly afterward, Shankar opened a small guesthouse in Hampi, which he named after his Austrian friend: Herman's Guest House. The relationship lasted for years: the Austrian friend traveled to Hampi almost every year, and if he could not, they would meet at a different place in India (e.g., Shankar would travel to the town of Gokarna, one day by bus from Hampi, when his friend was there with tourists) and often talked on the phone or Skype. Later on, Shankar was involved in another Austrian project, this time dedicated to organic farming. He used to say: "I like to learn from foreigners."[15]

Tourists also serve as signposts to help in making business decisions related to economic migration, because often—especially the ones who have been traveling for months in India—have a good understanding of the situation and can share their perspective as customers. Many seasonal migrants in Hampi ended there just because of tourists. One of them was Phurbu, who run a winter-season restaurant on the rooftop of the house owned by Geeta's mother-in-law for almost ten years. Earlier, while working in Goa, his buddy, also a Nepalese, married a tourist from France, and left to live there. When the couple's fried was planning a holiday in India, they put him in touch with Phurbu. They met, became friends, and when the Frenchman rented a motorcycle, he suggested he and Phurbu should go for a joint trip. On route they visited the remains of Vijayanagara kingdom, about which Phurbu—as a Nepalese—had never heard before. Following the trip Phurbu finished the season in Goa, and the next year he came to Hampi to set up his own business. Moreover, tourists also suggested "scenario B" strategies in case Hampi would be completely wiped out, helping people regain at least a little bit of certainty in this liminal situation. The owner of a T-shirt shop planned, in case his shop was to be demolished, to move with his family to Gokarna or Goa, or to a town in Tamil Nadu, where, as a tourist friend from the United States told him, "there are almost no shops yet, and no one is yet selling T-shirts" ("Just don't tell anyone!"—he warned me).

Often it was tourists who convinced my research partners to enter the tourism market at all, by providing them—sometimes by accident—with their first job. The trainee-guide in Hampi, already mentioned in chapter 3, had previously earned by offering short-term small loans (to fruit vendors or owners of tea stalls) and additionally leasing a few-room guesthouse in the village. One day, a tourist who rented a room there asked if he would show her around. The funny thing about it was that the man was new to Hampi—he

[15] Mr. Shankar Singh died from an illness in 2016. I chose not to use a pseudonym here in order to keep his memory alive.

came from a nearby big city—and knew little about the site and its history. He did not even know the way to the sunrise viewpoint, offering magnificent views over Vijayanagara's remains, so they got lost on the way and almost missed the main attraction. Nevertheless, the girl was very satisfied and paid him more than they agreed. Soon afterward, he decided to pursue that opportunity and joined the training for tourist guides. When we met, he was just about to receive his certificate.

Sometimes, business ideas that tourists shared with their Indian friends were more unconventional. In the early 1990s, an Australian journalist, whom I mentioned in chapter 3 and who had been coming to Hampi for years, advised his friend who was looking after Lakshmi, the temple elephant, to introduce double price system for Lakshmi's blessings: one for pilgrims and a different one for foreign tourists (in line with the principle of bottom-up leveling of global economic inequalities, discussed above). As a result, during my fieldwork Lakshmi collected one or two-rupee coins from Indians, but tourists had to pay in banknotes, that is, at least ten rupees per person. The elephant caretaker called this latter, more expensive service the "photo blessing." In fact, while the "real blessing" consisted in Lakshmi slightly touching the pilgrim's head with the tip of her trunk, when she received a banknote she kept her trunk on the donor's skull for much longer so that there was enough time to take a commemorative photo. Lakshmi collected fees herself using her trunk and, before conferring the blessing, handed over the money to her caretaker (or to me, when I sometimes filled in for him, which was one of my favorite activities and which caused sincere amusement among the pilgrims). I still do not know how Lakshmi differentiated tourists from Indians and that she could not be fooled by occasional attempts to trick her into accepting coins from the former (she took coins from tourists and handed them over to her caretaker, but she would not bless such a fake donor until she got a banknote; she could also change money if necessary). The elephant's caretaker claimed that she distinguished the two types of "customers" by smell, but my observations showed that Lakshmi took into account many other aspects, observing, for example, whether there would be some photo-up involved with people posing in front of cameras or smartphones—then she hold her trunk up for longer, but only provided she had received a banknote; if she received a banknote, but there was no hint of picture-taking, she gave a normal short blessing. The salary her caretaker received from the temple authorities for his job was small (about 100 dollars per month), even though this work required high qualifications (regarding feeding, cleaning, training, etc.). He thus earned money, together with his helper, mainly thanks to the tourist work done by Lakshmi. On average, the elephant collected 350 to 400 rupees (8 to 9 dollars) a day, and during the tourist season even up to 1,000 rupees. With these earnings the caretaker had to support his whole family—four children and a stay-at-home wife.

It was also inspiration by tourists—and their know-how—that brought about the first Korean restaurant in Hampi, already mentioned in chapter 3. Bhavani, whose story will be described below, together with her two brothers used to run a restaurant owned by their grandfather at the bazaar. Soon she noticed that Korean customers were reluctant to eat spicy Indian food and were often sick afterward, later sticking to a biscuit-and-banana diet. It was they who started to teach her how to prepare Korean dishes, which she gradually added to the menu (of course, in a modified, vegetarian version—in Hampi, a Hindu holy place, it is officially forbidden to serve any food containing meat, including fish, which is an important part of Korean dishes). Some of the ingredients not available in India, such as spices, were sent to her by her "Korean friends," while she herself learned how to make sauces, glass noodles and kimchi. With these skills and contacts, a few years later she managed to set up her own restaurant. Her daughter recalls: "It was a very famous restaurant. Koreans came to Hampi from Goa only because of my mother's cooking. One time, the whole bus from Benares came only to eat here. They did not even look at the monuments! They just had dinner and left." Once, an important guest from Japan, who came to a nearby city to visit the Indian energy and steel industry tycoon Jindal, was brought to the restaurant, since he wished to sample East Asian food in the Indian countryside:

> Our restaurant then had very modest interior design, the roof was leaking, customers were sitting on mats spread out on the floor. In addition, there had just been a power cut when they came. So, this VIP just sat there, in front of the restaurant, with a plate in his hand, like a beggar! And all his bodyguards stood around him, plus four police cars. The whole Hampi flocked to see it.

Knowledge and competences held by tourists regarding the tourism market—its standards, new technologies, marketing strategies—are constantly used to improve local services. Raj's "friend"—a tourist from France—designed a new menu for his restaurant including the caption "if you're satisfied, rate us on TripAdvisor and like us on Facebook." I myself often acted as a helpful "expert," like all other tourists who expressed their need to "fraternize with the natives." I was especially involved in proofreading, editorial and translation work: at Raj's, I proofread the description of his guesthouse and restaurant posted online (I removed not only all unnecessary spaces, but also the information that they offered "drinks"—instead of soft drinks—which would certainly be used against them, as officials and police could accuse them of supporting "immoral tourist practices"[16]). For Phurbu, I revised new business cards (one for a restaurant in Hampi, and the other for a restaurant in Kasol, Himachal Pradesh) before handing them over to a printing shop; at Norbu's request, I set up a Skype account for him, so that he could be in touch with "friends from abroad" and I explained to him how to use Facebook (which proved quite unfortunate, as Norbu has a habit of flooding his friends with devotional posts and videos); a painter from Rajasthan asked me to take a picture of one of his paintings and send it to his brother's email account, because the client was interested in buying and wanted to see it. Moreover, due to the growing number of Russian tourists, I acted as a translator (since "your languages sound the same"). The workers in the tourism sector, who were used to interlanguage communication, were sensitive to new, popular words, remembered them and then asked me about their meaning: what is "pasmatri" ["look"]?, what is "davai" ["come on"]?, how do you say "good" in Russian?

Finally, tourists contributed directly to local businesses, not simply by purchasing goods and services during their tourist trips, but by establishing business relationships initiated during such trips. This often took the form of creating markets in their countries of origin for goods produced in India, which—due to exchange rate differences between the Global North and Global South—could be a mutually beneficial solution. Many tourists were looking for an alternative way of earning money which would not involve corporate environment, while others were seeking an additional source of income. For instance, an Australian friend of Norbu and Dolma—a tour guide already mentioned in chapter 3—used to buy goods from them, which she then sold with profit upon returning home. The seamstress from Hampi sewed kurtas[17] ordered by tourists from Spain, who sold them on the Spanish market. Many local manufacturers and sellers dreamed of such a partnership (I myself was often asked what prices particular products could wield in my country and whether I thought there would be demand for them). Jay, a shopkeeper from Rajasthan, whose romantic relationship with a tourist from France was developing before my eyes, mentioned to her that if in her work—and she was a store space designer from Paris—she found "some nice place in some shop, he would send her bags[18] for sale." The woman also suggested that she would design Jay's store.

[16] Alcohol—as well as meat—is forbidden in Hampi.

[17] *Kurta* (Pers. *kurtā*) is a collective term for a long shirt or tunic, which takes very different forms depending on the region and occasion.

[18] Leather goods, especially bags, are Rajasthani specialties in the Indian tourism market, alongside jewelry and textiles.

Two stories—one from Dharamshala and one from Hampi—which I want to recall in more detail, illustrate the impact that instrumental and emotional support from tourists and their economic, social, and cultural capitals can have on the establishing and development of local businesses in the tourism sector. This, in turn, allows people who find themselves in a situation of structural subordination to overcome the constraints they face: to achieve economic self-sufficiency and social advancement.

Let's start with Dharamshala and Phuntsok, my peer (a man with very long hair, in cargo pants, keys on a snap hook). Phuntsok came to India in the late 1990s, then in his twenties, without education or professional experience. Like many young men, he could not stand the discipline of a government-run school for adult refugees and dropped out after six months. This is how he ended up in McLeod Ganj with twenty rupees in his pocket. A friend, who himself used to be *sarjor*,[19] lent him 1,000 rupees, which Phuntsok invested in a gas burner and a steaming pot. In this way, he began the then typical career of a Tibetan newcomer: he became a street vendor of *momo*. Every day, he would get up at 5:30 a.m., make dumplings and go round McLeod Ganj. His clients were "rich Tibetans," as he saw them at the time, "who could not be bothered to make *momo* themselves." It was then, "in the street," that he met a young American who started telling him: "Why are you doing this? It's such hard work, and the earnings are small" (a portion of five to seven big *momos* cost ten rupees [approximately zero to two dollar] at the time). However, Phuntsok had no alternative; it was not only about money, but also about his illegal status in India, which limited his options, including opportunities of a regular employment. At the time, the Indian authorities did not officially issue residential permits to new refugees from Tibet. Thus, the documents had to be acquired by "unofficial means" and using financial "incentives." At the request of an American friend, his mother came to his aid.

With the material support from the sponsor, Phuntsok was soon able to take up a job in the formal sector. He hired himself to wash dishes in a newly established Italian restaurant, located close to the main temple and the Dalai Lama's monastery (at that time there were no other Italian restaurants at

[19] New refugees try to support each other in a difficult position. Often institutional support—the one on the part of the government-in-exile, for example, the so-called Transit School for adult refugees discussed in the previous chapter—turns unsatisfactory. Moreover, the majority of refugees from Tibet have no relatives in India on whom they could count. Therefore, they create their own, kin-like bonds (they get to know each other most often in the aforementioned school or in monasteries) and try to look after themselves. This is how Kelsang's nomad tent-style coffeehouse was established (the one where jam sessions were held on "White Wednesdays"). It was founded by three Tibetan friends, who—apart from earning their living—had the ambition to implement projects that would support new refugees in the process of adapting to life in India. Also most of the founders and heads of the Tibetan NGOs I wrote about in the previous chapter are people with such background.

McLeod Ganj; moreover, it was a vegetarian-only venue, which was not very common in Dharamshala in the late 1990s). The restaurant was opened by a Swiss Buddhist, together with his wife, a French woman, and their Tibetan friend.[20] Among others, Richard Gere used to eat there when he visited Dharamshala.[21] For three years Phuntsok worked very hard. He would get up around 1:30 to 2:00 at night and start baking apple pies, shredding tons of vegetables and kneading pizza dough. However, he quickly started to get promoted: first as a cook's helper, then as a cook and finally as a chef. His status was viable enough for him to write to his sponsor that he no longer needed her support, because he was doing great. During the last six months when the restaurant was still open, he was already an official partner of the whole undertaking. When in 2002 the Swiss owner decided to leave Dharamshala, Phuntsok made a decision to buy the restaurant from him. Then the mother of an American friend came to his aid again and put in the seed money (to which Phuntsok added the savings he had managed to accumulate). He was to repay the rest in installments as the restaurant was profitable. For ten years, he worked virtually non-stop, but managed to keep the restaurant on the market. Shortly before we met, he had moved to the center of McLeod Ganj, where he rented a place in a Tibetan guesthouse (this is where we met—during one of my field trips, I literally lived in his restaurant as the door of my room looked out onto a terrace with tables). On that time Phuntsok was just keeping an eye on the business, as the restaurant had its own momentum. He employed newcomers (mainly as the front staff, that is, waiters) and Bihari workers in the kitchen. It was with him that Prakash found a job—I told his story in chapter 1—after the death of the restaurant owner who brought him from Bihar to Dharamshala. The street advert for Phuntsok's restaurant boasted that they made fresh tomato sauce every day and used genuine Swiss raclette cheese and Italian mozzarella. The restaurant was very popular with both tourists and Tibetans, thus creating another heterogeneous space in Dharamshala, where guests and hosts had the opportunity to meet. One of the pizzas was named after its Swiss founder and was composed of his favorite ingredients. Richard Gere, upon visiting McLeod Ganj, used to come to Phuntsok's restaurant to see how he was doing and eat the pizza named after the Swiss man.

---

[20] Switzerland is home to the largest Tibetan community in Europe.

[21] Richard Gere has been actively supporting the Tibetan cause for many years. He often visits Dharamshala, joining quite a large group of celebrity tourists. This presence is evident in the daily life of the diaspora: one of the popular movies of young refugee Tibetan cinematography has a telling title: *Richard Gere Is My Hero* (2007, India, dir. Wangchuk Tashi, Dorjee Tsultrim, produced by Tibet Motion Pictures & Arts). Sometimes in private homes of Tibetans, in the family photo album one can come across a photograph with a set inner table and a roll of toilet paper on it, with Gere siting by the table, surrounded by Tibetan dinner companions.

The second story illustrating the impact of tourist support on the development of local entrepreneurship in the tourism sector is the story of the owner of the first catering establishment in Hampi, opened for foreign guests in 1971 by Ram Singh,[22] a man in his nearly seventies on the time we met. Ram was born in a small village in Karnataka and came to Hampi at nineteen in 1967 in search of work. He comes from a poor family of farm laborers who, however, made sure that their son went to school. He combined his education with work, delivering milk to the villagers—without the involvement of all members, including children, the family would not have been able to survive. After completing tenth grade, he moved to a town near Hospet, where he found employment as a newspaper boy (delivering 600 newspapers a day, he received 60 rupees; a decent meal cost about 10 rupees at that time). Once, in one of the newspapers he saw an article about Hampi. He thought: "What a beautiful place, by the river, with a temple and such history! I would like to live there one day." So one day, on a public holiday, he just went to Hampi, using 100 rupees he borrowed from his employer for the trip (which was not enough anyway, so the last stretch of the road, 12 kilometers from Hospet to Hampi, he covered on foot). He spent a few hours in Hampi, liked it very much, but did not even stay overnight, because in the mid-1960s nobody rented rooms yet. Soon afterward, three months after the birth of her third son, Ram's mother died. Ram tried to help his middle brother to take up a job delivering newspapers, but he never went to school, so was not even able to read the recipients' addresses. However, he found employment as a helper in a local eatery, where he was provided with accommodation and food. Ram often visited his brother, which eventually became a bone of contention between his brother and the eatery's owner (who felt one brother handed unusually too big portions of food to the other). Eventually, they decided that together they would go to a "big city," where they would have a better chance on the job market. They went to Hospet, and from there—after two years and several misfortunes while working in local restaurants—to Hampi.

Initially, Ram headed to the ashram, where the *swami*[23] allowed him to stay for a few days and think about what he could do next (he did not have any money, since the money he earned earlier was just enough to survive). After about a week, the postmaster came to visit the *swami*, who told him about the "quite educated boy" looking for work. Together they came up with the idea that Ram could teach local children (although there was already a school with two or three classes in Hampi at the time, but with a mediocre

---

22 A real name.

23 Sanskr. *svāmī*. A honorary title given to Hindu leaders, many of whom run ashrams. Ashrams, in addition to performing religious functions, also carry out social activities and have often great political influence.

state-funded teacher who just "didn't care," so in reality the children had no schooling). The postmaster spread the word around the village that for a small payment of two rupees a week per student—these sums were to be paid toward Ram's salary—children could be sent to the ashram to study. Soon these lessons became popular and after a few months Ram had forty pupils; he taught mainly history and social sciences (he does not like mathematics or science, he was never good at them). However, he did it in an unusual way, using—completely intuitively—drama as an educational tool. To this end, he started to write plays himself: comedies and dramas he called "social drama" (e.g., about "stupid officials"). The residents liked his dramas so much that they proposed that he put on a play about Vijayanagara in which they would play parts. This is how Ram's first "mature" play was created.[24]

Over time, the number of students increased so much that the ashram became too small and with the help of the village council, one of the mandapas was adapted as a school building. Ram's social standing in Hampi was growing. He was on good terms with the neighbors and started being invited to festivals and family celebrations. One day, somebody brought a deaf girl, Shanta, to his school. Her parents hoped that Ram would teach her to speak. The girl was just sitting in class, but over time Ram became friends with her family, one of the oldest in Hampi. Four years later they got married (she was sixteen years younger) and they still are. They have a son and a daughter and several years ago became grandparents. This marriage allowed Ram to join the local community and an important family, and made it possible for Shanta to start her own family while staying with her parents (otherwise, as a bride she would have to move to her husband's and in-laws' house after marriage, but since Ram had nowhere to go, they lived in one of the houses owned by Shanta's family[25]). After the marriage, Ram's father-in-law forced him to give up the teaching job and instead help him look after the farm workers in the fields that the family leased from the temple. However, their paths quickly diverged. Ram had no inclination to agriculture, and felt that he was treated as if he was lesser man by the family (figure 5.1).

Earlier, while he was still working as a teacher, he met two foreigners, a Frenchman and a Belgian, who stayed in Hampi for a few months (one was a philosopher, the other cultural anthropologist, both were very interested in local culture). They became friends and Ram served as their "assistant-informant." It was they who gave him the idea to set up a catering business:

---

[24] And a new tradition was born in Hampi: staging plays at the Dasara Festival every year. One of the mandapas in the lower, uninhabited part of the bazaar served as a stage, and the actors were those residents of Hampi, who could read enough to learn a role from a script written by Ram. Overall thirteen plays were staged at the turn of the 1960s and 1970s.

[25] Ram belongs to the same caste as his wife's family—hence the marriage was possible by mutual consent—although he was treated like a poor relative, landless and in fact homeless.

Figure 5.1 Ram Singh with His Grandson, Hampi. *Source*: Photo by Natalia Bloch (2017).

"Hampi is such a beautiful place—they told him—and one day many tourists will come here." They decided to support the business, which they had in fact invented themselves, and provided tools and produce. They bought a fire burner (gas burners were not available at the time), glasses, tea bags and instant coffee powder.[26] They also invented the "appropriate local name," which they placed on a hand-painted sign: Trishul, or trident.[27] Initially, the *chai* shop was set up not in the village itself, but near the rocks on the way from the lower part of the bazaar to the Rama'a temple, where hippies most often lived at that time. The two foreigners did not want anything in return, only coffee and *chai* every morning and evening. Ram admits: "They gave me all the ideas. Everything they said became true. It was like magic, like a miracle." The three of them have never met again. Then, in 1971, both men were forty to fifty years old, so Ram suspects that they may already be deceased, but he often reminisces about them, sitting in his Trishul Restaurant.

[26] In India, such places are most often called *chai* shops, because they mainly serve *chai*—tea cooked with milk, sugar, and spices—and sometimes also snacks. Apart from Kerala, the southernmost state where coffee is grown and drunk, serving coffee in India is rather a response to the needs of foreign tourists.

[27] The symbolic weapon attribute of Shiva.

It was then that Ram brought his youngest brother to Hampi. It is him who today "prays for tourists every day" (their middle brother was sent to their home village to take care of their old, sick father, who contracted tuberculosis working in a tobacco factory, where he earned his living when he did not work in the fields). Ram's youngest brother worked as a farm laborer, and after hours served clients in a *chai* shop. With time, Ram started to develop his small business. He made a roofing out of a mat woven from coconut palm leaves, introduced simple meals and hired his brother "full-time," that is, during the season from 6:00 a.m. to midnight. At the same time, on land leased from his father-in-law, he built four modest huts with the capacity to sleep of two, which he rented out to travelers. After a few years, half of this land was presented to Ram and his wife by his father-in-law, while the second half—the one on which the huts stood—was transferred to Shanta's brother, Shankar, who later on built the already mentioned Herman's Guest House there. Ram then decided to move his catering facility to the village, all the more so as the Archaeological Survey of India ordered him to leave the site near the Rama's temple in preparation for listing Hampi as the UNESCO World Heritage site. Sometime later, in the late 1980s, when the village was already becoming popular with tourists, the Nepalese appeared in Hampi. They came from Pune, a city located more than 500 kilometers north, where a German, one of Osho's students, had set up the first European bakery in India, called simply the German Bakery.[28] The Nepalese he hired learned the art of baking "German" cakes, breads and rolls and spread this know-how to the entire tourism sector in India.[29] In Hampi, they started with Ram's restaurant, which they rented and where they built a large oven in the background. One of the Nepalese became Ram's son-in-law years later. The couple ran the Trishul Restaurant during the winter season (figure 5.2) and moved to Leh in Ladakh for the summer to continue working in the same business.

However, during my fieldwork Ram still opened the restaurant even when the season was over and he did not have many customers, and the menu was limited to just few dishes (mostly cooked by his daughter-in-law). It is hardly surprising, after all, that he literally lived in a restaurant, that is, in the back of the restaurant, and had to go through it to get out of his house. This is how we met. In April the temperature in Hampi reaches over forty degrees Celsius in the shade. This turned everybody—including me, although I have no such inclination—into very early risers. It was impossible to sit in the room, as the fan was only moving the burning hot air, so I was looking for space for my

[28] The man later moved his business to Goa (D'Andrea 2007, 187).

[29] The characteristic feature of these bakeries was a chill-out atmosphere, favored by mattresses replacing chairs and world music oozing from the speakers. To this day, cafés and restaurants run by Nepalese in India follow this model.

Figure 5.2 Trishul Restaurant, Hampi. *Source*: Photo by Natalia Bloch (2013).

morning field note–writing sessions. At 6:00 a.m., I was often the only client, before other tourists came back after watching the sunrise. I used to hang out there with my laptop every day and, little by little, Ram started telling me his story, in one hour to one-and-a-half hour-long "episodes."[30] He finished with: "Still people address me as 'Sir.'"[31] Immigration to Hampi gave him a chance for social advancement, and his business in the tourism sector, sparked by inspiration and material support of friendly tourists, allowed him to gain a stable economic position, which made it possible for him to educate his children (his daughter has a bachelor's degree in economics). He is proud that people respect him.

The most far-reaching form of cooperation between tourists and workers or small entrepreneurs in the tourism sector is joint business. Raj was very jealous of a friend who, together with an ex-girlfriend, an Italian girl, opened a small guesthouse in Goa. "He walks dressed in Italian jeans and doesn't have to worry about anything"—said Raj, while before he started to lease the guesthouse on the other side of the river (she was his client, this is how they met and became a couple), he used to drive an auto-rikshaw, and even earlier, as a young boy, to graze cattle ("he didn't even go to school," complained Raj, who himself was quite well educated). Another guesthouse, in

[30] Ram was an unusual interlocutor, extremely careful with details and, most importantly, with regard to chronology, which was very rare in my research in Hampi. Then I discovered the reason for his excellent memory—for many years he has been keeping a diary in which he recorded all the more and less important events in his life and that of the village.

[31] The use of this English honorary title in modern India—especially among Indians themselves—as a sign of respect is part of the colonial legacy.

Kerala state, was run by a former tour guide from Poland, who, while working with Polish tramping tourists in India, met a Kerala-born man, whom she married. Jay was planning to open a restaurant in Manali (in the state of Himachal Pradesh, where he moved for the summer season), as a joint business with an Israeli friend. They hoped to cater to the needs of long-term tourists from Israel usually residing in this Himalayan hill station. From the perspective of tourists, such undertakings are an example of a phenomenon described in the literature as "lifestyle entrepreneurship in tourism" (Ateljevic and Doorne 2000) or "travel-stimulated entrepreneurial migration" (Snepenger et al. 1995). It refers to those tourists who transform themselves into migrants and set up small businesses in the field of tourism. Often they employ their cultural capital and social networks to attract fellow nationals as clients. Motivated by the desire to escape stress, overworking and the fast pace of life in neoliberal economies of their countries of origin, they try to achieve their own well-being in places which they identify with the idea of slow life. They choose tourism sector because it seems familiar to them (after all, they have recently been tourists themselves) and accessible (it does not require any specialist qualifications or large financial investment, especially when using the greater purchasing power of money from the countries of the Global North). Their ventures are not aimed at maximizing profit, but only at satisfying the needs necessary to achieve a satisfactory quality of life, and often the growth of such business is intentionally limited, to such extent that researchers describe it as "non-entrepreneurship" (Shaw and Williams 1998). These migrants "deliberate reject a market ethos" and the idea of development based on economic growth, redefining the capitalist framework of the tourism sector (Ateljevic and Doorne 2000, 379). In this sense, their motivations coincide with the attitude displayed by some tourism workers, who dissociate themselves from the strive for market profitability and opt for the non-economic benefits of working in the sector (such as tranquility, the possibility to live in a beautiful surrounding and establish close relations with "clients"), which I discussed in chapter 3.

This kind of ventures, which I had the opportunity to observe during my field research, are often the result of love relationships between tourists and tourism sector workers. This was the case of the Kashmiri-French couple and their clothing line from Dharamshala, the Tibetan Kelsang and the Japanese Tamiko and their Tibetan-Japanese café, or the Japanese woman and the auto-rickshaw driver from Andhra Pradesh who together set up a small guesthouse for Japanese tourists in Hampi. The last couple never even considered living in Japan or even moving to an Indian city: "There, I would only chase the same thing as [I did] in Kyoto: trying to be so brilliant, so self-confident. Here, I have found peace, time goes by slowly, you can think, live according to how the sun rises and sets, and in Japan [there is] only work, work, work, over and over again, just thinking about clients, stress." Sometimes the cause-and-effect relationship is the other way round: the tourist is looking for an alternative lifestyle in India and proposes a joint venture to a local person he or she met and with whom they formed a close relationship. In both cases, both parties benefit, especially since having a local partner is often the only way to legalize economic activity in India. Furthermore, in doing business in countries where many things are done informally using local networks, the instrumental support of a competent local person not only facilitates, but is often indispensable for the whole venture.

WOMEN AND TOURISM: CONTACTS WITH TOURISTS AS AN EMANCIPATORY FORCE

The status of women in relation to men was often presented by female scholars as inferior, which in feminist studies was initially considered to be a universal rule. This was particularly true of the restrictions imposed on women with regard to their sexuality and of economic oppression, which were often theorized with the help of psychoanalysis and Marxism (Rubin 1975). The causes of the unequal power relations between women and men were sought within cultural interpretations and social structures. Sherry B. Ortner (1974) formulated the fundamental question "is female to male as nature is to culture?," and Michelle Zimbalist Rosaldo (1974) pointed to domestic-public opposition as a source of lower status of women. The women's reproductive functions—the fact that it is they who give birth and raise children—were to place them closer to the order of nature than culture and inscribe them in the space of domesticity. Scholars used Claude Lévi-Strauss' (1969) concept of the exchange of women between men as the universal origin of social organization. On the other hand, claims of universal male domination were criticized for their Eurocentric generalizations and challenged by women anthropologists conducting research in various cultural contexts (e.g., Leacock 1981; Strathern 1988; see also the entire volume edited by Matthiasson 1974). It was also argued that such representations were often the result of patriarchal optics adopted by (male, Western) researchers (and often also women researchers), that is, the Euro-Atlantic androcentrism (e.g., Slocum 1975; Moore 1988, ch. 2). It was pointed out that in communities perceived as egalitarian, whether they were hunter-gatherers or nomads, women did not occupy a secondary position, but participated together with men in the process of raising children and ensuring survival, or both parties' roles were complementary and equally prestigious. It was the development of modern societies, their increasing stratification and pushing women into the private sphere that was to lead to their subordination. Forced export of modernity by colonialism was to create (or strengthen) the subaltern position of women in non-European societies (Channa 2004; Leacock 1975).

This is why postcolonial critics have so often conceptualized a colonized woman as doubly subordinated—both by European racism and patriarchy, often shared by colonizing and colonized men. Indeed, British administrators largely sanctioned the social order created by the indigenous male elites, thus perpetuating the domination of the high caste and stiffening the caste category itself through censuses, "ethnographic" studies and codifications (Dirks 2002). At the same time, the replacement of community goods with individual property greatly shook the position of women and their property rights, all the more since these rights were guaranteed only to men, which in turn led to an increased demand for male children (Channa 2004; see also Etienne and Leacock 1980). Scholars pointed out that it was a series of distortions made by colonialism that pushed Indian women into a subaltern position.

Although the subordination of women served colonizers as evidence of the "barbarism" of colonized men, and colonized men treated any "modernizing" actions dedicated to women and introduced by the colonizers either as an attack on traditional values or as reforms that they themselves would like to introduce, they all shared the view that it was up to them to speak on behalf of women.[32] This became a pretext for Gayatri Chakravorty Spivak to ask a crucial question for postcolonial studies: "Can the Subaltern Speak?" (1993). Her seminal essay made a figure of an Indian widow burned with her deceased husband's body—that is, *sati*—a topical example of subalternity. Women were also the "prototype" for Edwin Ardener's "muted group" theory (1975a, 1975b), mentioned in chapter 4. Issues of power relations in the context of gender are therefore a dimension which cannot be overlooked in sociocultural studies of any phenomenon. In chapter 1, I have already addressed this topic while discussing patterns of spatial mobility among workers and small entrepreneurs in the tourism sector. In the next two sections I will discuss the issue of relations between tourism and gender.

Spivak's pessimistic response to the question she asked met with criticism that her choice of *sati* as a symbol of subjugation cannot be regarded as representative of all women or even of women in India, because it is a conceptual and social category that "comes into being only when the subject dies," so it is the most perfect example of silenced subject (Loomba 2015, 231). First of all, just as it is difficult to speak of one "Indianness," it is not possible to make generalizing conclusions about the "Indian woman," because the roles and position of women (as well as men) in India vary greatly depending on social class,

[32] Later they were joined by women, but these were women of the higher castes, from which the Indian middle class was to emerge (see Sarkar and Sarkar 2007).

caste, ethnicity, religion, occupational group or age, and their intersectionality. Therefore, it is necessary to always locate our perspective in the specific context of our research. Secondly, Indian society was and is undergoing dynamic transformations as a result of various global processes (invasions, colonial modernization projects, postcolonial socialist programs and, finally, globalization) that entail changes in the status and potential of particular groups, including women (Channa 2004). One of these processes is the development of tourism, and especially its informal sector, which offers women a number of opportunities, mainly because it involves many jobs and activities that do not require formal qualifications (Ashley, Roe, Goodwin 2001).

Hampi, as a village which emerged as an outcome of mobility and was constantly fueled by the influx of migrants, is socially and culturally heterogeneous. Despite numerous internal divisions within this community—not only ethnic, religious and related to caste, but also between "old" and "new" residents, wealthier and poorer, or supporters of different political factions—the local social boundaries and hierarchies seemed to be quite fluid. At the same time, however, the families that have lived in the village for three generations are characterized by quite a high degree of social conservatism, despite many years of exposure to liberal attitudes and behavior of foreign (and more recently middle-class Indian) tourists. This is evident, for example, in the still dominant model of arranged marriages—"love marriages" are rare and when they occur, they are met with strong opposition from the relatives. Women are not allowed to marry again after their husbands' death. Widows remain at the mercy of the family of the deceased spouse. In this sense, family is a powerful weapon that can act as an excluding force. Here it is tourism that helped many of the women whom I met during my fieldwork to overcome the abovementioned limitations, primarily through the possibility of economic independence. As a rule, a woman living in a rural community—for example, in one of the villages around Hampi—has little opportunity to earn money unless her parents decide to provide her with good education, for which they must have adequate resources, since the quality of teaching in governmental schools is very low. The wages for agricultural labor do not allow for financial independence and can only serve as a contribution to the household budget. Tourism, on the other hand, provides such opportunities, also due to the involvement of tourists in the development of local entrepreneurship, a phenomenon I described in the previous section.[33] At this point I would like to emphasize that I do realize that the project of women emancipation through economic independence has been criticized as an universalist approach of Western feminism—white, heterosexual, secular, and middle-class (Butler

[33] How marriage is perceived as an "escape attempt" from a social destiny (Cohen and Taylor 1992) is discussed in the last section of this chapter.

1990; Kishwar 2005; Bobako 2013)—and that the empowerment programs inspired by neoliberalism have been seen as producing a *homo oeconomicus* and thus maintaining the capitalist *status quo* (see Sharma 2008, 26, 30–61). Moreover, in the Indian context, as shown in research by Anna Romanowicz (2017), the NGOs implementing empowerment projects are managed almost exclusively by middle-class women, which often leads to the reproduction of existing class divisions. However, in this particular context, the possibility of procuring economic capital through the presence of tourists is desirable by women because it gives them the chance to at least partially liberate themselves from the control exercised by their families. But it is not just that—contact with tourists also gives women the opportunity to acquire alternative social networks and to gain cultural capital, on which they can then build a sense of satisfaction and self-esteem. Paradoxically, this is possible due to the lack of an institutional context, which would impose the top-down meanings of empowerment.

Such emancipatory role of tourism is illustrated by the story of my landlady, Bhavani.[34] When we first met, Bhavani was almost forty years old, had been a widow for sixteen years and a grandmother for four years. She told me: "I am an Indian, but I don't like many things in India. Sometimes I ask God: 'Why don't you just kill all the men, then everything would be much simpler.'" Her marriage to her mother's younger brother was arranged when she was fifteen. An old, black-and-white photograph hanging on the wall of the Bhavani's room shows them all: her as an infant, him as a ten-year-old, her mother, father, and her mother's other brother. Such marriages are practiced in the southern states of India, including Karnataka, where they are legal. The rationale behind it is to keep both the daughter and the property within the family, because the dowry stays with the girl's grandparents, who are also her in-laws. However, Bhavani was not convinced by these arguments; she could not look at her uncle as her husband. She also blamed him for the fact that their only son was born with a number of congenital faults: he could not hear or speak, had a serious visual impairment, and epilepsy, as a result of which he was delayed in his psychophysical development (he started to hold his head on his own when he was two years old, and could only walk when he was seven). Besides, "family or not family," as Bhavani used to say, "all Indian men are the same: they need us only as slaves."

When after the marriage she moved to Hampi, she was shocked: "I saw a refrigerator, and a phone for the very first time. In my village, there was not even electricity. We used to go to sleep at 7 p.m. when it was getting dark." In Hampi, they lived in a mandapa, and after ten months of marriage she gave

[34] Bhavani was very aware of her role as a partner in the research process, so I am telling her story under her name.

birth to a daughter. Soon after, she started working in the Welcome Restaurant run by her grandfather at the main bazaar (the restaurant developed from a small *chai* shop set up in the early 1980s). It was 1992, and there was a growing influx of tourists in Hampi. She liked this work, although it was not easy: "Sometimes I didn't even see daylight. I cooked all day, and when I would finally go outside, there were already stars in the sky. We opened at 6 a.m., and closed before midnight." She prepared up to 200 meals a day, one of her brothers was taking care of the supplies, and the other helped in the kitchen. She worked even when she was pregnant for the second time and spent only the last two weeks of her pregnancy at home. She learned to cook for tourists from tourists themselves. Above, I have already described how Bhavani became the first Korean cuisine specialist in Hampi. That was not all she could do, though. Italians showed her how to prepare "real spaghetti" (she made noodles by herself, as she had a special utensil for this purpose), green pesto and cheese "just like parmesan." She also learned English. Initially, she was not able to communicate with her foreign guests at all, although she went to school and completed several grades, but in her village the whole education was in Kannada language. To this day she cannot write or read English (that is why we never wrote to each other, we always called).

At that time her husband was starting his own hotel and restaurant across the river. He had a knack for business. He had set up the first telephone booth in Hampi for international calls, had brought the first sound system to organize techno music parties, and through his extensive networks catered for the first groups of tourists coming from Goa. He was rarely at home. He did not accept his son's disability and avoided contact with him. Like many people in Indian villages, he believed this was a bad *karman*:

> People were afraid to touch him! My husband never took him on his lap, never! So, with the exception of my brother and mother, nobody helped me take care of him. And even [using] the toilet was a challenge, because you need two hands to pull the pants up ["with sari it's easy!," Bhavani joked]. So I used to hang him over my arm, like a towel, and washed his bottom. When [I was] taking a shower, I took him with me to wash him. I breastfed him until he was 5, because he didn't want to eat anything else. I lost my left nipple when once during feeding he had an epilepsy attack—I could not open his mouth.

Even if the husband would show up at home every now and then, he was most often drunk. Bhavani's grandfather/father-in-law told him then to go to bed in a room he and grandmother slept, in order to protect his granddaughter/daughter-in-law from having forced sex with his son. Bhavani did not want to have any more children as she was afraid that they would be just as sick: "It is because of the same genes that we should not have children." She took the son for medical consultations, camped out in the hospital corridors, and even went to a special center in Mysuru, where she and her son learned sign language for a month. One of the tourists, a physiotherapist, showed her some exercises she did after the restaurant was closed. She then weighed about forty kilograms. For breakfast she ate only rice, and for dinner—a banana and sleeping pills, otherwise she could not sleep. In the morning, in order to wake up, she drank a strong espresso—she still does it today. It is one of the few houses in Hampi where you drink coffee in the morning, not *chai*.

People were gossiping about her a little, that she was working for a living, and she should not, because she is Lingayat.[35] But she did not pay attention to it, because work allowed her to forget about everyday worries and gave a new meaning to her life. All the more, the restaurant was very popular among tourists. Bhavani had one major advantage over the competition—all other restaurants were run by men and she was a woman. I noticed on many occasions that Bhavani was very aware of the benefits that gender provided her in the tourist business. First of all, she attracted tourist women traveling alone who wanted to befriend her because they either felt safe in her company, that is, not exposed to the advances of local men, or on the contrary, who were looking for a trusted person to talk about their Indian loves. Secondly, the image of an Indian woman oppressed by a patriarchal society who needs to be saved (see Loomba 2015, 153–171) is still very much present in tourism imaginaries as part of the colonial legacy. By defining selected practices toward women in conquered societies as backward or even barbaric, the colonizers were able to legitimize their power through the category of a civilizing mission. Within the framework of "social reforms," women were unveiled in the Arab countries (see, for example, Fanon 1967) and the practice of *sati* was abolished in India (Spivak 1993). Spivak captured the essence of these policies in a phrase: "white men saving brown women from brown men" (Spivak 1993, 93). In reality, however, "white men" often only strengthened these practices by making colonized men defend what was under attack, identifying women with the essence of traditional culture and anti-colonial nationalism (Chatterjee 1989; cf. Rabinow 1977 in the context of North Africa).[36] In this

[35] *Liṅgāyat*—followers of Shiva, quite numerous in Hampi (and throughout Karnataka). Currently they are waging a political struggle for recognition as a separate religion, not only a sect within Hinduism. Lingayat position themselves high in the social hierarchy (they like to compare themselves to Brahmins).

[36] All the more, because practices which were often fought by colonial "reformers" were misunderstood or concerned only a particular group or area. For instance, *sati*, as Spivak herself writes (1993), was not common in the entire subcontinent, as suggested by colonial representations, but it referred mostly to women from Bengal, and those of high birth. It was connected, paradoxically, with their high social position resulting from the right to inheritance, which some family members tried to limit with the use of the ritual of *sati*, thus referring to the widow's affection for her

way, female bodies became a field of ideological struggle between colonizing and colonized men, and women's voices remained muted. Paradoxically, many reformers in postcolonial governments also saw modernity as the chance to end the oppression of women. Therefore, especially in the eyes of tourists assuming the attitude of "friendly moral ethnocentrics," as Krzysztof Podemski (2005, 206–207) calls them, Bhavani embodied the socio-economic progress of India, which had to be supported.

Bhavani became a widow after nine years of marriage—her husband, like many other men in Hampi, died "from drinking." He spent the last months of his life in bed, as an addict: "When you poured water into his mouth, he did not drink. We had to mix it with whisky! We had to carry him in the shower, like a baby. I had two completely dependent men to look after: a husband and a son." When he finally died, she was relieved, but on the other hand, widowhood excluded her from many areas of social life. She was not invited to family celebrations, she could not participate in ceremonies, including her own daughter's wedding, which she funded and organized by herself. She could not even wear jasmine flowers in her hair or bracelets on her hands. She regretted that she was not a Lambani; then she could marry again, "because those women are equal to men" (her daughter was of the same opinion, she spoke of women from lower castes with jealousy: "Less education, more freedom"). Bhavani believed that her higher social status did not favor her; on the contrary, it was a limitation. The Indian anthropologist Subhadra Mitra Channa points out that this can also be interpreted as part of the colonial legacy. The British tried to "modernize" Indian women born into higher castes following the "Victorian prototype," bound by the precepts of Puritanism. In contrast, uneducated Indian women dressed much more freely and talked about sex issues without restraint, often joking (Channa 2004).[37] This is why "friends," as Bhavani calls the tourists with whom she had formed closer relationships, were a substitute for the missing social and emotional support of relatives: "Friends give me more happiness than the family." Although, certainly, if it were not for her brothers who were very supportive of her life decisions, it would be more difficult for her to survive.[38]

Six years later, when her father-in-law died and brother-in-law decided to take over the business, Bhavani had to leave Welcome Restaurant. She took

---

deceased husband. Only after identifying this practice with "Indianness," as a result of colonial condemnation, it became more widespread. In this sense, it was British colonialism which "produced" the Indian practice of *sati*.

[37] Channa recalls that her mother, educated by British missionaries, had to wear a long petticoat under her sari, which in the hot climate of Bengal caused her great physical discomfort.

[38] The relationship between brother and sister is—after the relationship between son and mother—the most important of the male-female relationships in Indian societies. Sometimes it is extended to unrelated men, for example, friends of a brother, which create a safe network around the woman and give her freedom to interact with men (Channa 2004).

the children and moved to semi-detached small house located in the center of Hampi, which her husband got from his father and her grandfather (who in turn had bought the land with the money earned through the restaurant; the other half of the house went to her husband's other brother). When she left, she took one sheet and 150 rupees with her, while the daily medicine for her son cost 100 rupees. She was trying to get some state support, she had all the required documents stating her child's disability, but the officials had been sending her away, telling her to "come back tomorrow." The same happened with the widow's pension. In the end she gave up and decided she preferred to earn money herself. She started sewing clothes for sale, even ten sari blouses a day. But the money she earned in this way was small, her son's needs—big, she also had to start collecting money for her daughter's dowry. Besides, she missed contact with tourists. So, she decided to start her own restaurant on the rooftop of the house she lived in. For this she needed a loan: "For fifteen years I lived [on the bazaar] in front of the bank, and I had never been inside! . . . When I went in [to borrow money] for the first time, I was not able to sign, because my hand was shaking so much. . . . It took me an hour to give six signatures." She started all over again, but ever since she worked at the Welcome Restaurant, she already had contacts and know-how, was able to keep simple accounts, so she quickly managed to make a good name for herself. The brothers, as usual, came to help. However, she is still annoyed that when she goes to the bank or buys something in installments, there is always the expectation that a man from her family will sign the paperwork: "And this is me who runs the business, I make the money!"

Contrary to the doctors' pessimistic prognosis, Bhavani's son grew up and slowly developed, to the point where she even managed to send him to school in Hospet. He was thirteen years old when, one day, he stopped in the street to tie his shoe. A school bus was honking for him to clear the road, but he was unable to hear it. He got hit and suffered a heart attack as a result: "When my brother and I entered the hospital room, his body was already covered with a white cloth. I closed my eyes and opened them after two months. I don't remember anything." She sank into herself. She neglected her daughter, who decided to quit her college and get married. The marriage arranged by the family lasted less than five months. The girl ran away from her husband to Hampi, but was already pregnant. She gave birth to a daughter, who was four years old, when we first met. This was another reason why Bhavani felt socially excluded. Today the three of them live in one household (figure 5.3).

Her beloved Welcome Restaurant was demolished in 2011—just like other businesses and homes at the bazaar. She had to close her own restaurant after four years, because her husband's brother's family did not want to rent their part of the rooftop anymore. For the next few years, she ran restaurants in rented places—one of which was so famous that *Lonely Planet* wrote that it

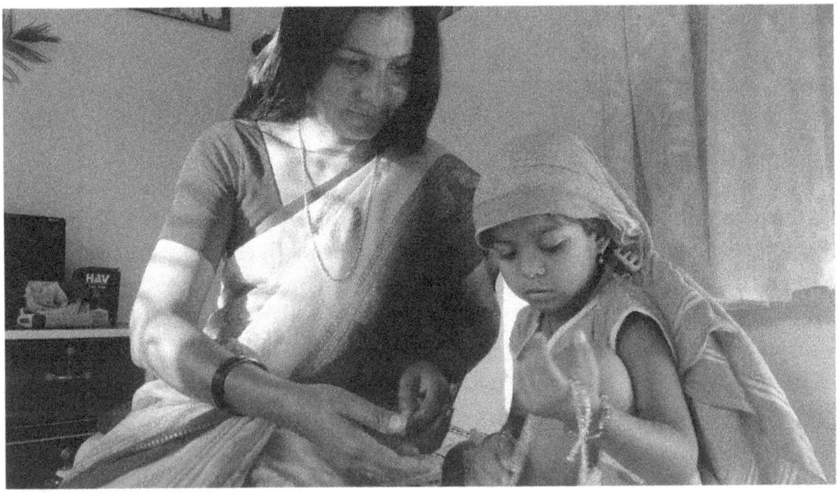

Figure 5.3 Bhavani with Her Granddaughter, Hampi. *Source*: Photo by Natalia Bloch (2017).

served great Korean and Japanese food. Finally, she decided to take another loan from the bank to renovate two rooms on the first floor of her part of the house: both with bathrooms, air conditioning and hot water. She, her daughter and granddaughter lived on the first floor, and the rooms were rented to tourists, from which they made a living. In her business, Bhavani focused on regular clients, who arrived often every year, for two or three weeks, sometimes even longer. Some of them were friends from the times she had worked in a restaurant or friends of friends, or friends of friends of friends. Also in this business, Bhavani's gender often played to her advantage, because tourist women who traveled alone often wished to stay in a guesthouse run by a woman.[39] One of her regular clients was a Korean woman who, when she first came to Hampi, was just over twenty years old and made her first independent trip. Because she was very trusting, she often fell prey to the swindlers, so when she finally came across a woman working in the tourism sector, she immediately clung to her. Since then, for eight years, she was coming to Hampi every year for a month. She spent most of her time with

[39] More and more initiatives have been emerging in India and Nepal to respond to this gender-based tourist need for safety. In Dharamshala, Tibetan women opened a guesthouse called Ladies Venture and dedicated to women traveling alone (however, due to the declining number of foreign tourists, mainly in the wake of widely reported in the world media rape cases in Indian metropolises, the guesthouse gave up his "women only" character). In Pokhara, a popular tourist destination in Nepal and a starting point for treks to and around Annapurna, 3 Sisters Trekking Adventure was established, which was the first Nepalese trekking agency founded by women and training women as guides.

Bhavani, watching TV together, and when Bhavani went shopping, she kept an eye on the house. She called her "mother." Bhavani also hosted Korean tourists referred by her brother—after she withdrew from the catering business, he opened his own restaurant with Korean food, using the accumulated cultural capital.

Bhavani was very proud of her two rooms. She consulted with many tourists on what alteration to make, what to improve, and so on. She knew that "Europeans" appreciated cleanliness and did not like bright colors, so she painted the walls beige. She also bought a thick, decent mattress, very expensive, because she often hosted elder guests, already retired, and for them the comfort of sleep was very important. For those who could not sleep or wanted to relax before bedtime, there was a TV set. There was also a big mirror in the room, because if men wanted to shave and women to do make up, there was better light there than in the bathroom. Bhavani was bursting with pride when she learned from tourists that she was in the Korean and German editions of the *Lonely Planet* guide. However, she recalled the times when she run a restaurant with a sentiment. She still sometimes cooked at her brother's place when "old friends" showed up and asked about her and said that they missed her food. She was moved by the fact that people remembered her, that they called on different occasions or sent photos from her restaurant. She complained a little—or rationalized—that today's tourists are different, "not so friendly anymore," that they shout from the table: "hey, clean it up," "as if I were a cleaning lady." Meanwhile, Bhavani positioned herself as a partner in her relations with tourists (see chapter 3). She recalled that her mother, when she needed ten rupees, had to ask her father, because she had absolutely no money at all, and yet, as Bhavani strongly emphasized, she had worked hard in the field. Luckily, Bhavani lived in Hampi, where—as she said—"most women work for a living, each one sewing something, selling bananas or coconuts. It is much better here than in the villages."

In the Tibetan diaspora, on the contrary, the rhetoric of gender equality is omnipresent. Refugees not only build their global image in opposition to the significant Chinese Other—referring to "Western" values in the process of inventing traditions (Hobsbawm 1983)—but also use this mechanism to differentiate themselves from the host society. They therefore perceive caste and the subaltern position of women in terms of moral ethnocentrism as a sign of Indians' backwardness. By creating positive self-representations of the traditionally egalitarian community,[40] Tibetans appear as civilized Asians,

[40] In order to Tibetanize this tradition, on a rhetorical level, Buddhism is used as an opposition to hierarchical and masculinized Hinduism and the practice of polyandry is recalled, interpreted as evidence of the high status of Tibetan women in opposition to patriarchal polygyny, which is largely a product of idealization (cf. Gyatso and Havnevik 2005).

winning the support of international women's rights organizations for the Tibetan cause (Mountcastle 1997, 247). Although these declarations were not always in line with actual practices,[41] certainly the harsh living conditions in the mountainous climate forced the Tibetan communities to adopt a kind of gender egalitarianism, at least in relation to the sphere of work. While politics and religion, especially in central Tibet, were areas of male domination, in agriculture, pastoralism, and commerce there were no clear dividing lines and women worked to support households equally with men (cf. Aziz 1978). In Lhasa, the capital of Tibet, many women ran their own stores and stalls (Bell 1928, 157). This model is still practiced, as I have already mentioned in chapter 1. The presence of Tibetan women in the tourism sector—or more broadly in trade and services—is visible and widely accepted. I even heard Tibetan men say that women are more effective in establishing relationships with clients and bringing about transactions (women themselves, in turn, sometimes claimed that this was an excuse for men's laziness). As a result, Tibetan women run businesses—stalls, shops, restaurants, and guesthouses—not only together with male partners (like Dolma and Norbu in Hampi and Spiti or Dolma's sister and her husband in Goa), but also on their own, even when they have husbands, who often take over housekeeping and childcare responsibilities (like Dolma's second sister or Dolkar, both running stalls in McLeod Ganj). Moreover, it seems that exile has only strengthened their independence in this respect—the need to adapt to the economic realities of the host country forced refugees of both sexes to earn for living.

As far as marital arrangements are concerned, in pre-PRC rule in Tibet, they were very diverse and flexible. There were different forms of relationship between man and woman (Grent 2002, 115), divorces were allowed, since Tibetan marriage was not particularly religious in nature, and widows remarried; even having illegitimate children was not always perceived as a problem (Subba 1990, 107). Today, the diaspora is clearly dominated by monogamous "love" relationships, and marriages are concluded quite late, even after people turn thirty, while women rarely give birth to more than two children.[42] At the same time, the nationalist Puritanism characteristic of the diaspora largely forces women, as guardians of Tibetan values, to take care of their moral conduct, which is also part of the image politics of the diaspora's elite. As a result, single mothers had been in a way swept under the carpet for

---

[41] For instance, women employed in the exile administration are expected to wear the so-called traditional Tibetan dress, while men can dress "in Western style."

[42] While in Hampi I was seen as a pretty old mother-to-be (Bhavani, two year elder than me, was already a grandmother then; I even began to lie about my age when being asked about it), in Dharamshala my emancipated girlfriends worried whether my motherhood would not affect my professional career.

a long time, and their difficult social situation was only brought to light by campaigns launched by NGOs.

Tendill Sangmo,[43] whose story I wish to describe here, is such a mother. She was born in Uttarakhand.[44] Her grandparents originated from Ngari, a region in western Tibet, and traded across border with India. Before the Chinese rule was introduced in Tibet that border had been very fluid, so they had finally settled on the Indian side and Sangmo's parents were already born in India.[45] There are about ten Tibetan families and one gompa[46] in their village, in addition to several families scattered around the surrounding villages. At home, the elder generation spoke Tibetan, but the younger in Hindi; the elder women dressed in Tibetan *chubas*, while the younger in the North Indian *salwar kameez*.[47] Sangmo attended an Indian school, she cannot read or write in Tibetan (she once confessed to me: "I did not have this privilege"[48]), and she only learned Tibetan when she arrived in Dharamshala. The cultural practices of the *gyakar Khampa* are a mixture of many influences, but two things are clear evidence of their sustained distinctiveness. First of all, they are Buddhists, although, as Sangmo claimed, "we think like Indians," referring bitterly to her upbringing as a girl. She was not supposed to laugh, talk or sing out loud, or play with boys; she was so frustrated with such treatment that whenever her parents left, she used to beat up the neighbors' son (his mother did not intervene because the boy was very naughty). Secondly, they follow endogamy (Sangmo was outraged when she found out that her cousin married an Indian man; of course, many of her colleagues had Indian boyfriends while studying in college, but "then fortunately they married Tibetans").

Sangmo's parents traded in plastic bracelets, metal rings, and incense. They did not even have a stall and laid out their modest assortment on a sheet of plastic spread on the ground. Teenage Sangmo earned money by selling things she knitted herself (later, when having her own bakery, both she and her daughter knitted while waiting for the customers—the daughter was making a headband for her mother, and Sangmo was knitting a cardigan for her

43 A real name.

44 A state bordering with Himachal Pradesh, Tibet (today in the People's Republic of China) and Nepal, being largely a cultural borderland.

45 They are thus *gyakar Khampa*—that is, "Indian Khampas"—whom I mentioned in chapters 2 and 4. *Gyakar Khampa* function on the margins of both the diaspora (as a not-Tibetan-enough holders of Indian citizenship) and the Indian Himalayan communities (there they are not-native-enough and have a limited right to land ownership). For refugees—including those born in India—they are often invisible and they themselves tend to keep a distance from the Tibetan refugees' identity politics.

46 Wylie: *dgon pa*—a Buddhist temple.

47 A feminine garment popular in the northern states of India and Nepal (but also in Bangladesh or Pakistan), consisting of wide, tapering pants and a tunic (sometimes also a scarf, called a *dupatta*).

48 But Sangmo's younger siblings—brother and sister—were enrolled in a Tibetan boarding school, as was Dolma. This shows that Tibetan education is valued in this community.

daughter). Then she made some money by singing two songs on the radio because the local DJ liked her voice (he even promised to arrange a scholarship for professional voice training, but when her father found out about it, he deemed it not appropriate for a girl to sing and that was the end of her career). She got married young and her marriage was arranged. Her husband, a Tibetan, came from a wealthy family, but was addicted to alcohol and drugs and did not treat her well. Violence became part of the deal very early and did not stop even when Sangmo was pregnant. She tried to commit suicide several times and finally decided to run away, initially alone, leaving her daughter in her husband's home. She survived with the help of relatives who gave her shelter and a local influential women who gave her job. Almost two years later, after the divorce (she says: "I was the first Tibetan-Indian woman who exercised her right to divorce"), she decided to leave the town along with her daughter and settle as far away as possible from her ex-husband's family and her own mother, who did not accept her decision. It was 2009.

She chose Dharamshala. As with many new refugees, it appeared to her as a place offering many opportunities. She knew someone through whom she met another person, thanks to whom she found a house run by one of the leading Tibetan activists. He took her in. She did not have to pay any rent, her contribution to the home budget was cooking (she laughed that before she got there, the crew of activists had fed almost exclusively on instant soups). Then she came across the NGO founded by the Tibetan man and his Korean wife, already mentioned in this chapter, which supports single parents by giving them employment and running a free nursery. Since Sangmo knew a little English from the course she took in Uttarakhand, she was offered a job at a café run by the organization (most of the women who were supported by this NGO were refugees, so as they spoke Mandarin at best, but no English). However, the salary she was receiving was so low that she could not afford to rent a room for her own in McLeod Ganj, especially since she had to pay her daughter's kindergarten fee. So, she still lived with activists near Lower Dharamshala, but she already had enough money for food, bus tickets, and inexpensive clothes. The commuting was, however, tedious.

The café, apart from being an additional source of income for the NGO and a place to sell handicraft made by women employed in the organization, also had a marketing function.[49] Located in one of the main streets of McLeod Ganj, it was supposed to promote the NGO and its activities among tourists and thus attract volunteers and, above all, sponsors (the same function was performed by walks organized once a week during the tourist season for tourists

[49] Other Tibetan NGOs do the same, however, in the case of this foundation, which relies less on volunteers and more on sponsors and handicraft sales, a café combined with a shop and an information center was of particular importance.

and babies from the nursery, and combined with a picnic at the main temple). Sangmo started her career at the café by helping in the kitchen (and burning cakes), but was quickly promoted to a managerial position, until she was able to run the whole place by herself. After less than three years of working for the organization, she decided to leave. She wanted to become independent and earn better money in order to match her daughter's growing needs. She tried different jobs: selling beauty products door-to-door and coach tickets online, making handicrafts, working as an accountant and, meanwhile, baking something to order. Eventually, she realized that she liked the latter the most and knew it best, so she took a loan and decided to try her hand at the tourist business. She rented a tiny place with barely three tables in the basement in the center of McLeod Ganj and opened a bakery there, which she named after her daughter. Woeser Bakery quickly began to gain popularity thanks to the skills she acquired in her work for the NGO, and especially the skills she learned from tourist-volunteers, whom she befriended: from a Canadian woman—how to prepare a handmade cappuccino (because she could not afford a coffee maker), from an American woman—how to bake muffins, and so on.[50]

She worked hard. She used to show up in the bakery at 8:00 a.m. in order to prepare fresh cakes, open at 10:00 a.m., close at 8:00 p.m., and then do cleaning. All that took almost thirteen hours a day. But unlike most seasonal small businesses in the tourism sector, she took one day off per week. She usually spent this time in the company of her tourist friends, making trips or going out to restaurants. Tourists also provided her with new recipes, like the one for "Russian tea biscuits" from a British friend (later a Russian lady customer asked her why they were "Russian," but she liked them very much anyway). Sangmo also searched for recipes on the Internet, and then modified them in her own way. She introduced a few vegan cakes, because more and more tourists asked about them (but also in response to a growing clientele from the south of India who does not eat eggs). Sangmo's bakery was a "heterogenous space" (Edensor 1998), so typical for McLeod Ganj—she hosted urban middle-class Indians (Sangmo was one of those Tibetan businesswomen who had plenty of Indian clients), tourists from the so-called Global North (especially long-term volunteers) and the Tibetans themselves (including monks and elder Tibetan women dropping by for a cup of tea). From the latter, Sangmo never took money, but toward tourists she was ruthless—there were no discounts even for friends (and prices were not low).

From the times when she worked at the NGO's café, Sangmo had extensive contacts on Facebook, which she skillfully used to promote her business.

[50] Paradoxically, launching her own business was not treated by the NGO's leadership as its beneficiary's empowerment, but as a disloyalty and unfair competition. She was even asked not to bake the same cakes that were served in the NGO's café.

Actually, she was online all the time: in the bakery, the laptop was on the counter, while Sangmo did not let go of her smartphone. She read with satisfaction emails and posts from tourists who wrote that they missed her cakes. One day, when we sat together at her shop, she posted on Facebook: "describe me in one word." The answers appeared immediately: "determined," "independent," "strong," "inspiring," "vibrant." They arrived from different parts of the world: from an Indian journalist from Mumbai, from a nice American tourist, from a French volunteer, from an Australian woman who invited her to her wedding. Tourists from Indian metropoles had been encouraging her to open a bakery in Delhi. Sangmo burst with pride, and I could clearly see that it gave her a lot of self-confidence. Just like in 2012, when she won first place in the ranking of restaurants in McLeod Ganj at TripAdvisor (and she is still ranked as the best café in McLeod Ganj). Tourists' comments read as follows:

> "The Tibetan woman who runs this [bakery] is a delight and she knows how to bake!! Of the many baked goods sampled at various restaurants and street stands, hers were the best! She seems to have 'western' recipes, i.e., they taste like real baked goods. Always fresh and a nice variety, it is definitely the best place for cakes, cookies, and the like" (a tourist from the United States). "This woman and her young daughter Woeser are passionate about bringing mouthwatering delicacies to those of us who will appreciate them. Sangmo started this business on a shoe string and she is hard working and inspiringly determined to succeed as a business woman. Whatever you do, don't miss this hidden treasure. You won't be disappointed!" (a tourist from the United States). "And we have joined the fan list of this quaint little bakery located just below . . . !! Run by a passionate and very warm young lady, we had some finest carrot cake, chocolate cake and lemon bars" (an Indian tourist). "The bakery has a real buzz, come all times off year and you will be impressed with the delicious display of mouthwatering sweet delights, interesting people who come through the doors and the delightful young woman who runs the show" (a tourist from the United Kingdom).[51]

After a few months of operation, a British woman turned up at the café. She ran a fitness club in Mumbai, and in McLeod Ganj, which was by comparison a very quiet place, she rented a big, nice apartment. However, since she often left for business, she offered Sangmo to live with her in exchange for housekeeping, cooking and taking care of her dog. Sangmo herself, talking about the British lady, used the term "my roommate," positioning herself as a partner in this relationship. This allowed her to survive the period of business start-up, so when the "roommate" decided to move out of McLeod

[51] https://www.tripadvisor.in/Restaurant_Review-g1092107-d3930797-Reviews-or340-Woeser_Bakery-McLeod_Ganj_Dharamsala_Kangra_District_Himachal_Pradesh.html.

Ganj after two years, Sangmo could afford to rent a room on her own. She could also afford to send her daughter to a Tibetan boarding school located two-hour drive from Dharamshala—the school had a good reputation and in this way, Sangmo could take care of her business during the week.

She was proud of her financial independence also because her husband "hasn't even earned thousand rupees in his life!," living off his family's money. For her self-esteem, it was also important when her uncle along with a few friends from her hometown visited her. They came to McLeod Ganj for sightseeing. The British woman was not at home, so Sangmo welcomed them as if they were at her home (she showed me a photo they took together on the terrace of the apartment with a beautiful view of the mountains in the background). In the evenings she took them to various restaurants: Japanese, Korean, and Italian, which made a great impression on them. McLeod Ganj seemed to them as a cosmopolitan metropolis: "You can't even get rice and *dal*[52] here!—they were shocked—and apples are twice as expensive as ours!" When they found out how much Sangmo paid for renting her tiny premises, they could not believe it either: "You can rent a whole house in our town for that!" When they returned to their hometown, they shared impressions. Soon afterward, Sangmo started receiving invitations to weddings and other family celebrations from all those who had previously socially shunned her (one of them even used a honorific form of address). She usually did not go to these events, because she was busy at the café, but sent expensive gifts to further emphasize her success. She also started to act as a donor, working for an NGO set up by Australian former tourists to India, which supports educational opportunities of Indian children from Himalayan villages.

Sangmo was not in a relationship, as she claimed, she had no time for dates. But she started traveling—every year she took her daughter to a new destination on a tourist map of India, enjoying the ability of being a tourist. When we saw each other last, she was just starting a new small business: manufacturing political souvenirs for tourists, that is, crochet-made bookmarks which read "I love Tibet" and "I love momo." As a parting gift, she gave me a pin with a wish: "I hope to go back to Tibet."

LOVE IN THE TIMES OF TOURISM: RACE AND GENDER

The stories told so far in this chapter show that contrary to popular belief, the emancipatory potential of tourism triggered by establishing close

[52] A staple meal—consisting of rice and thick lentil sauce—consumed in the northern states of India.

relationships between tourists and members of the host communities, does not have to take on the character of love relations. Although, of course, this dimension in its various aspects—sexual, emotional, or formal—is also present in tourist encounters.[53] During my fieldwork, I discovered that each of my research partners had at least one family member or friend who was in a relationship with a female or male tourist. Frequently, such a relationship resulted in couple's emigration to the tourist's country of origin (although not always, as I discuss in this section). Take, for example, a Tibetan from Nepal, the owner of one of the seasonal restaurants in Hampi. In my conversations with him, the topic of "buddies who left" often recurred: one, with whom he worked in a restaurant in Goa, met a tourist from South Korea who turned out to be very wealthy and influential—they got married and left (later he also brought his brother to Korea); another, also a Nepalese, married a German woman—much older—and left; yet another emigrated to Australia in exactly the same way. Raj, the owner of a family-run guesthouse, often told me about "the boys from Hampi who left"—to Italy, Australia, the Netherlands, or France. He thought that "they were very clever, they knew what was going on here and they left as soon as they could." He often returned to this topic, expressing a strong sense of jealousy. Eventually, he admitted that he also wanted to leave, but his family did not let him go—his marriage with a decent Indian woman was arranged and soon he became a father. Nevertheless, such stories ignited hopes, as they made dreams very real. All the more so because marrying a citizen of the host country is, in most of the migration policies of the Global North countries, the surest way to obtain a visa and then legalize further stay and work.[54] Ultimately, it is the fastest way to become a citizen. The phenomenon of migrating abroad through tourism is therefore an example of one mobility stemming from another—tourism-driven—mobility, the former being not only spatial but also social (although social advancement through a relationship with a tourist does not have to involve emigration, which I describe below).

More often these relationships involve tourist women, which is at odds with the classical approach to tourism with the prefix "sex," usually described as a continuation of colonial power, being implemented through commodification of female bodies. In such an interpretative framework, a white, often older man from the wealthy North exploits sexually women from the poor South, thus benefiting from global economic inequalities (see,

[53] See Singh (2019) for his broad analysis of love in the context of tourism (and beyond).

[54] Of course, the system tries to seal itself against the so-called paper marriages. For instance, in Poland, the credibility of such relationships is verified by conducting separate interviews with spouses about the partner and the relationship itself (and comparing the consistency of responses), conversations with neighbors to confirm whether the couple really lives together, and visits of the district policeman with the same purpose.

for example, O'Connell Davidson 1996). And even if the tourist is not a man, she still plays such a role in the relationship: the "nations [of Third World countries] are forced into the 'female' role of servitude, of being 'penetrated' for money, often against their will; whereas the outgoing, pleasure seeking, 'penetrating' tourists of powerful nations are cast in the 'male' role" (Graburn 1983, 441). However, the dynamics of these relationships, as I could often observe, was much more complex and escaped such clear-cut conceptualizations reducing intimate relations, which occur as a result of tourism, to binary categories of domination and subordination. Certainly, more relationships between tourist women and local men were fostered by the fact that the tourism sector in India is still dominated by male workers, and a relatively high number of single traveling tourists are women (this was the case at least until media started reporting on gang rapes in India). If we take into account the long-term nature of many of these travels and the considerable size of "heterogeneous spaces" that tourism in India offers, all this creates numerous opportunities for closer relationships. A characteristic feature of such relationships was a visible difference in the social status of the partners. Single tourist women were usually middle-class, twenty or thirty plus, university graduates (students more often travel in a group of friends). Men with whom they fell in love most often did not have the opportunity to obtain formal education due to structural conditions that had forced them to enter the labor market early in order to make a living.

Let's consider an example of Jay, a salesman from Rajasthan, who traded in clothes in Hampi during the winter season, and moved to Manali, Himachal Pradesh, for the summer. Jay had been earning his living since he was twelve. At first, for seven years he sold coconuts for temple offerings in his native, holy town of Pushkar. He was brought to Hampi by an acquaintance, who ran a shop across the river: he cooked for others, made tea, cleaned up. Once he got familiar with the place, he decided to open his own small shop, especially since his brother run this kind of business in Manali. During my last field trip to Hampi, Jay met Sophie. Sophie was thirty-two years old (thus she was seven years elder than Jay), lived in Paris and designed store spaces. She then had just arrived in Hampi and was looking for cheap accommodation. He asked her if there was anything he could do to help and recommended a place. Gradually, Sophie visited his shop more and more often, spending more and more time there. They sat together, waiting for customers and chatting. I also saw them in the evenings, in a restaurant run by a Tibetan man from Nepal. They even went to Hospet to see a movie in the cinema and do some shopping. The guys from the street started smiling ambiguously when they talked about them. Then, Sophie was supposed to leave Hampi in two or three days, but those "two to three days" dragged on. Eventually he had to go home for the holiday and she also continued her trip and sightseeing in

India, but she called him every day and they talked long time. Then he came back and so did she. She was supposed to leave again in two or three days and put it off again. Eventually she left for Mysuru, a popular tourist destination in Karnataka, but after two days called Jay to join her. He hesitated for a while, but in the end, he went—a befriended Rajasthani shopkeeper offered to take care of his store, and it was already the end of the season anyway. From Mysuru they continued southward to Varkala, in Kerala state, to a beautiful, romantic beach. They rented a room together. Jay became a tourist. After returning to Hampi, he closed the shop and together they headed north to Rajasthan. They planned to stop by in Rishikesh and then take a plane to Delhi, because she had to return home. He was to take her to the airport and say goodbye, and from there, go straight to Manali for the summer season. This way, they could spend more than a month with each other. They agreed to meet next year—Sophie would then design a decent store for Jay, and he asked her to look around to see if there is a place in Paris where he could sell his merchandise. He was thinking about going to Paris.[55]

There are more such stories. The relationship between an Australian woman and a young auto-rickshaw driver from Hampi lasted for three tourist seasons already. The man had lost his family house in a mandapa as a result of the bazaar's demolition, and lived with his parents and younger brother in a makeshift shack encampment. The Australian woman did her best to help him through this difficult time. There was also a relationship between a young woman from Japan (a holder of a degree in Japanese literature) and another auto-rickshaw driver from Hampi. The woman was on her tour around the world, when she came to Hampi for two weeks, but stayed for three months. Then she continued her journey, even returned to Japan, but kept in touch with her Hampi love by phone. Eventually, she came back for good and they got married. On their honeymoon, she took him to Agra so that he could see the Taj Mahal for the first time. A few years later their son was born. They made a living running a two-room guesthouse for Japanese and Korean tourists. He was from a polygynous family, his father, who originated from a neighboring state, took two sisters from Hampi as his wives. The man's mother decided to leave her husband, prone both to alcohol-addiction and violence, and returned to Hampi with her three sons. Her brothers, however, were not particularly eager to help the woman who abandoned her husband, so the boys, who were still underage at the time, had to go to work. They lived in a mandapa near the bazaar, hired themselves to help in restaurants during the season, and after the tourists left, worked as farm laborers. After the evictions and demolition,

[55] He did not go. They separated, she got pregnant while traveling to Peru. But Jay went to Europe. Two seasons later he met a Swedish tourist, who took him to her country. He met her parents, they traveled around Scandinavia and got engaged.

they moved to a neighboring village and the man took a loan from the bank to buy an auto-rickshaw. He started to drive tourists and in this way met his future Japanese wife.

The stories quoted above show that tourism creates great opportunities for young men from low social strata. In what other context would Jay—a petty seasonal salesman who completed no more than six or seven (he does not remember exactly how many) grades of elementary school—have the opportunity to meet and attract an educated Frenchwoman? Tourism in this sense is a space facilitating encounters. But what makes both parties of the tourist contact decide to go beyond the service provider and customer arrangement, and take the risk of building a close relationship despite the differences(s)? My research shows that men involved in the informal tourism sector build their attractiveness in the eyes of foreign tourist women using other capitals than economic and social ones, namely, (a) an appearance perceived as attractive; (b) a different kind of masculinity as compared to the one dominant in the tourist women's countries of origin, and, paradoxically, (c) the subaltern position. Let us take a closer look at these.

In the relationships discussed in this study, the physical difference seems to be attractive for both sides of the contact, something that I witnessed on numerous occasions. Of course, this perception of physical attractiveness is firmly rooted in the cultural constructions of the category of race (Loomba 2015). In India, the cult of white skin is still very much alive, especially with regard to women. Television, billboards, and press are full of adverts of cosmetic products that brighten the skin tone (see Parameswaran and Cardoza 2009). Critics point out that all major female stars of mainstream Indian cinema, that is, Bollywood, have a very bright complexion (and are often the faces of advertising campaigns for skin lightening creams). Matrimonial adverts, in addition to information about caste, age and education, often also mention "fair complexion." Bright skin is, therefore, synonymous not only with beauty, but also with high social status (Mishra and Hall 2017). Ronald E. Hall (1995) calls this phenomenon the "bleaching syndrome" and argues that it is common in many postcolonial countries in Asia and Africa. How deeply internalized is this perception of race by women themselves became evident in the spontaneous statements I had so often heard. My girlfriends from Hampi, describing other women, used to say: "she is so pretty, she has such fair skin" (cf. Hunter 2002 on American women of African and Mexican origin). In the southern states of India, whose inhabitants usually have a darker skin tone than their fellow citizens in the north, the bride's face, hands and sometimes even their entire body is covered with a yellow turmeric paste, which, according to the participants at the weddings I attended, looked beautiful, as "even dark skin becomes fair then." Once, during a lazy afternoon when we watched TV in Geeta's house, I chatted with her elder daughter, who was then just over twenty. We started talking about marriage. I asked her which relationship she thought was better—a "love"-one or arranged. She instantly replied that arranged marriage was better, because love often makes a girl make a bad choice: "she can choose a poor boy or [and this was said with obvious contempt] a black one!" Sangmo, too, when she spoke with loathing about her cousin, who had married an Indian man, added: "And she had such fair skin, more than I do! And she married this black Indian man. I couldn't believe it!" Such associations are, of course, part of the colonial legacy (de L'Estoile 2008) and the mimicry mechanism (Bhabha 1984), that is, the desire of the colonized to become like a colonizer, and thus to overcome their sense of inferiority. For white is the color of power—the colonizers who put themselves in the position of civilizing superiority had fair skin. However, for tourist women, the darker skin tone was the sign of health, vitality and sensuality, as opposed to the metrosexual pallor of men from their native countries. Although positively valued by them, such dark skin perception is also the result of colonialism. The colonial discourse, founded on the binary opposition of nature and culture, attributed the non-European Other to the world of nature and thus to instincts and sexuality (see Loomba 2015).

In the case of male tourists searching for intimacy in the countries of the Global South, researchers point out that sexual gratification is often not their leading motivation at all (Oppermann 1999, 254–255; Sacramento 2015), as these men could achieve it in other ways (such as using paid sex services in their native country or simply picking up a woman). However, the need for a different kind of femininity—more adoring, tender, and devoted—from that currently prevailing in the men's countries of origin plays an important role here.[56] Many men are unable to position themselves within a new model of gender relations, in relationships in which the woman, as a result of emancipatory processes, is an equal economic partner, expressing high demands and pursuing her own goals outside the relationship (Giddens 1992). According to Samuel Veissière, "many Northern men do experience some form of existential disempowerment brought by the erosion of patriarchy and 'traditional' family values in the Northern public sphere, and seek to reaffirm their domination . . . in the Global South" (2011, 31). Many have experienced repeated disappointment in attempts to establish an intimate relationship and difficulties in building a lasting relationship and are afraid to be exposed to further humiliation (Sacramento 2015; O'Connell Davidson 2001; Piscitelli 2004). Guided by Orientalist imagination, they look for the lost ideal of femininity, a woman who holds a man and the house she creates for him at the very center

---

[56] Researchers of the prostitution phenomenon point out that men who pay for sex are also often concerned with "love" and that they feel disappointed due to insufficiently emotional, purely commercial attitudes of prostitutes in Western societies (Kruhse-MountBurton 1995).

of her own existence—whether she be Thai, Brazilian, or Ukrainian. Women, on the other hand, skillfully recognize these needs and respond to them, seeing them as a chance to improve their own economic and social situation. Many studies show that, frequently, it is the privileged party—a tourist from the Global North—who is exploited in these relationships, when he becomes a regular sponsor of the whole family of a woman or one of several "fiancé-sponsors" (Cohen 1993b). It is therefore important to be careful in interpreting the sexual relationships in the area of tourism exclusively in terms of unilateral exploitation, since the dynamics of such relationships tends to be much more complex.

The same mechanism works the other way round. The single tourist women, whom I met during my fieldwork, saw Indian and Tibetan men as devoted, emotional, and sensual—in opposition to indecisive, insecure and reserved men from their countries of origin. Disappointed by their previous relationships with men "who don't know what they want," men they had to strive for, they were delighted to be adored, admired, cared for, especially in a foreign environment (cf. Herold et al. 2011; Dahles and Bras 1999). Moreover, just as the position of tourist women from the "better world" (and the associated capitals) is attractive to men from the Global South, the women themselves—often "new moral tourists" motivated by the desire to save the world (Butcher 2003)—see the disadvantaged situation of their potential partners as creating a moral obligation. This is particularly evident in the context of the Tibetan diaspora, where refugee status is identified with the romantic image of a persecuted freedom fighter who has been deprived of his homeland and separated from his family by the communist regime. Here, tourists assume the role of missionaries, capable of saving those whom they see as victims.

What drives love in the times of tourism is a literary genre that could be described as "travel love story." The autobiographical book *Namma. A Tibetan Love Story* by the British author Kate Karko (2000) is already a classic piece in this category. Karko, with a BA degree in cultural studies and a designer, met her future Tibetan husband, a refugee from Amdo,[57] when he was selling *momo* on the side of the road leading to the main temple in McLeod Ganj. After the wedding, they left together for London, where Tsedup tried to make money by playing supporting roles of Native Americans in movies. A large part of the book is a romantic account of the couple's visit to Tibet—which took place after Tsedup obtained his British

---

[57] One of the three historic provinces of Tibet, located in the Northeast. The vast majority of new refugees who fled to India in the 1990s originated from Amdo and neighboring province of Kham, which—contrary to the central province of Tibet, U-Tsang—were directly incorporated to the Chinese provinces (U-Tsang officially has a status of autonomous region).

citizenship—where Karko became part of her husband's nomadic family. In 2006, the book *Mantras & Misdemeanours. An Accidental Love Story* was published, authored by the New Zealand tourist, Vanessa Walker, with a foreword by the Dalai Lama himself (who treated this book, in accordance with the refugee identity regime, as a story about young Tibetans who "have the responsibility . . . to preserve and promote Tibet's rich spiritual heritage and its unique culture"). Walker also met her future husband in Dharamshala. He also was a newcomer from Tibet—in this case from Kham—and on top of that, he had under his belt a two to three year stay in Sera monastery in southern India. Walker had already noticed the scale of the phenomenon:

> I . . . couldn't fail to notice over the years the extraordinary number of western women, with all the accoutrements of sanity—careers, social lives, beliefs, intelligence—who would take the trip to McLeod Ganj and return with a Tibetan husband, sometimes within weeks. What was it about this place and these people? My own close friend wrenched herself away from a happy, secure relationship in Australia to learn Tibetan for a year and by the time I caught up with her in McLeod Ganj ten months later had succumbed to the charms of a former monk with dodgy English. I heard of a Canadian lawyer who visited for a week, met a Tibetan man and had her nuptials just days later. Was there something in the water? (Walker 2006, xii) . . . For a moment I step out of our private relationship and see us as a cultural phenomenon; after all, I am one of seemingly hundreds of western women in this town who have a Tibetan boyfriend. Part of me is embarrassed that I might be seen as just another western woman who's picked up a Tibetan man along with Buddhism and a packet of prayer flags . . . because that's how I've judged other people in the past. (Walker 2006, 69–70)

What made her go for him? What was this "something"? Above all, it was the refugee poverty. The book contains a description of very basic conditions in a room rented by Choying, which probably did not differ from the one where Pasang lived and to which he took candidates for his sponsors. In addition, there was a tragic life story, especially the escape from Tibet, accepted with stoic (if not Buddhist) attitude. There was also the ability to cook, which was supposed to be alien to Australian men (it all started with Choying inviting Vanessa to his room for a homemade dinner, thus satisfying her desire for "hot authenticity"). Add to this his charming timidity, compared to the more offensive approach of Indian men, and his "Western" look (hip-hop pants, sports shoes), which gave Choying's Otherness a familiar character. The couple went to the tourist's country of origin, where their son was born. Choying took up a job in the factory, like other husbands of Vanessa's friends who went with them to Australia and New Zealand. They were planning to travel to Tibet together, after Choying would acquire New Zealand citizenship.

Seven years later, during my field research, yet another book was published: *Dharamsala Days, Dharamsala Nights: The Unexpected World of the Refugees from Tibet*, self-published by a Canadian woman using the pseudonym Pauline MacDonald (2013). Despite the apparent academic look (the author calls herself a researcher and dubs the process by which she collected materials—research, which could be described as a covert participant observation) and the exposing character of the book that caused a lot of controversy in Dharamshala, as it targeted the idealized global image of the diaspora, the book itself is a love story about a volunteer tourist and a newly arrived refugee from Tibet. It is in his defense that the author accuses the diaspora elites and the India-born refugees of neglecting the refugees from their homeland, who are the main protagonists of her book. Speaking from a position of moral superiority by referring to the category of human rights, MacDonald argues that foreign tourists are fulfilling the role of the Tibetan government-in-exile, offering refugees who have been left to fend for themselves "food, shelter, visas, sex, and, sometimes, enduring love" (with sexuality being strongly emphasized here, as opposed to the other publications mentioned above, which presented a more romantic picture of love). The book presents Dharamshala as a "heterogeneous space," where visitors and locals (though also visitors) can party together, consume cheap Indian whiskey and enter into sexual relationships. In this sense, one can expect that the book will not so much change the perception of Dharamshala and the internal divisions within the refugee community, but fuel touristic imaginations of the need for tourist women coming from the "better world" to save individual refugees.[58]

Every time I returned to Dharamshala it turned out that some of my friends had "disappeared"—I later found out that within a year they met a girl, got married and left abroad. Young Tibetan men, who were looking for contact with foreign tourists visiting McLeod Ganj, became a permanent part of the social reality of the town. This phenomenon was noted, among others, by the young refugee Tibetan cinematography. A movie with a telling title *We're no Monks. A Struggle for Identity*,[59] which could be read as a manifesto of the young generation of Tibetans living in exile, shows young refugees—also those born in India—"hunting" for tourist women brought every day by an interstate bus, who are then shown sitting with young Tibetan men in local

[58] Also, a Japanese woman who married a Hampi auto-rickshaw driver published a book *My Life. Is It Happy or Not?*, which is a printed version of her travel blog. The book is written in Japanese and it is very likely that it will also raise the hopes of Japanese tourist women to find love in the times of tourism.

[59] A feature movie, dir. Pema Dhondup, prod. Clear Mirror Pictures, Wilderness Films India Limited, 2004.

cafés. This phenomenon arouses ambivalent feelings in the community. On the one hand, the conspicuous Tibetans hanging out with their *Inji* partners in McLeod Ganj do not fit well—according to the diaspora's elite, especially the elder generation—with the global image of either a spiritual Buddhist or a modest refugee, and are thus incompatible with the diaspora's identity politics. On the other hand, most of the people I know, although they do not officially approve such practices, realize that the situation of newcomers from Tibet is extremely underprivileged, both legally and economically. In such situation, the relationship with a foreign tourist—in addition to acquiring a sponsor (which is often the case)—is frequently the only way to "escape" from their social destiny (Cohen and Taylor 1992). As Sangmo used to say: "They don't approve, but when it comes to visas, then they approve!" Such "escapes," therefore, can be interpreted as manifestation of agency, which consists in skillfully recognizing the available opportunities within unfavorable circumstances resulting from the subalternity.

However, such relationships are not always synonymous with emigration to the tourist's country of origin. Sometimes, although much less often, the couple tries to make a living in India by starting a business in the tourism sector together, as I have already described in the section of this chapter dedicated to the support received by the local entrepreneurship. Erik Cohen (1971) made the same observations with regard to young Arab men living in Israel. By establishing relationships with tourist women from European countries, they were looking for a way out of their disadvantaged position, that is, little or no chance of economic advancement, political agency or satisfaction of sexual needs. This last aspect is of particular importance in the case of (at least seemingly) conservative societies, where the sphere of sexual relations is highly regulated and often only accessible through marriage. However, for young men at lower ranks of the social hierarchy this is often a distant and uncertain vision.

Of course, not only men enter into love relationships with tourists, although certainly they are much more visible, which gives the impression that such an arrangement prevails. As Sangmo noted: "Guys like to show off, they hang out with Western girls as if they were medals on their chest. They brag, and the girls do not. Girls have boyfriends, but they don't advertise, just as they do with Tibetan boyfriends." The fact that a relationship with a white Western woman is seen as a sign of prestige for men from the Global South, while for women it is more often something embarrassing and socially obscure, is quite common (Herold et al. 2011). Moreover, in countries where the presence of women in the public space is limited, there are also limited possibilities to make contact of this kind (Tucker 2009, 451). Nevertheless, tourism, at least to a certain extent, expands these opportunities, although the close relationships between male tourists and women are sometimes so invisible that I often became aware of their existence only after a long time of doing fieldwork. Finally, more often than in the case of relationships between tourist women and local man, these are legalized.

What does marriage with a tourist give women? It certainly allows to break away from the rigid framework of a conservative society. Above all, it is a chance to escape from social destiny available to women whose marital status is not accepted (divorcee or single mothers) or underprivileged (widows), and/or women originating from a low caste or a marginalized ethnic group (which does not mean that a high status at birth automatically benefits women, as demonstrated by Bhavani's example). Men who decide to have such relationships are generally aware of their role in them. Such couples are more likely to settle in India because Indian women who have no formal education are usually more afraid of moving to an unfamiliar world than men. Moreover, the mere fact of being married and the financial support received from a foreign husband make those women able to function independently of both the family and community. A white, wealthy man does not appear here as a neo-colonizer satisfying his sexual desires with the use of global economic inequalities. On the contrary, I often had the impression that he agreed to be "used" (although sometimes he had a vested interest in it, as I demonstrate below).

Let's examine the relationship between Hans and Sahana. Hans first came to Hampi after the demolition of the main bazaar, in which Sahana's family, who belonged to a low caste of tailors, lost their home and business—a small clothing store. Looking for any possibility to earn money, Sahana began selling coconut water and *chai* near the temple, placing herself at the bottom of the self-employment ladder in the informal tourism sector in India.[60] This is how she met Hans, a young tourist from Germany. They started spending time with each other, and when Hans returned to Hampi after a few months, they got married. They rented a small house in a village some distance from Hampi, so as she could break away from the network of family entanglements, that is, a large family living in a small rented apartment (although in reality he was the one who was more entangled, being the recipient of numerous requests for financial support from Sahana's relatives). No one knew exactly how Hans made a living, although the life he led in the Indian province did not require much money. He knew about electricity and IT, sometimes hired himself for more complicated installations in luxury tourist

[60] Both of these microbusinesses do not require any special investments or skills: fresh coconuts are bought from a farmer and sold to tourists as a cold drink (it is enough to cut the tip of the nut with a machete so that the customer can drink the refreshing water from inside the nut with a straw); to make *chai*, all you need is a kettle, granulated black tea, milk, sugar, some spices and plastic cups.

resorts, sometimes he did IT jobs remotely. Anyway, after the wedding, Sahana no longer had to sell coconuts in front of the temple.

Poor families in India often accept such relationships, also because they allow them to avoid paying dowry. This is all the more so as collecting such large amounts of money often entails incurring obligations to the entire family, or, worse still, the need to use the services of a money lender. Even though dowry has been forbidden in India, it is still practiced by various social groups, although nowadays mainly in the form of expensive gifts that the bride's family gives to the groom and his family.[61] Because it takes such a form, which does not involve money directly, it is difficult to detect and punish. Moreover, the marriage itself—including all the events preceding the main ceremony, to which all the distant relatives, neighbors and important people in the village must be invited—is very expensive. The inability to raise funds for this purpose generally limits the possibility of getting married, or even eliminates such chances at all. Meanwhile, a non-Indian husband not only does not expect this kind of gratification, but often makes a financial contribution to the family budget himself. And a "non-poor" husband, combined with him being "not black"—as Geeta's daughter said—is a prerequisite of a successful relationship. A tourist husband—most often because of his wealthy status resulting from global economic inequalities—satisfies both these conditions.

The second marriage that was concluded in Hampi during my fieldwork was Gabriel marrying Puja, which is a clear example of Puja's agency and her open pursuit of her own interests. Puja met Gabriel during his second tourist stay in Hampi, when she was eighteen. He was then thirty-five. She immediately caught his attention: modern, self-confident, wearing jeans and a T-shirt, she did not resemble the other local young women (Gabriel: "A girl wearing jeans in Hampi! Can you imagine that?!"). They made contact easily. Puja was a child of the tourism informal sector. She never went to school, started working as a ten-year-old, initially selling postcards to tourists—many careers in tourism started that way. At that time, before the digital photography era, it was a great business. She gave all the money she earned back to her parents, herself getting just a few rupees of pocket money. With time she

[61] The groom, depending on his caste and economic status, can expect expensive clothes, a golden watch, a motorcycle and even a car. At this point, it is worth emphasizing that in many lower social strata in India dowry was absent in the past. In the case of castes performing service activities (like laundry and ironing), the value of a woman resulted from her skills and work she did. Paradoxically, it was modernity that contributed to the popularity of dowry, so strongly associated today with the "Indian tradition." Social advancement related to the education of men and the departure from caste professions resulted in women losing their previous role. Moreover, the search for an educated wife—however less educated than a man—made ground to the argument that after marriage she would be financially supported by her husband, for which he was entitled to a dowry (Channa 2004; see also Kapadia 1995).

showed tourists around Hampi, in return asking them to buy postcards, but they started to pay her for the tour. After a few years she became the only female guide in Hampi. An Italian man who lived on the other side of the river taught her English and cheered her on, while an older, deceased guide taught her the history of Hampi. She got used to contacting tourists, with whom she quickly established close relationships. As for Gabriel, it all started when she asked him to take her on a trip around India with him. She wanted to see the world, to be a tourist. At that time, however, they were only friends, and Puja confessed to him that she fell in love with a local boy. Of course, it was a secret relationship, and when she got pregnant, the boy disappeared. What happened next is unknown. People say that she decided to have an abortion, she said she gave birth, but the baby was given to relatives in a distant village and died soon after.

Anyway, Puja lost her chance to get married. It was then that she asked Gabriel to marry her, which he agreed to. He knew that Puja had a mercantile approach to this marriage, but as a social worker, who had for years been involved in helping drug addicts, he was sensitive to situations of social exclusion. He wanted to take her with him to Italy, but she kept postponing her trip to the city to deal with the passport formalities, using various excuses. Finally, it became clear that she did not want to leave. She did not mind that Gabriel lived permanently in Europe. It was enough for her that she was formally married and thus socially "protected" (from potential attempts at exploitation, solicitation, insulting), and could count on her husband's financial support. All the more so because she came from a very poor family of the Valmiki ethnic group, which is considered by the state as a Scheduled Tribe. Her mother, like Sahana's mother, was deceased and her father was addicted to alcohol. She had eight siblings, herself being the third daughter, with two elder sisters married and living in their own houses, while the younger siblings were still children. She and her father were therefore the only breadwinners of the family. They lived—just like the auto-rickshaw driver having a relationship with an Australian woman—in a tent on the outskirts of a resettlement site to which displaced Hampi residents were to be moved. Before the bazaar's demolition, in which they lost their home, Puja's parents used to cut down the grass for the temple elephant and later on sell coconut water. But there was something else. Puja did not fit in with the local pattern of femininity—people were gossiping that she was "boyish": she dressed a bit like a man, did not put on makeup, wandered around by herself, sat with the boys at the *chai* stall after dark, chewed betel and spit. She was offensive, sometimes even aggressive. She got used to being an object of minor mockery, often striking preemptively. Probably, even if it was not for the story with the illegitimate child, she would not have a chance to find a husband in such a small community as Hampi, who would accept her lifestyle. Relationship with a foreign man was therefore strategically the best option in her situation.

In general, however, men who settled permanently in the Hampi area—built houses, planted trees, and had children—were married to Lambani women (those whom Bhavani envied as free and equal with men[62]). For the Lambani, who still function on the fringes of the mainstream society, such a relationship meant a social advancement and, at least in terms of expectations, the source of economic capital.[63] Moreover, a relatively more independent position of women in this group, when compared to Hindu women functioning in the caste system,[64] gave them more freedom in establishing contacts with tourists. I know of cases when a romance between a Lambani woman working in the informal tourism sector and a tourist was an excuse for a woman to end a failed marriage with a Lambani man (despite, for instance, having three small children with him). At the same time, I also know that mothers were bringing their daughters to foreign, white men as candidates for wives. These relationships, in contrast to the other discussed in this section, were characterized by a rather large age difference between spouses. Also, some of these men, having been widowed, remarried a Lambani woman.[65]

The relationships between women and tourist men should also be considered in the context of how these women perceive local men and their own positioning vis-à-vis them. Puja, herself professionally independent and willing to be present in public spaces, rebelled against the fact that "women are kept indoors," that "they cannot do what they want." When I pointed out, for example, that all the *chai* stalls around us were run by women, Puja triumphed: "Well, that's it! They are all single mothers, they must somehow earn money to support their children. But at least they are not dependent on husbands." Bhavani spoke plainly: "I would not recommend any Indian man as a husband. I know that girls from the West like Indians, when I was still running a restaurant, they came to me as I was a woman and told me all about it. But I always told them: 'Go, enjoy, have fun, travel for two or three months, but that's it, then forget about him, no marriage!'" She believed that all such marriages were unhappy and that they all ended up with a divorce, because "Western" women were not fit to be "slaves," "they were free, independent," while Indian men "prefer to have wives from the countryside, because they want them to see their husbands as gods and not question any of

62 I elaborated on this group in chapter 1.

63 I was told a story of a Lambani woman's family, who—after the couple's breakup—demanded a high compensation from the man, threatening him with violence.

64 See chapter 1 on the division of gender roles in mobile business models.

65 Most of the cases of premature deaths in Hampi I know are, as people say, "deaths caused by drinking," possibly due to cirrhosis or other health damage caused by the abuse of cheap, poor-quality alcohol. In the case of the Lambani, the equality also works in this respect.

their actions." As a proof of this claim, she mentioned the widespread practice of the so-called nonresident Indian men, that is, those living and working in the Global North countries, who look for wives in India, obviously through arranged marriages. Bhavani, however, considered the opposite solution, that is, an Indian woman marrying a "Western" man, as ideal: "Without yelling, without beating, giving freedom, doing what you want, traveling . . ." According to this logic, all women who managed to enter into such a relationship were, according to Bhavani, "lucky." In these representations, the Indian man was strongly Orientalized: he appeared primitive, impulsive, dominant and disrespectful of women's rights, while the man from the "West" was his complete opposite and therefore a kind of salvation from a subaltern situation. In the case of Tibetan men, opinions about them are not so unfavorable, which may in part be due to the different status of women in the Tibetan community and the diaspora's narrative of gender equal rights, which I discussed at the beginning of this chapter. Nevertheless, Tibetan women also complained, though mostly, that young Tibetan men today were reluctant to settle permanently and that they only got married (with "Western" women) because they were motivated by the possibility of obtaining a visa.

The dynamics and reciprocity of these relationships are enhanced by the fact that tourists, mentioned so far mainly in the context of satisfying other kinds of needs (emotional, ambitious, etc.), can also benefit "tangibly" from them. It is mainly about those male and female tourists who—for various reasons, although mostly related to lifestyle—decided to change their status to immigrants and settle down in India. Marriage—as I already mentioned in the previous chapter, in the context of migration to "greener pastures"—is the easiest way to legalize one's stay (all the more so because these tourist-turned-migrants, living with funds from their country of origin or developing small business in the informal sector, are not able to obtain the right to stay on the basis of employment). This leads to a situation where a tourist determined to settle in India is willing to pay for the possibility of getting married or formally remains in an unsuccessful relationship due to the legal status it gives him or her (an example here is the Swedish woman, Nelli, who, despite the actual years of separation from her Rajasthani husband, never divorced him). Another important benefit of marrying an Indian citizen is the acquired right to buy land and/or property. Most of the men I know who settled in Hampi and its surroundings "fell in love" with the site—its magic, almost surrealistic landscape and elusive atmosphere—and, as one of them openly admitted, "were looking for a way to stay here." Hampi came first, then there were women. For many of these men, settling down in India was an escape from a life dominated by work and competition. The low cost of living in this part of the Global South made these projects viable, especially since most of these people earned money in one of the currencies of the Global North (playing on the stock market, renting a proprietary apartment, receiving an inheritance, performing a remote IT job, or exporting locally manufactured products). So maybe Gabriel is not "used" at all? When we saw each other last time, he started to promote responsible tourism to India, including Hampi, recruiting clients from among his friends in Italy. Perhaps his future is also connected with India, and his marriage to Puja will make it possible?

Conclusions

While writing this book, I was guided by the desire to overcome fatalism resulting from a momentous impact of Foucault's conceptualizations on social studies of tourism (Cheong and Miller 2000), the desire to liberate ourselves from the "'Foucault effect' which formatted our thinking about the world in terms of knowledge/power, giving no hope of liberation from a world ruled by power relations" (Domańska 2008, 34).[1] My goal was, following the Sherry B. Ortner's (2016) postulate, to overcome the "dark tendencies" dominant in anthropology since the 1980s, to interpret the world through the prism of omnipresent power, exploitation, and inequalities. This optics has led us to a dead end: a paralyzing criticism that sealed us against such categories as empathy, solidarity, care, reciprocity, or closeness, that is, disabled an "anthropologies of aspiration and possibility" (Appadurai 2016, 2013). The postulate to move beyond the "suffering subject" was also formulated in the field of social studies of tourism, where a critical turn sparked work on the "academy of hope" (Ateljevic et al. 2007). I would like this book to be part of the above calls to go beyond the existing approaches by showing the emancipatory potential of tourism.

However, I wish to strongly emphasize that this undertaking is not about ignoring the broader contexts of power. Agency or closeness do not abolish complex relations of dominance and subordination, both of a discursive and very tangible nature related to global economic inequalities, but they are capable of challenging them (Pritchard et al. 2011, 945–946). The anthropological reclaiming of the subaltern voices is not a matter of "slid[ing] into romanticized notions of free choice, individualism, and the apotheosis of

[1] Ewa Domańska quoted a largely forgotten essay by Jean Baudrillard, who as early as in 1977 urged us to "Forget Foucault!"

intimacy" (Herzfeld 2005, 49). I do not want the numerous examples showing tourism as a source of individual empowerment or as a platform for the self-representations of marginalized groups, which I have recalled in this book, to serve as a tool for relativizing various abuses produced by the tourism industry, but I wish them to show how people, at grassroots levels, are able to overcome them and build meaningful relationships across the difference. Social practices escape the neat, carefully constructed categories (Bushell and Staiff 2012, 261), and "the population living in the areas visited by tourists retains the power of judgment and is able to make a spectacular use of it in a very intelligent way" (Wieczorkiewicz 2008, 167).

In other words, I want this book to go beyond the perspective of tourism as a source of economic exploitation, perpetuating social inequalities and unwanted cultural change. This approach leads to intellectual closure and ethical paralysis, leaving little conceptual field for alternative uses of tourist contact (cf. Smith 2006, 41), such as the acquisition of cultural capital or the strengthening of economic and social position through contacts with tourists. This essentially anti-colonial optics, imposing a dichotomous interpretative framework, is in fact "moralizing and elitist" (Routledge 2001, 237) and results in perceiving those on the other side of the tourist contact as passive victims of someone else's ideas and needs. This kind of victimization is discursively prejudicial, because it takes away from those it puts in the subaltern role their power to judge, decide, and act. The only action the subaltern can take is resistance, often carried out in secret in the form of cunning or guile (hence the anthropologists' tendencies to romanticize deception in relations between the subalterns and the privileged, see Simoni 2017). This is an outdated narrative, which reduces the tourist encounter to the framework of power imposition versus subordination/resistance. I think that its potential has already been exhausted.

At the same time, by stigmatizing tourists as hedonistic, superficial, and ignorant, we are like the Western European elites from the early days of tourism, who longed for times when travel was the privilege of the rich before Cook's customers flooded Europe with what they perceived as uneducated, vulgar, and pitiful (Hamilton 2005). Thomas Cook, a "father" of organized, collective travel, was an English Baptist pastor. In 1841, as part of the anti-alcohol movement, he prepared a one-day railroad trip for 570 of his charges, for which participants had to pay one shilling. The idea was to distract laborers from alcohol consumption in smoky factory pubs by providing them with an alternative way of spending their free time and their hard-won right to holidays within the new industrial reality of urbanizing England. The train offered the possibility of mass transportation, which significantly reduced travel costs, making them affordable for laborers. This created a kind of class conflict: the elites defended access to the Scottish hills and the Swiss Alps, comparing Cook's tourists to a herd of animals, whose leader first goes in front, then on the side, and then at the very end, "just like a shepherd dog," to which Cook was supposed to respond that a lack of education does not necessarily mean that tourists do not understand the places they visit; why, he asked, should not the wonders of nature be accessible to everyone? (Hamilton 2005). The development of tourism is, therefore, linked to the struggle for the rights of the lower social classes. It is a kind of democratization of travel (see Withey 1998), which we often forget, denying tourists the ability to authentically experience travel.

Of course, the framework which tourism creates for interpersonal encounters is a challenge if we wish to establish meaningful rapport. There are social and economic inequalities that both parties have to overcome, which often, despite their best intentions, is not easy. The inner struggle with the perception of the Other—but also of oneself—is shown in the reflections of Kristin, a tourist-volunteer in her twenties, who was one of the protagonists in this book:

> When you leave Germany, you think: "Oh, I'm so well prepared, I'm so open-minded, I'm so progressive, politically rather left-wing. I know so many people and the encounter with a different culture will be no problem for me. I'm so tolerant, super-tolerant." And then you come here and you think: "How I hate it!" . . . [They] either try to marry you or to pimp you a cab to the airport. . . . And they all are taking pictures of you. This is so 100% impolite, unmannered, and uneducated. And all these things come to your mind, but you think: "No, no, no! You can't generalize in this way. . . . You can't be that racist, you can't even think like that."

Such feelings often reinforce the need for familiarity and evoke a defensive reaction, which locks tourists in a safety bubble created by other tourists (backpackers, tramping tourists, volunteers, etc.). Paradoxically, this problem often affects people who see themselves as open to other cultures and affirming cultural diversity. Focusing on Otherness makes the tourist gaze fail to see all those elements that connect "us" with "them." Moreover, all the Otherness is then interpreted in cultural terms. Kristin did not allow herself any protection against practices she did not accept, for fear that this would be seen as her lack of openness and tolerance. In this way, however, "touting" was not interpreted by tourists as a local tactic for dealing with global economic inequalities, and "chatting up" a result of a misunderstanding of the cultural codes of clothing or the way female tourists behave, but as part of an "Indian culture" (often referred to as "mentality"). Since culture itself is understood in an essentialist manner as immanent, these characteristics are seen as inalienable and typical of the whole group. The above mechanism also works the other way round. Tourist women's way of being is often interpreted by Indian men as an obvious proof of sexual permissiveness, treated of course as an integral part of an immoral "Western culture." Moreover, tourist women often allowed local men much more than they would have done in their country of origin, not because they were looking for romance or sensual pleasure in tourism, but because they "did not want to offend them [i.e., the Others]." This apparent openness—manifesting as no reaction to supposedly accidental touching—was in turn interpreted by Indian men as sexual openness. As a result, such contact only led to the consolidation of Orientalist-Occidentalist imaginaries.

Moreover, the difficulties in establishing close relations on the grounds of tourist encounters are due to structural factors, that is, the very nature of the trip in which tourists participate. This is shown especially by the research carried out among the participants of the tramping tours, who—despite their declared willingness to "fraternize with the natives" and pointing out the Other as the main goal of their travel—in practice were largely focused on their travel companions, with whom they spent most of their time, often moving from one place to another within the framework of a tight schedule of an organized trip. Nevertheless, even they were able to experience an interaction with the people they met, which they felt as authentic.

The research that has been the basis of this book allowed me to reach many stories which neither support the argument of "distanced friendship" between tourists and workers in the informal tourism sector, nor confirm the exclusively "apparent participation" of the former. This is a story about the grassroots, informal encounters which take place through tourism and the possibilities that arise from them. It is "in the sticky materiality of practical encounters," as Anna L. Tsing (2005, 1) writes, that fruitful "frictions" occur: "A wheel turns because of its encounter with the surface of the road; spinning in the air it goes nowhere. Rubbing two sticks together produces heat and light; one stick alone is just a stick" (Tsing 2005, 5). It is a story of "encounters across difference" (Tsing 2005, 3) which happen between people coming from different cultural, social, economic, and political contexts, about how mutual perceptions are shaped, how self-representations and relationships are constructed, and what implications these encounters have for individual stories of the parties involved. These encounters would never have happened if it had not been for the space created by international tourism, and especially the informal sector which gives the local population, migrants, and refugees a democratic access to tourists, while at the same time allowing them to move beyond the market-driven service provider-customer scheme. Such a bottom-up view demonstrates that it is very simplistic to view the relationship between tourists and those who form the communities they visit solely through the prism of domination and subordination. Tourism in the postcolonial context is certainly embedded in power relations, but people play them differently, for instance, by attempts to position themselves as partners and build relations based on reciprocity. This does not mean abolishing the hegemonic system, but it does constantly undermine it (Kanemasu 2008, 120, 127):

> To be a tourist, but also a tourism service provider, is to be mobile and transient and to become involved, even if only superficially, in the worlds and lives of others. While tourism often stands for the commoditization of a uni-dimensional culture, the exoticization and eroticization of contact with the Other, along with cosmopolitanisms constructed on the foundation of colonialism and orientalism, it can also foster interpersonal relationships which involve genuine intercultural exchanges. (Salazar 2012, 15–16)

On the pages of this book I have shown how tourist encounters take place on the most grassroots level of direct human relations, albeit embedded in a wider, global, postcolonial framework. I have made tourists an object of "native" gaze, looking at how "they" see "us." The map of the world drawn in this way from the perspective of those who are situated on the global "pleasure peripheries" (Turner and Ash 1975) shows that, contrary to the popular beliefs of tourists themselves, they are not perceived solely through the prism of marketability or the Occidentalist optics demonizing "the West" (Buruma and Margalit 2005). Paradoxically, although we are talking about relations built within the economic sector, wealth, and thus potentially the greatest profitability, were not at all the main determinants of the positive valuation of individual groups of tourists. These were the open attitude and cultural sensitivity. Workers and small entrepreneurs of the tourism sector, who became the protagonists of this book, skillfully diversify tourists, recognizing the heterogeneity of this group, often much better than official agencies responsible for promotion and development of tourism. Certainly, the informal nature of the sector, which creates space for communication on an individual level (Macleod 2004, 85), and such personalized contact allow people to avoid generalizing judgments.

The growing presence of domestic tourists also reveals the fluidity of the Self—Other dichotomy, as the Indian Self often appears to be even more different than a foreign tourist. Similarly, the perception of economic inequalities, especially in the context of the post-economic crisis tourists from Europe, shows that workers in the tourism sector are aware of global processes and changes. Above all, however, the overlapping of domestic and foreign tourism largely cancels out the colonial character of the latter, since the availability of tourist spaces for both Indians and foreigners refutes the accusations of the enclavic nature of the "peripheries of pleasure." This is of particular importance in the postcolonial context within the mountain resorts set up by the British and perceived in the past as "white enclaves"—their use by Indian citizens on a par with European tourists is therefore symbolic (Richter 1989, 106). Moreover, domestic tourists come from a rapidly growing middle class and often appear to be more affluent than Europeans, especially since the latter are mostly backpackers, the biggest group of foreign tourists in these spaces.

My research also showed that workers and small entrepreneurs in the tourism sector are very skillful in their use of cultural resources, thus responding to the tourist need for "cool authenticity." In this context, I see the widespread critique of commodification as an attempt—referring to colonial nostalgia—to freeze the Other in pre-modernity. This is not only a paternalistic denial of other people's right to change but also a denial of the right to dispose of their own cultural capital, which is often their only capital (previously appropriated by colonialism and anthropology). In this sense, the positive valorization of local cultures under the influence of tourist interest can be read as opposed to the mimicry of the colonial period as understood by Homi Bhabha (1984), when the colonized wanted to imitate the colonizers in the hope that by adopting their culture and rejecting their own—perceived as inferior—they would become equal to their "masters." The commodification of culture not only means access of marginalized communities to the market economy (Azarya 2004) but is also a source of pride, a tool to maintain or revitalize local traditions, to retell themselves, to create or strengthen a sense of identity (Comaroff and Comaroff 2009). The performances produced for tourism can be affirmative in practice and serve as a basis for the construction of positive self-identification. It is a "culture work" that abolishes the differences between what is materially viable and what is beneficial symbolically (Adams 1996, 219).

Of course, cultural production for the tourist market is a field where different meanings clash and where the right to cultural heritage and economic benefits are negotiated, which is especially relevant in the context of Tibetanness—represented by Buddhism and the Dalai Lama—as a desirable attraction. Hence, people whose ethnic identity is much more complex exposes its Tibetan element in self-representations created within the framework of tourism; hence, the Tibetanization of space or ethnic employment policy is used as a popular practice of authentication. Refugees themselves see the production and trade of Tibetan artifacts as a means of cultural exchange, whose primary goal is to gain international support for the Tibetan cause, in accordance with a modern reinterpretation of the historical "patron-priest" relationship. In this way, Tibetans position themselves as custodians who generously share the unique Tibetan culture and in return expect political support of their struggle for recognition and the right to self-determination.

However, it is becoming more and more difficult for workers and small entrepreneurs in the tourism sector to appeal to the "cool authenticity" of products in the face of the global capitalism. It is much easier to authenticate the services and create a sense of "hot authenticity" by providing opportunities for multisensory experience and establishing direct human relations. All the more so because such forms of tourism as tramping or backpacking create this kind of desire (hence the "live" contact with culture in workshops or cooking classes and joint meals). Both authenticities are socially produced, and we should treat them as cognitive categories with which people organize the world. The question is therefore not whether or not something is authentic, but how certain sights, objects, and experiences become authentic for people.

The field research which I conducted in Hampi and Dharamshala clearly contradicts the conclusions drawn by Kristin Lozanski (2010) that tourist industry workers are "invisible" to tourists as they remind them of the commodified nature of their travels. My observations show that—despite the context of the service provider–consumer arrangement which does not favor the tourist need for authenticity—these workers are very successful in building meaningful relationships with tourists, who also did not seem to be particularly bothered by the above context. Moreover, the high intercultural competence of these workers—acquired precisely through their contact with tourists—did not so much affected their authenticity as it facilitated establishing the relationships, even though Lozanski claimed that the mobility and cosmopolitanism of Others also make them "invisible" to tourists. Workers and small entrepreneurs, with whom I worked, positioned themselves as partners and employed the principle of reciprocity in their relations with tourists, thus overcoming the distance between the service provider and the client (and the power relations inherent in the tourism industry) at the level of grassroots contacts. This was achieved, among other things, through chatting, which was an expression of opposition to commercialized, dehumanized contacts reduced exclusively to monetary transactions. Moreover, perceiving oneself as a helpful friend or an expert in one's own culture allowed tourism workers to build a sense of self-esteem as experts in the culture desired by tourists (which at the same time increased the "cool authenticity" of goods and services offered). On the one hand, chatting with tourists was an opportunity for the hosts to get information about life in other countries and to get in touch with foreigners. The tourists themselves, on the other hand, driven by the need to "fraternize with the natives," were eager to enter into such relations, and the business context of these encounters did not prevent them from feeling them as authentic.

The exchange of gifts and favors, which accompanies the relations built in this way, gives them the reciprocal character, binding both parties and enriching them. This exchange often involves photographs, gifts, parcels (sent after tourists return to their countries of origin), hospitality (including tourists inviting over their befriended tourism workers), or other favors (e.g., tattooing in exchange for sightseeing). Changing practices of photographing, and especially the popularity of selfies, make taking pictures no longer a distancing and objectifying activity. The division of roles between the photographer and the poser becomes blurred, and the act of taking photos itself often becomes a way of establishing relationships. Moreover, the growing number of domestic tourists makes foreign tourists themselves the object of their own gaze, which is often interpreted as an exchange. Finally, the availability of relatively cheap smartphones and the prevalence of Internet messengers change the inherently temporary nature of the relationships built in this way, allowing them to last beyond the limits of holiday time, which forces us to redefine the category of "distanced friendship" (Wieczorkiewicz 2008, 207). All the more so because this virtual-visual closeness is inclusive in nature, including in the contact also people who are illiterate.

Direct contact with tourists enabled by the informal sector allows workers and small entrepreneurs to strengthen their own intercultural competences and act as cosmopolitan natives, that is, to construct cosmopolitan identities while staying put in a place, beyond spatial mobility with which such identities are usually associated. This allows us to conceptually overcome the dichotomy of rootedness in territory/culture and cosmopolitan openness (Appiah 2006), so deeply grounded in the way guests and hosts in tourism studies are perceived (Glick Schiller and Salazar 2013, 187). The fact that the workers of the tourism sector become participants in the world of mobile meanings and ideas can also be interpreted as traveling-in-dwelling (Clifford 1992, 108, 1997), that is, enjoying the cultural diversity of the world without having to move physically. Due to this kind of cultural exchange tourism workers do not perceive their relations with tourists as one-sided and themselves as inferior within these relationships. Emphasizing their own high competence in the "tourist culture" is often accompanied by a narrative of responsibility for tourists—their well-being and safety, especially in relation to tourist women. My research partners positioned themselves not only as experts in satisfying tourist needs but also as caring guardians, thus responding to the tourist need for safety and familiarity in a socially and culturally unfamiliar reality. Daniel Heuman (2005) points out that such aspects of relationship as care, exchange of gifts and respect, which are present in small-scale tourism activities, are features of the so-called traditional hospitality.

It is also characteristic that, just like in the ideas about particular groups of tourists, also in terms of building relations with them, workers and small entrepreneurs of the informal tourism sector cut themselves off from the capitalist nature of tourism through the category of tranquility. Often they were motivated by other-than-economic motivations and did not treat profit maximization as the main determinant of their decisions and strategies. Just as profitability was not the key criterion for their preferences related to the types of tourists—with post-economic crisis "Europeans" as their favorite guests—profitability was not the key criterion for choosing a place of business. Especially mobile workers and small entrepreneurs in Hampi deliberately gave up their jobs in Goa, indicating as reasons the activities of the local mafia, the widespread demand for commissions, corrupt police, and greedy and dishonest residents. Tranquility was the most important criterion for their satisfaction.

At the community level, this book shows how groups that found themselves in a subaltern position—as a result of exile (in the case of Dharamshala) or evictions (in the case of Hampi)—try to use their own attractiveness in the field of tourism and direct contacts established with tourists to resist oppression and carry out their political struggle. The tourists, seen in this context as representatives of the powerful "West," become allies in this struggle—the recipients of self-presentations and demands and then their carriers. Here tourism becomes a means to gain global moral and political support. Apart from the effectiveness of this strategy in terms of actual change in the situation of both communities, the actions of Tibetan refugees and displaced Hampi residents show that tourism can be a platform for the recovery of the subaltern voices, as they are heard and positively received by tourists effectively engaged in political activities for both communities. Although the language used for this purpose derives from an international dictionary (of human rights, the right to self-determination or cultural heritage), it has been efficiently appropriated by "muted groups" (Ardener 1975a, 1975b) to overcome their disadvantaged position.

In Dharamshala the political appropriation of tourism has already taken an institutionalized form and is being implemented in several areas. The first is the extensive production of "political souvenirs" with the "Free Tibet" logo—by purchasing, wearing, and giving them to others, tourists make the political goals of the Tibetan diaspora audible. And even if they do not buy such souvenirs, they are still recipients of their messages, because the tourist space of Dharamshala is saturated with political declarations. Thus, the production for the tourist market has in fact two recipients: tourists themselves and also members of the host community within the project of nation-making in exile. It is accompanied by a rich offer of educational events (documentary film screenings, meetings with former political prisoners, "awareness talks," etc.). For several years now, political reality tours have also been organized, with programs including encounters with local activists, visits to social institutions, talks with the victims of torture, and talks even with the head of the government-in-exile; their participants are also directly involved in organizing political events. Educating the dominant group about the situation and aspirations of the dominated group by the underprivileged themselves is indicated as one of the ways to extract muted narratives (Obre 1995). Wherein, the expectation of political support can be seen as a reinterpretation of the historical "patron-priest" relationship—refugees offer to the world the ancient heritage of Tibetan Buddhism, expecting in return the support for their cause (Klieger 1992). The tourists, perceived as representatives of powerful political players, have been involved in this peculiar relationship of reciprocity, which in turn allows Tibetans to position themselves as equal partners in this symbolic exchange and not as poor refugees begging for help.

Hampi, in turn, is an example of how the elites of a postcolonial state, in the name of protecting material cultural heritage, silence the voices of certain groups of citizens, stigmatizing them as illegal encroachers who economically exploit the national treasure. Paradoxically, it is the tourists, often from former colonial metropolises, who have become allies of the displaced inhabitants in their struggle for their own vision of "living heritage" as opposed to "dead monuments," the right to housing and the right to satisfy their livelihood through work in tourism. Unlike in Dharamshala, however, in Hampi a transnational solidarity-based activism for justice is only just being born, and it is the initiative of the tourists themselves. It is the tourists who appear in the media, organize social media campaigns, make films, and write petitions to international organizations and the Indian government to stop the evictions from the village. In this sense—and again in opposition to Lozanski's claims—they are part of the local community's struggle for a bottom-up pro-poor tourism that has allowed the people of Hampi to turn from poor farm laborers into small entrepreneurs benefiting—both materially and immaterially—from the presence of tourists. The main idea behind pro-poor tourism is to strengthen competences and support local entrepreneurship, something which should be done by the state in collaboration with the private sector and grassroots organizations. In the case of Hampi, not only was there no such support—the inhabitants themselves developed the whole tourist infrastructure and service sector—but their activities were stigmatized by the authorities as illegal. The remaining residents are in danger of losing their small businesses and the already-displaced inhabitants were not offered any real alternative for meeting their families' needs.

Tourists are also engaged in direct actions for the social goals of both communities. This is done primarily through volunteering within the nongovernmental sector, often developed thanks to the tourists themselves and their capital. In Dharamshala, they are small, grassroots organizations focused on strengthening refugees' competences in foreign languages and new technologies. These skills enable participation in the virtual world, promotion of Buddhism, direct access to tourists, and mobility (e.g., further emigration to the countries of the so-called Global North). The scale of this phenomenon is so large that it has led to an oversupply of volunteers. Their motivations are very different, and the experience itself is perceived differently depending on the type and time of work performed, the so-called beneficiaries, and the biographies of tourists themselves. What they have in common, however, is their declarative disconnection from their mass alter ego, who only "consumes" places visually (Urry 1995), the desire to satisfy the need for "hot authenticity" by establishing direct rapport with the residents, albeit in a secure institutional framework, and the need for self-development or time to plan for the future. Paradoxically, it was the most committed volunteers who turned out to be the most critical and who more often than others adopted the attitude of "friendly moral ethnocentrics," leaning toward rationalization and modernization projects in the spirit of civilizing mission (Podemski 2005).

Many studies have addressed volunteer tourism and attempts to construct a "new moral tourist" (Butcher 2003), especially in the postcolonial contexts. Its critique includes, among others, continuing the colonial civilizing mission, paternalistic construction of the "ideal victim," imposing the idea of modernization on the rest of the world, maintaining global economic inequalities (since it is poverty in a depoliticized and aestheticized form that is the main attraction for tourists), transforming questions about the institutional causes of injustice into questions about individual morality, leakage to institutions that have specialized in mediating the recruitment of volunteers for local organizations which do not have such opportunities, that is, in short, perpetuating the neoliberal capitalist order on which the global North-South division is founded. Furthermore, my research in Dharamshala revealed many drawbacks of this kind of tourism: lack of cultural competence of volunteers, their random selection, the fact that they are increasingly young and lack didactic experience, and finally a big turnaround of both volunteers and their students. However, such criticism leads us nowhere, as I wrote at the beginning of this section in the broader context of critiques concerning the whole area of tourism. It is based upon generalizations and does not allow us to see other, bottom-up benefits. Autonomy in terms of organization—direct acquisition of volunteers and independent target setting by local organizations—reduces the leakage and hegemonic nature of development projects. Poverty does not have to be the main attraction of this form of tourism (and in the case of Tibetan refugees it is not). Even if volunteers have selfish motives or are not competent enough, their presence may bring other benefits (e.g., volunteers are also recipients of political messages of the diaspora). The organizations which rely on the work of volunteers have become a platform for talking about social problems not addressed by the diaspora and about the economic self-sufficiency of individuals, so far perceived as an aspiration unworthy of a refugee.

Above all, however, volunteer tourism, and especially language classes with tourists arranged by nongovernmental organizations create spaces for direct, informal contact with foreigners. This is particularly important for those who are not employees of the tourism sector and have limited access to foreigners and the potential benefits it brings. What is more, these are spaces which allow to establish close relationships—mainly with so-called conversation partners—which, when developed into friendship or even love, may in the future bring in a real change in the life of individuals (such as obtaining emotional, instrumental, or financial support, the possibility to emigrate to "greener pastures," etc.). In Hampi, which has not yet built stable, institutional forms of volunteer tourism, the political commitment of tourists and their direct actions for the local community intertwine, just as grassroots social initiatives intertwine with work in the tourism sector. Contacts with tourists and the opportunities they offer in terms of social, cultural, and economic capital, inspire the inhabitants—who are skilled in moving between the charity and tourism—to undertake prosocial activities. Volunteer tourism, reality tours, pro-poor tourism, or transnational solidarity–based activism (i.e., the active support by tourists of local communities whose human rights, including political and economic ones, have been violated) are forms of tourism which, according to Freya Higgins-Desbiolles (2010), have the potential to bring justice. Or, at least, they give a sense of moral support to those who find themselves in a situation of subalternity and hope that their voice will be heard.

Personal relationships developed with tourists can create opportunities for individual empowerment. The stories I told in the last chapter demonstrate the social, cultural, or economic capital that is accumulated through relationships with tourists—in the form of diversification of social networks, intercultural competences, or material support—has the potential to overcome the limitations faced by people belonging to groups in a situation of structural subordination because of their gender, social position, or political status (such as women in patriarchal systems, people from the lowest social strata, or newly arrived refugees). Democratic access to heterogeneous tourist spaces makes it possible to obtain informal sponsorship, support for local business development (in the form of know-how, investments, or just inspiring ideas), or to find a partner/spouse. All this can result in a real change in the quality of life through economic improvement and/or social mobility. In this perspective, developing contacts with foreign tourists may be an emancipatory strategy for women (especially single mothers, divorcees, widows, women from marginalized communities or from very poor families who cannot afford a dowry) and a chance for social advancement for young men, finding themselves at the bottom of the social hierarchy. The more so as the informal sector offers a number of jobs and activities that do not require formal qualifications and relationships with tourists make it possible to acquire alternative social networks and cultural capital, on which in turn a sense of satisfaction and self-esteem is built. The relationship between tourism and gender should not be read only through binary categories of domination and subordination. The dynamics of such relationships are much more complex, and love has the potential to overcome structural limitations. Often it is also an impulse to start one's own business in the tourism sector or to emigrate abroad, which in both cases may involve social mobility.

Skillful building of close relations with tourists shows, therefore, that the so-called hosts are not passive objects of the tourist gaze, subject to external domination, but "active agents responding to the opportunities available to them" (Lozanski 2010, 751). These relationships are often based on the principle of reciprocity, bringing various benefits to both parties, both tangible (earnings, discounts, good quality goods, gifts, sponsorship, investments, the possibility of emigration, or legalization of stay, etc.) and intangible (emotional support, self-esteem, cultural competences, satisfaction of sexual needs, sense of security, etc.).

In conclusion, while conducting a fieldwork, my attention was often drawn to the discrepancy between the way in which critics of tourism presented it as a continuation of colonial relations and the way in which it was perceived by the workers and small entrepreneurs of the informal tourism sector in Hampi and Dharamshala. The latter saw tourists as allies in an unequal political struggle—against the occupier, their own government, or international regimes—and, on a more individual level, as an opportunity to improve their living conditions or advance in society. The tourists themselves, on the other hand, were fully aware of the framework in which they and the others were operating, and yet were able to build close, reciprocal relationships with the people they met during their trips, contrary to the widespread concepts of a blinding desire for authenticity making tourist industry workers "invisible" to tourists. The most striking example was the involvement of tourists in the development of local entrepreneurship in the area of tourism—through inspiration, motivation, or material support—aiming to overcome economic inequalities from the bottom up. These tourists understood that tourism serves not only to satisfy their cultural needs but also to secure the livelihoods of those who act as hosts. Feeling a moral obligation based on reciprocity, they diminished the distance between themselves and their hosts and engaged in both instrumental and emotional support. In this way, they redefined the moral basis of the tourist encounters and the power relations inscribed in them.

Such reinterpretation of the host-guest relations requires, of course, certain structural conditions. First of all, it is certainly supported by the diversified nature of the tourism sector in India. Tourism in India does not have the classical form of three "s" tourism (i.e., sun, sand, and sea), covering a wide range of attractions (cultural, historical, ethnic, religious, natural, etc.) and activities (trekking, visits to national parks and nature reserves, participation in festivals, pilgrimages, volunteering, participation in self-development courses, etc.). Moreover, a large share of individual and long-term tourism within the sector certainly fosters the establishment of close relations between the two sides of the tourist contact. Secondly, equal access to tourist spaces for foreign and Indian tourists—and generally highly developed domestic tourism—means that there is no resentment and consequently sabotage of the development of enclavic tourism, as was the case in Jamaica or the Philippines (Richter 1989, 106). Thirdly, it is the informal sector, which still dominates in many destinations, serving the so-called budget tourists, that allows for the development of unmediated relations between tourists and workers/small entrepreneurs in the tourism sector. This sector provides relatively broad, democratic access to tourists, increasing the possibilities of participation in the benefits it generates by reducing leakage, as money spent by tourists goes directly to local communities (Wilson 1997b). Moreover, flexibility of small entrepreneurs in terms of time management and the blurring boundaries between the public and domestic spheres within the family businesses allow people to go beyond the rigid framework of client–service provider relationship. This, in turn, allows for very different actors to accumulate intercultural competences necessary for working with tourists, which makes many more people into intermediaries benefiting from access to tourists and their capitals (cultural, social, or economic).

At the same time, the relations between guests and hosts described in this book should also be read through the lens of transformations happening within tourism itself. The tourist's involvement in political struggle or direct actions for the benefit of the local community is part of the project of "moralizing tourism" (Butcher 2003), that is, the attempts to give tourism a deeper meaning, to make it useful in a wider sense than just as a source of material benefits for individual service providers. Moreover, I agree with the proponents of pro-poor tourism that the idea itself does not necessarily require creation of alternative forms of tourism (e.g., ecotourism). Often it is enough to transform the mainstream tourism in such a way that it brings stable benefits to those who are on the other side of the tourist encounter and can be a space for developing meaningful relationships. It is certainly crucial to increase participation, that is, the possibility to offer goods and services to tourists directly by "small players," and to give up the attempts of the state and large business to overcontrol the relations that hosts establish with tourists. In both sites analyzed in this book, I could observe bottom-up initiatives which were conducive to individual and community empowerment without the intervention of the state (or precisely because of its lack), while in Hampi such intervention served to suppress such grass-roots initiatives.

Unfortunately, as I have already indicated in the introduction to this book, in the last two decades the Indian authorities have been clearly striving for increasing regulatory measures or even limiting the scope of both the informal sector and low-budget tourism, favoring a more luxurious type of tourist services, which the government agencies seem to think is what wealthy clients (and the high-end hotel lobby as well as politicians involved in it) expect. In its recent tourism development strategy, the Indian Ministry of Tourism planned to focus on "niche tourism," involving such "products" as luxury cruises, medical tourism, wellness tourism, golf tourism, polo tourism, and conference tourism for businesses (Ministry of Tourism 2017, 5). Such tourism is to be standardized and developed to international standards. It is to be operated by "trained manpower" (e.g., training program launched to prepare "delivery boys" to serve food "with skill, style and civility")[2] to satisfy "quality standards" in terms of "cleanliness, hygiene and upgraded facilities and practices," while the development of the industry is to entail "large investments" and "world class infrastructure" (Ministry of Tourism 2017, 6–7, 14, 18, 22, 39, 52–53). The Ministry declared it would follow a stakeholder collaboration model, but stakeholders here are understood to be not local communities or even village or town councils, but various central and state government agencies and powerful umbrella and industry organizations, such as the Federation of Indian Chambers of Commerce and Industry, the Federation of Hotel & Restaurant Association of India, or the Travel Agents Association of India (Ministry of Tourism 2017, 14–15). Neither are small entrepreneurs in the informal sector—that is, the protagonists of this book—members of such organizations nor are nongovernmental organizations working for equal access to tourism, such as nationwide Equations, which supported Hampi residents in their struggle. According to the Ministry of Tourism, all the above changes will increase employment, increase economic development, and contribute to poverty eradication (2017, 14, 18, 20). Nevertheless, the example of Hampi shows that the narrative of standardization, professionalization, and the need to improve the quality of services is often used to the detriment of small entrepreneurs in the informal sector and tends to have the opposite effect to the goals enumerated above. Arun Saldanha and Anthony D'Andrea have conducted research on the informal tourism sector in Goa and both agree that the ban on outdoor events actually wiped out the international fans of the Goa trance scene, which affected

---

[2] Rickshaw drivers, boatmen, and street vendors from Benares have been covered by the "sensitization program" in order to "achieve an improved tourist specific ambience" (Ministry of Tourism 2017, 53).

mostly the local people for whom these tourists were the main source of income, that is, women selling chai at events, young men renting motorcycles, mechanics fixing them, owners of Internet cafés, beach bars, bakeries, and small guesthouses (Saldanha 2007, 184, 202; D'Andrea 2007, 196). Unfortunately, this might be a scenario for both Hampi and Dharamshala.

References

Abu-Lughod, Lila. 1990. "Writing Against Culture." In *Recapturing Anthropology: Working in the Present*, edited by Richard G. Fox, 137–162. Santa Fe: School of American Research Press.

Adams, Kathleen M. 1984. "Come to Tana Toraja, 'Land of the Heavenly Kings': Travel Agents as Brokers in Ethnicity." *Annals of Tourism Research* 11(3): 469–485.

———. 1999. "Taming Traditions: Torajan Ethnic Imagery in the Age of Tourism." In *Converging Interests: Traders, Travelers and Tourists in Southeast Asia*, edited by Jill Forshee, Christina Fink, and Sandra Cate, 249–261. Berkeley: University of California Press.

Adams, Vincanne. 1992. "Tourism and Sherpas, Nepal: Reconstruction of Reciprocity." *Annals of Tourism Research* 19(3): 534–554.

———. 1996. *Tigers of the Snow and Other Virtual Sherpas: An Ethnography of Himalayan Encounters*. Princeton: Princeton University Press.

Adler, Judith. 1985. "Youth on the Road: Reflections on the History of Tramping." *Annals of Tourism Research* 12(3): 335–354.

Agar, Michael. 1980. *The Professional Stranger: An Informal Introduction to Ethnography*. New York: Academic Press.

Ageenko, Dmitry S., George S. Papazyan, and Andrey Apukhtin. 2013. "Portrait of a Russian Tourist. Myths and Realities." *European Researcher* 39(1–2): 128–130.

Aitchison, Cara. 2001. "Theorizing Other Discourses of Tourism, Gender and Culture: Can the Subaltern Speak (in Tourism)?" *Tourist Studies* 1(2): 133–147.

Aitken, Carmen, and Colin Michael Hall. 2000. "Migrant and Foreign Skills and Their Relevance to the Tourism Industry." *Tourism Geographies* 2(3): 66–86.

Albers, Patricia C., and William R. James. 1983. "Tourism and the Changing Photographic Image of the Great Lakes Indians." *Annals of Tourism Research* 10: 123–148.

———. 1988 "Travel Photography: A Methodological Approach." *Annals of Tourism Research* 15: 134–158.

Alvesson, Mats, and Kaj Sköldberg. 2000. *Reflexive Methodology: New Vistas for Qualitative Research*. London: Sage Publications.
Amit, Vered. 2000. "Introduction: Constructing the Field." In *Constructing the Field: Ethnographic Fieldwork in the Contemporary World*, edited by Vered Amit, 1–18. London and New York: Routledge.
Anderson, Vivien, Richard Prentice, and Kazumasa Watanabe. 2000. "Journeys for Experiences. Japanese Independent Travelers in Scotland." *Journal of Travel & Tourism Marketing* 9(1/2): 129–151.
Ang, Ien, and Jon Stratton. 1995. "The Singapore Way of Multiculturalism: Western Concepts and Asian Cultures." *Sojourn: Journal of Social Issues in Southeast Asia* 10(1): 65–89.
Appadurai, Arjun. 2013. *The Future as Cultural Fact: Essays on the Global Condition*. London and New York: Verso.
———. 2016. "Moodswings in the Anthropology of the Emerging Future." *Hau: Journal of Ethnographic Theory* 6(2): 1–4.
Appiah, Kwame Anthony. 2006. *Cosmopolitanism: Ethics in a World of Strangers*. New York: W.W. Norton.
Aramberri, Julio. 2001. "The Host Should Get Lost: Paradigms in the Tourism Theory." *Annals of Tourism Research* 28(3): 738–761.
Araoz, Gustavo F. 2011. "Preserving Heritage Places Under a New Paradigm." *Journal of Cultural Heritage Management and Sustainable Development* 1(1): 55–60.
Ardener, Edwin. 1975a. "Belief and the Problem of Women." In *Perceiving Women*, edited by Shirley Ardener, 1–17. London: Malaby Press.
———. 1975b. "The 'Problem' Revisited." In *Perceiving Women*, edited by Shirley Ardener, 19–27. London: Malaby Press.
Ashley, Caroline, and Dilys Roe. 2002. "Making Tourism Work for the Poor: Strategies and Challenges in Southern Africa." *Development Southern Africa* 19(1): 61–82.
Ashley, Caroline, Dilys Roe, and Harold Goodwin. 2001. *Pro-Poor Tourism Strategies: Making Tourism Work for the Poor—A Review of Experience*. London: Overseas Development Institute, International Institute for Responsible Tourism and Centre for Responsible Tourism.
Ateljevic, Irena, Annette Pritchard, and Nigel J. Morgan. 2007. "Editors' Introduction: Promoting an Academy of Hope in Tourism Enquiry." In *The Critical Turn in Tourism Studies: Innovative Research Methodologies*, edited by Irena Ateljevic, Annette Pritchard, and Nigel J. Morgan, 1–11. Oxford: Elsevier.
Ateljevic, Irena, Candice Harris, Erica Wilson, and Francis Leo Collins. 2005. "Getting 'Entangled.' Reflexivity and the 'Critical Turn' in Tourism Studies." *Tourism Recreation Research* 30(2): 9–21.
Ateljevic, Irena, and Stephen Doorne. 2000. "'Staying Within the Fence': Lifestyle Entrepreneurship in Tourism." *Journal of Sustainable Tourism* 8(5): 378–392.
Azarya, Victor. 2004. "Globalization and International Tourism in Developing Countries: Marginality as a Commercial Commodity." *Current Sociology* 52: 949–967.

Aziz, Barbara Nimri. 1978. *Tibetan Frontier Families. Reflections of Three Generations from D'ing-ri*. Durham: Carolina Academic Press.

Bachórz, Agata. 2013. *Rosja w tekście i doświadczeniu. Analiza współczesnych polskich relacji z podroży* [*Russia in Text and Experience: Analysis of Contemporary Polish Travel Narratives*]. Kraków: NOMOS.

Banaszkiewicz, Magdalena. 2012. *Dialog międzykulturowy w turystyce. Przypadek polsko-rosyjski* [*Intercultural Dialogue in Tourism: The Polish-Russian Case*]. Kraków: Wydawnictwo Uniwersytetu Jagiellońskiego.

Banerjee, Poulomi. 2015. "No More Vodka on the Beach: Russian Tourists Shy Away from Goa." *Hindustan Times*, April 12, 2015. https://www.hindustantimes.com/india/no-more-vodka-on-the-beach-russian-tourists-shy-away-from-goa/story-qy6BohmoTcdl1ibWFAqxoJ.html.

Barretto, Lita. 2010. "Russians Face Backlash in India's Goa." *AFP*, March 8, 2010. org/web/20101012155358/https://www.google.com/hostednews/afp/article/ALeqM5j5QIbFrGQ-jtg-uUMxlJcbpto0ug.

Barthes, Roland. 1972. *Mythologies*. Translated by Annette Lavers. New York: Hill and Wang.

Baudrillard, Jean. 1977. *Oublier Foucault*. Paris: Galilée.

Bell, Charles. 1928. *The People of Tibet*. Oxford: Clarendon Press.

Bhabha, Homi K. 1984. "Of Mimicry and Man: The Ambivalence of Colonial Discourse." *October* (*Discipleship: A Special Issue on Psychoanalysis*) 28: 125–133.

Bhardwaj, Surinder Mohan. 1973. *Hindu Places of Pilgrimage in India: A Study in Cultural Geography*. New Delhi: Thomson Press.

Bhargava, Yuthika. 2017. "New 'Incredible India 2.0 Campaign' on Anvil." *The Hindu*, June 24, 2017. http://www.thehindu.com/business/new-incredible-india-20-campaign-on-anvil/article19141347.ece.

Bhattacharyya, Deborah P. 1997. "Mediating India: An Analysis of a Guidebook." *Annals of Tourism Research* 24(2): 371–389.

Bianchi, Raoul V. 2000. "Migrant Tourist-Workers: Exploring the 'Contact Zones' of Post-Industrial Tourism." *Current Issues in Tourism* 3(2): 107–187.

Binder, Jana. 2004. "The Whole Point of Backpacking: Anthropological Perspectives on the Characteristics of Backpacking." In *The Global Nomad. Backpacker Travel in Theory and Practice*, edited by Greg Richards and Julie Wilson, 92–108. Clevedon: Channel View Publications.

Bishop, Peter. 1989. *The Myth of Shangri-La: Tibet, Travel Writing and the Western Creation of Sacred Landscape*. Berkley: University of California Press.

Bissell, William Cunningham. 2005. "Engaging Colonial Nostalgia." *Cultural Anthropology* 20(2): 215–248.

Bloch, Natalia. 2008. "The Young Tibetan Exile Today: A Democrat, Environmentalist, Vegetarian? Globalization and Negotiating Identity in the Situation of Displacement." In *Exploring Home, Neighboring and Distant Cultures*, edited by Lech Mróz and Aleksander Posern-Zieliński, 239–258. Warszawa: Wydawnictwo DiG.

———. 2010–11. "Between Vishnu and Mao Zedong: The Maoist Movement in Multi-Ethnic Nepal." *Ethnologia Polona* 31–32: 169–189.

———. 2011. *Urodzeni uchodźcy. Tożsamość pokolenia młodych Tybetańczyków w Indiach* [*Born Refugees: The Identity of the Young Generation of Tibetans in India*]. Wrocław: Wydawnictwo Uniwersytetu Wrocławskiego.

———. 2014a. "Constructing Borders within Diaspora. 'Born Refugees,' Newcomers and Bargaining Tibetan Identity." In *Middle Grounds, Ambiguous Frontiers and Intercultural Spaces*, edited by Aleksander Posern-Zieliński and Lech Mróz, 135–154. Poznań: Instytut im. Oskara Kolberga.

———. 2014b. "Kolonizatorzy, turyści, antropolodzy. Dziedzictwo kolonialne w turystyce i kolonialna nostalgia w antropologii" [Colonizers, Tourists, Anthropologists. Colonial Legacy in Tourism and Colonial Nostalgia in Anthropology]. *Konteksty* LXVIII(1): 187–195.

———. 2016a. "Overheating with Authenticity: Between Familiarity and Otherness in Multisensory Experiencing of India by Tramping Tourists." *Déja Lu* 4. https://www.waunet.org/wcaa/dejalu/issue4.phtml.

———. 2016b. "O pożytkach z przypadków szwendania się i nomadologii. Autorefleksyjny reportaż antropologiczny" [On the Advantages of Serendipity, Nomadology and Hanging-Out: Self-Reflexive Anthropological Reportage]. In *Auto/biograficzne aspekty praktyk poznawczych* [*Auto/Biographical Aspects of Cognitive Practices*], edited by Marcin Kafar, 63–87. Łódź: Wydawnictwo Uniwersytetu Łódzkiego.

———. 2016c. "Evicting Heritage: Spatial Cleansing and Cultural Legacy at the Hampi UNESCO Site in India." *Critical Asian Studies* 48(4): 556–578.

———. 2017a. "Barbarians in India: Tourism as Moral Contamination." *Annals of Tourism Research* 62: 64–77.

———. 2017b. "Taxonomic Panic and the Art of 'Making Do' at a Heritage Site: The Case of Hampi UNESCO Site, India." *Anthropological Notebooks* 23(3): 19–44.

———. 2018. "Making Tourists Engaged by Vulnerable Communities in India." In *Anthropology of Tourism in Central and Eastern Europe: Bridging Worlds*, edited by Sabina Owsianowska and Magdalena Banaszkiewicz, 181–197. Lanham: Lexington Books.

———. 2019. "Refugees as Donors: 'Rich' Tibetan Refugees, Evicted Indian Slum Dwellers and a Smart City." *Journal of Refugee Studies*. DOI: 10.1093/jrs/fez106.

Blundell, David. 2005. "The Traveling Seminar: An Experiment in Cross-Cultural Tourism and Education in Taiwan." *NAPA Bulletin* 23(Special Issue: Tourism and Applied Anthropologists): 234–251.

Bobako, Monika. 2011. "Płeć, rasa, seksualność w kolonialnych ekonomiach władzy" [Gender, Race, Sexuality in Colonial Economies of Power]. *Nowa Krytyka* 26/27: 83–112.

Boissevain, Jeremy. 1996a. "Ritual, Tourism and Cultural Commoditization in Malta: Culture by Pound?" In *The Tourist Image. Myths and Myth Making in Tourism*, edited by Tom Selwyn, 105–120. Chichester: Wiley.

———, ed. 1996b. *Coping with Tourists: European Reactions to Mass Tourism*. New York and Oxford: Berghahn Books.

Boorstin, Daniel J. 1962. *The Image: A Guide to Pseudo-Events in America*. New York: Atheneum.

Bourdieu, Pierre. 1986. "The Forms of Capital." In *Handbook of Theory and Research for Sociology of Education*, edited by John G. Richardson, 241–258. New York: Greenwood Press.

Boyne, Steven, Fiona Carswell, and Derek Hall. 2002. "Reconceptualising VFR Tourism. Friends, Relatives and Migration in a Domestic Context." In *Tourism and Migration: New Relationships Between Production and Consumption*, edited by Colin Michael Hall and Allan M. Williams, 241–256. Dordrecht: Kluwer Academic Publishers.

Brettell, Caroline B. 2002. "The Individual/ Agent and Culture/Structure in the History of the Social Sciences." *Social Science History* 26(3): 429–445.

Britton, Stephen G. 1982. "The Political Economy of Tourism in the Third World." *Annals of Tourism Research* 9(3): 331–329.

Bruner, Edward M. 1989. "Of Cannibals, Tourists, and Ethnographers." *Cultural Anthropology* 4(4): 438–445.

———. 1995. "The Ethnographer/Tourist in Indonesia." In *International Tourism: Identity and Change*, edited by Marie-Françoise Lanfant, John B. Allcock, and Edward M. Bruner, 224–241. London: Sage Publications.

———. 1999. "Abraham Lincoln as Authentic Reproduction: A Critique of Postmodernism." *American Anthropologist* 96(2): 397–415.

———. 2005. *Culture on Tour: Ethnographies of Travel*. Chicago and London: The University of Chicago Press.

Buchowski, Michał. 2005. "Ku odpowiedzialnej antropologii" [Towards a Responsible Anthropology]. *(op.cit.,)* 1(22): 6.

Bulag, Uradyn E. 2012. "Seeing Like a Minority: Political Tourism and the Struggle for Recognition in China." *Journal of Current Chinese Affairs* 41(4): 133–158.

Burman, J. J. Roy. 2010. *Ethnography of a Denotified Tribe: The Laman Banjara*. New Delhi: Mittal Publications.

Burns, Georgette Leah. 2004. "Anthropology and Tourism: Past Contributions and Future Theoretical Challenges." *Anthropological Forum* 14(1): 5–22.

Burszta, Wojciech J. 2001. "Wólka, kozy i 'kocanka'" [Wólka Village, Goat, and Strawflower]. In *Asteriks w Disneylandzie. Zapiski antropologiczne* [*Asterix at Disneyland. Anthropological Notes*], edited by Wojciech J. Burszta, 176–181. Poznań: Wydawnictwo Poznańskie.

———. 2008. Przeznaczenie: kultura. Cztery uwagi o antropologii globalnego turyzmu [Destination: Culture. Four Notes on the Anthropology of Global Tourism]. In *W kręgu humanistycznej refleksji nad turystyką kulturową* [*In the Circle of Humanistic Reflection on Cultural Tourism*], edited by Marek Kazimierczak, 17–27. Poznań: Akademia Wychowania Fizycznego.

Buruma, Ian, and Avishai Margalit. 2005. *Occidentalism: The West in the Eyes of its Enemies*. New York: Penguin Books.

Bushell, Robyn, and Russell Staiff. 2012. "Rethinking Relationships: World Heritage, Communities and Tourism." In *Routledge Handbook of Heritage in Asia*, edited by Patrick Daly and Tim Winter, 247–265. London and New York: Routledge.

Butcher, Jim. 2003. *The Moralisation of Tourism: Sun, Sand . . . and Saving the World?* London and New York: Routledge.

Butcher, Jim, and Peter Smith. 2010. "'Making a Difference': Volunteer Tourism and Development." *Tourism Recreation Research* 35(1): 27–36.

Butler, Judith. 1990. *Gender Trouble: Feminism and the Subversion of Identity*. London and New York: Routledge.

Calkowski, Marcia S. 1997. "The Tibetan Diaspora and the Politics of Performance." In *Tibetan Culture in the Diaspora*, edited by Frank J. Korom, 51–57. Wien: Verlag der Österreichischen Akademie der Wissenschaften.

Callanan, Michelle, and Sarah Thomas. 2005. "Volunteer Tourism." In *Niche Tourism: Contemporary Issues, Trends and Cases*, edited by Marina Novelli, 183–200. Oxford: Elsevier.

Chang, T. C. 1997. "From 'Instant Asia' to 'Multi-Faceted Jewel': Urban Imaging Strategies and Tourism Development in Singapore." *Urban Geography* 18(6): 542–562.

Channa, Subhadra Mitra. 2004. "Globalization and Modernity in India: A Gendered Critique." *Urban Anthropology and Studies of Cultural Systems and World Economic Development* 33(1): 37–71.

Chatterjee, Partha. 1989. "Colonialism, Nationalism, and Colonized Women: The Contest in India." *American Ethnologist* 16(4): 622–633.

Cheater, Angela P., ed. 1999. *The Anthropology of Power: Empowerment and Disempowerment in Changing Structures*. London and New York: Routledge.

Chen, Susan T. 2012. "When 'Exile' Becomes Sedentary: On the Quotidian Experiences of 'India-Born' Tibetans in Dharamsala, North India." *Asian Ethnicity* 13(3): 263–286.

Cheong, So-Min, and Marc L. Miller. 2000. "Power and Tourism: A Foucauldian Observation." *Annals of Tourism Research* 27(2): 371–390.

Choedon, Yeshi. 2018. "The Unintended Consequences of India's Policy on Citizenship for Tibetan Refugees." https://idsa.in/policybrief/unintended-consequences-of-india-policy-on-citizenship-for-tibetan-refugees-ychoedon-230218.

Chon, Kaye Sung, and Tsutomu Inagaki, eds. 2000. *Japanese Tourists: Socio-Economic, Marketing and Psychological Analysis*. New York: The Haworth Hospitality Press.

Clifford, James. 1992. "Travelling Cultures." In *Cultural Studies*, edited by Lawrence Grossberg, Cary Nelson, and Paula Treichler, 96–116. London and New York: Routledge.

———. 1997. *Routes: Travel and Translation in the Late Twentieth Century*. Cambridge and London: Harvard University Press.

Cohen, Erik. 1971. "Arab Boys and Tourist Girls in a Mixed Jewish-Arab Community." *International Journal of Comparative Sociology* 12(4): 291–304.

———. 1972. "Towards a Sociology of International Tourism." *Social Research* 39(1): 164–189.

———. 1973. "Nomads from Affluence: Notes on the Phenomenon of Drifter-Tourism." *International Journal of Comparative Sociology* 14(1–2): 89–103.

———. 1979. "A Phenomenology of Touristic Experiences." *Sociology* 13(2): 179–201.

———. 1982. *The Pacific Islands from Utopian Myth to Consumer Product: The Disenchantment of Paradise.* Aix-En-Provence: Centre des Hautes Etudes Touristiques.

———. 1985. "Tourist Guides: Pathfinders, Mediators, and Animators." *Annals of Tourism Research* 12(1): 5–29.

———. 1988. "Authenticity and Commoditization in Tourism." *Annals of Tourism Research* 15(3): 371–386.

———. 1993a. "The Heterogenization of a Tourist Art." *Annals of Tourism Research* 20(1): 138–163.

———. 1993b. "Open-Ended Prostitution as a Skillful Game of Luck: Opportunities, Risk, and Security Among Tourist-Oriented Prostitutes in a Bangkok Soi." In *Tourism in South-East Asia*, edited by Michael Hitchcock, Victor T. King, and Michael J. G. Parnwell, 155–178. London and New York: Routledge.

———. 2004. "Backpacking: Diversity and Change." In *The Global Nomad: Backpacker Travel in Theory and Practice*, edited by Greg Richards and Julie Wilson, 43–59. Clevedon: Channel View Publications.

———. 2006. "Pai: A Backpacker Enclave in Transition." *Tourism Recreation Research* 31(3): 11–28.

———. 2007 "Youth Tourists in Acre. A Disturbance Becomes a Lifelong Preoccupation." In *The Study of Tourism: Anthropological and Sociological Beginnings*, edited by Dennison Nash, 50–59. Amsterdam: Elsevier.

Cohen, Erik, and Mark Neal. 2012. "A Middle Eastern Muslim Tourist Enclave in Bangkok." *Tourism Geographies* 14(4): 570–598.

Cohen, Erik, and Robert L. Cooper. 1986. "Language and Tourism." *Annals of Tourism Research* 13(4): 533–563.

Cohen, Erik, and Scott A. Cohen. 2012. "Authentication: Hot and Cool." *Annals of Tourism Research* 39(3): 1295–1314.

———. 2015. "Beyond Eurocentrism in Tourism: A Paradigm Shift to Mobilities." *Tourism Recreation Research* 40(2): 157–168.

Cohen, Scott A. 2010. "Reconceptualising Lifestyle Travellers: Contemporary 'Drifters.'" In *Beyond Backpacker Tourism. Mobilities and Experiences*, edited by Kevin Hannam and Anya Diekmann, 64–84. Bristol: Channel View Publications.

Cohen, Stanley, and Laurie Taylor. 1992. *Escape Attempts: The Theory and Practice of Resistance in Everyday Life.* London and New York: Routledge.

Cole, Stroma, and Jenny Eriksson. 2010. "Tourism and Human Rights." In *Tourism and Inequality: Problems and Prospects*, edited by Stroma Cole and Nigel J. Morgan, 107–125. Oxford: CABI Publishing.

Coles, Tim, and Dallen J. Timothy, eds. 2004. *Tourism, Diasporas and Space.* London and New York: Routledge.

Comaroff, John L., and Jean Comaroff. 2009. *Ethnicity, Inc.* Chicago and London: The University of Chicago Press.

Conklin, Beth A., and Laura R. Graham. 1995. "The Shifting Middle Ground: Amazonian Indians and Eco-Politics." *American Anthropologist* 97(4): 695–710.

Conran, Mary. 2006. "Beyond Authenticity: Exploring Intimacy in the Touristic Encounter in Thailand." *Tourism Geographies* 8(6): 274–285.

———. 2011. "They Really Love Me! Intimacy in Volunteer Tourism." *Annals of Tourism Research* 38(4): 1454–1473.

Cooper, Malcolm J. 2002. "Flexible Labour Markets, Ethnicity and Tourism-Related Migration in Australia and New Zealand." In *Tourism and Migration: New Relationships Between Production and Consumption*, edited by Colin Michael Hall and Allan M. Williams, 73–86. Dordrecht: Kluwer Academic Publishers.

Crick, Malcom. 1989a. "Representations of International Tourism in the Social Sciences: Sun, Sex, Sights, Savings, and Servility." *Annual Review of Anthropology* 18: 307–344.

———. 1989b. "Shifting Identities in the Research Process: An Essay in Personal Anthropology." In *Doing Fieldwork: Eight Personal Accounts of Social Research*, edited by John Perry, 24–40. Sydney: The University of New South Wales Press.

———. 1992. "Ali and Me: An Essay in Street-corner Anthropology." In *Anthropology and Autobiography*, edited by Judith Okely and Hellen Callaway, 173–189. London and New York: Routledge.

———. 1995. "The Anthropologist as Tourist: An Identity in Question." In *International Tourism: Identity and Change*, edited by Marie-Françoise Lanfant, John B. Allcock, and Edward M. Bruner, 205–223. London: Sage Publications.

Croall, Jonathan. 1995. *Preserve or Destroy: Tourism and the Environment*. London: Calouste Gulbenkian Foundation.

Crouch, David, Rhona Jackson, and Felix Thompson, eds. 2005. *The Media and the Tourist Imagination: Converging Cultures*. London and New York: Routledge.

Cutter, Susan L., Bryan J. Boruff, and W. Lynn Shirley. 2003. "Social Vulnerability to Environmental Hazards." *Social Science Quarterly* 84(2): 242–261.

Dahles, Heidi, and Karin Bras. 1999. "Entrepreneurs in Romance: Tourism in Indonesia." *Annals of Tourism Research* 26(2): 267–293.

D'Andrea, Anthony. 2007. *Global Nomads: Techno and New Age as Transnational Countercultures in Ibiza and Goa*. London and New York: Routledge.

Dann, Graham. 1996. "The People of Tourist Brochures." In *The Tourist Image: Myths and Myth Making in Tourism*, edited by Tom Selwyn, 61–82. Chichester: Wiley.

Das, A. K. 1996. "Preliminary Findings on Divine and Royal Themes in the Chariot Festival at Hampi." In *Vijayanagara: Progress of Research, 1988*–1991, edited by D. V. Devaraj and C. S. Patil, 210–215. Mysore: Directorate of Archaeology and Museums.

Das, Veena. 2010. "Engaging the Life of the Other: Love and Everyday Life." In *Ordinary Ethics: Anthropology, Language, and Action*, edited by Michael Lambek, 376–399. New York: Fordham University Press.

Datta, Saikat. 2007. "Molotov on Mandovi: The Russian Mafia Has Barged into Goa, Running Its Arms and Drug Trade Freely." *Outlook, India*, January 29, 2007. https://www.outlookindia.com/magazine/story/molotov-on-mandovi/233732.

Davies, Charlotte Aull. 1999. *Reflexive Ethnography: A Guide to Researching Selves and Others*. London and New York: Routledge.

De Boeck, Filip. 1996. "Postcolonialism, Power and Identity: Local and Global Perspectives from Zaire." In *Postcolonial Identities in Africa*, edited by Richard Werbner and Terence Ranger, 75–106. London: Zed Books.

de Certeau, Michel. 1984. *The Practice of Everyday Life*. Berkeley: University of California Press.

Deitch, Lewis I. 1989. "The Impact of Tourism Upon the Arts and Crafts of the Indians of the Southwestern United States." In *Hosts and Guests: The Anthropology of Tourism* (2nd edition), edited by Valene L. Smith, 223–236. Philadelphia: University of Pennsylvania Press.

de Kadt, Emanuel, ed. 1979. *Tourism: Passport to Development? Perspectives on the Social and Cultural Effects of Tourism in Developing Countries*. New York: Oxford University Press.

de L'Estoile, Benoît. 2008. "The Past as It Lives Now: An Anthropology of Colonial Legacies." *Social Anthropology* 16(3): 267–279.

Dirks, Nicholas B. 2002. *Castes of Mind: Colonialism and the Making of Modern India*. Delhi: Permanent Black.

Dirlik, Arif. 1994. "The Postcolonial Aura: Third World Criticism in the Age of Global Capitalism." *Critical Inquiry* 20(2): 328–356.

Domańska, Ewa. 2008. "O poznawczym uprzywilejowaniu ofiary (Uwagi metodologiczne)" [On the Cognitive Privilege of the Victim (Methodological Notes)]. In *(Nie)obecność. Pominięcia i przemilczenia w narracjach XX wieku* [*(Non)Presence: Omissions and Silence in the Narratives of the 20th Century*], edited by Hanna Gosk and Bożena Karwowska, 19–36. Warszawa: Elipsa.

Dorjee, Tenzin. 2013. "Why Lhakar Matters: The Elements of Tibetan Freedom." *Tibetan Political Review*, January 10. https://sites.google.com/site/tibetanpoliticalreview/articles/whylhakarmatterstheelementsoftibetanfreedom.

Douglas, Mary. 1977. "Deciphering Meal." In *Food and Culture: A Reader*, edited by Carole Counihan and Penny Van Esterik, 36–54. London and New York: Routledge.

Dreyfus, Georges. 1998. "The Shuk-Den Affair: Origins of a Controversy." *Journal of the International Association of Buddhist Studies* 21(2): 227–270.

Dryden, John. 1672. *The Conquest of Granada by the Spaniards*. In Two Parts. Acted at the Theater-Royall. London: T.N. for Henry Herringman.

Duval, David T. 2003. "When Hosts Become Guests: Return Visits and Diasporic Identities in a Commonwealth Eastern Caribbean Community." *Current Issues in Tourism* 6(4): 267–308.

Dzieduszycka, Ewa. 1912. *Indye i Himalaje. Wrażenia z podróży* [*India and the Himalayas: Impressions from the Journey*]. Lwów and Warszawa: Skład Główny w Księgarni Polskiej B. Połonieckiego and E. Wende i Spółka.

Eadington, William R., and Milton Redman. 1991. "Economics and Tourism." *Annals of Tourism Research* 18(1): 41–56.

Echtner, Charlotte M., and Pushkala Prasad. 2003. "The Context of Third World Tourism Marketing." *Annals of Tourism Research* 30(3): 660–682.

Edensor, Tim. 1998. *Tourists at the Taj: Performance and Meaning at a Symbolic Site*. London and New York: Routledge.

———. 2001. "Performing Tourism, Staging Tourism: (Re)producing Tourist Space and Practice." *Tourist Studies* 1: 59–89.

Edwards, Elizabeth. 1996. "Postcards: Greetings from Another World." In *The Tourist Image: Myth and Mythmaking in Tourism*, edited by Tom Selwyn, 197–221. Chichester: Wiley.

Enloe, Cynthia. 2000. *Bananas, Beaches and Bases: Making Feminist Sense of International Politics*. Berkley: University of California Press.

Eriksen, Thomas H. 2001. "Between Universalism and Relativism: A Critique of UNESCO Concept of Culture." In *Culture and Rights: Anthropological Perspectives*, edited by Jane K. Cowan, Marie-Bénédicte Dembour, and Richard A. Wilson, 127–148. New York: Cambridge University Press.

Esman, Marjorie R. 1984. "Tourism as Ethnic Preservation: The Cajuns of Louisiana." *Annals of Tourism Research* 11(3): 451–467.

Etienne, Mona, and Eleanor B. Leacock, eds. 1980. *Women and Colonization: Anthropological Perspectives*. New York: Praeger.

Evans-Pritchard, Deirdre. 1989. "How 'They' See 'Us': Native American Images of Tourists." *Annals of Tourism Research* 16(1): 89–105.

Fabian, Johannes. 1983. *Time and the Other: How Anthropology Makes Its Object*. New York: Columbia University Press.

Fahmi, Wael Salah. 2008. "Global Tourism and the Urban Poor's Right to the City: Spatial Contestation within Cairo's Historical Districts." In *Tourism Development Growth, Myths and Inequalities*, edited by Peter M. Burns and Marina Novelli. Wallingford: CABI International.

Fanon, Frantz. 1963. *The Wretched of the Earth*. Translated by Constance Farrington. New York: Grove Press.

———. 1967. "Algeria Unveiled." In *A Dying Colonialism*, translated by Haakon Chevalier. New York: Grove Press.

Ferro, Marc. 1997. *Colonialism: A Global History*. London and New York: Routledge.

Fillitz, Thomas, and A. Jamie Saris. 2013. "Authenticity Aujourd'hui." In *Debating Authenticity: Concepts of Modernity in Anthropological Perspective*, edited by Thomas Fillitz and A. Jamie Saris, 1–24. New York and Oxford: Berghahn Books.

Fischer, Edward F. 2014. *The Good Life: Aspiration, Dignity, and the Anthropology of Wellbeing*. Stanford: Stanford University Press.

Fogelson, Raymond D., and Richard N. Adams, eds. 1977. *The Anthropology of Power: Ethnographic Studies from Asia, Oceania, and the New World*. New York: Academic Press.

Forster, Edward Morgan. 1924. *A Passage to India*. New York: Harcourt, Brace and Company.

Foucault, Michel. 1980. *Power/Knowledge: Selected Interviews and Other Writings, 1972–1977*. Translated by Colin Gordon, Leo Marshall, John Mepham, and Kate Soper. New York: Pantheon Books.

———. 1981. "The Order of Discourse." Translated by Ian McLeod. In *Untying the Text: A Post-Structuralist Reader*, edited by Robert Young, 51–78. Boston, London, and Henley: Routledge and Kegan Paul.

———. 2002. *The Archaeology of Knowledge*. Translated by A. M. Sheridan Smith. London and New York: Routledge.

———. 2003. *The Birth of the Clinic: An Archaeology of Medical Perception.* Translated by A. M. Sheridan Smith. London and New York: Routledge.

Franklin, Adrian. 2003. *Tourism: An Introduction.* London: Sage Publications.

Freitag, Tilman G. 1994. "Enclave tourism development: For whom the benefits roll?" *Annals of Tourism Research* 21(3): 538–554.

Fritz, John M., and George Michell. 1991. *City of Victory: Vijayanagara, the Medieval Hindu Capital of Southern India.* New York: Aperture Press.

———. 2009. *Hampi: A Story in Stone.* Mumbai: Eminence Designs.

———. 2012. "Living Heritage at Risk: Searching for a New Approach to Development, Tourism, and Local Needs at the Grand Medieval City of Hampi." *Archaeology.* http://www.archaeology.org/issues/53–1211/letter-from#art_page1.

———. 2014. *Hampi Vijayanagara.* Mumbai: Jaico Publishing House.

Fröbel, Folker, Jurgen Heinrichs, and Otto Kreye. 1980. *The New International Division of Labour: Structural Unemployment in Industrialized Countries and Industrialization in Developing Countries.* Translated by Pete Burgess. New York: Cambridge University Press.

Gandhi, Leela. 1998. *Postcolonial Theory: A Critical Introduction*: New York: Columbia University Press.

Gaonkar, Dilip Parameshwar. 2002. "Toward New Imaginaries: An Introduction." *Public Culture* 14(1): 1–19.

Ghodsee, Kristen. 2005. *The Red Riviera: Gender, Tourism, and Postsocialism on the Black Sea.* Durham: Duke University Press.

Giddens, Anthony. 1992. *The Transformation of Intimacy: Sexuality, Love, and Eroticism in Modern Societies.* Cambridge: Polity Press.

Gillen, Jamie. 2011. "Off the Record: Segmenting Informal Discussions into Viable Methodological Categories." In *Fieldwork in Tourism: Methods, Issues and Reflections*, edited by Colin Michael Hall, 199–208. London and New York: Routledge.

Glick Schiller, Nina, and Andrew Irving, eds. 2014. *Whose Cosmopolitanism? Critical Perspectives, Relationalities and Discontents.* New York and Oxford: Berghahn Books.

Glick Schiller, Nina, and Noel B. Salazar. 2013. "Regimes of Mobility Across the Globe." *Journal of Ethnic and Migration Studies* 39(2): 183–200.

Goffman, Erving. 1961. *Asylums: Essays on the Condition of the Social Situation of Mental Patients and Other Inmates.* New York and Toronto: Anchor Books.

Golde, Peggy, ed. 1986. *Women in the Field: Anthropological Experiences.* Berkeley: University of California Press.

Golinowska, Karolina. 2013. "Oblicza etniczności w dobie globalizacji" [Faces of Ethnicity in the Age of Globalization]. *Kultura–Historia–Globalizacja* 14: 103–116.

Goodwin, Harold. 2008. "Pro-Poor Tourism: A Response." *Third World Quarterly* 29(5): 869–871.

Graburn, Nelson H. H. 1983. "Tourism and Prostitution." *Annals of Tourism Research* 10(3): 437–442.

———. 1984. "The Evolution of Tourist Art." *Annals of Tourism Research* 11(3): 393–413.

Gramsci, Antonio. 1996. *Prison Notebooks* (Vol. 2). Translated by Joseph A. Buttigieg. New York: Columbia University Press.

Gravari-Barbas, Maria, and Nelson H. H. Graburn, eds. 2016. *Tourism Imaginaries at the Disciplinary Crossroads*. London and New York: Routledge.

Greenwood, Davydd J. 1977. "Culture by the Pound: An Anthropological Perspective on Tourism as Cultural Commodization." In *Hosts and Guests: The Anthropology of Tourism* (1st edition), edited by Valene L. Smith, 129–138. Oxford: Blackwell.

Gregory, Derek. 2001. "Colonial Nostalgia and Cultures of Travel: Spaces of Constructed Visibility in Egypt." In *Consuming Tradition, Manufacturing Heritage*, edited by Nezar AlSayyad, 111–151. London and New York: Routledge.

Grent, Nellie. 2002. "Polyandry in Dharamsala: Plural-Husband Marriage in a Tibetan Refugee Community in Northwest India." In *Tibet, Self, and the Tibetan Diaspora: Voices of Difference*, edited by Paul Christiaan Klieger, 105–138. Boston: Brill's Tibetan Studies Library.

Gudykunst, William B., ed. 2003. *Cross-Cultural and Intercultural Communication*. Thousand Oaks: Sage Publications.

Guha, Ranajit. 1982. "On Some Aspects of the Historiography of Colonial India." *Subaltern Studies* 1: 1–8.

Gupta, Akhil, and James Ferguson. 1992. "Beyond 'Culture': Space, Identity, and the Politics of Difference." *Cultural Anthropology* 7(1): 6–23.

———, eds. 1997. *Anthropological Locations: Boundaries and Grounds of a Field Science*. Berkeley: University of California Press.

Gupta, Pamila. 2009. "Goa Dourada, the Internal Exotic in South Asia: Discourses of Colonialism and Tourism." In *South Asia and Its Others: Reading the "Exotic,"* edited by V. G. Julie Rajan and Atreyee Phukan, 123–148. Cambridge: Cambridge Scholars Publishing.

———. 2014. "Frozen Vodka and White Skin in Tourist Goa." In *Tourism and the Power of Otherness: Seductions of Difference*, edited by David Picard and Michael A. Di Giovine, 95–109. Bristol: Channel View Publications.

Gutowska, Aleksandra. 2015. "Going Beyond Poverty Tourism? The Narratives of Kibera Residents and Their Meanings Constructed within the Frame of Slum Tours." *Folia Turistica* 37: 99–122.

Gyatso, Janet, and Hanna Havnevik, eds. 2005. *Women in Tibet: Past and Present*. London: C. Hurst & Co.

Haldrup, Michael, and Jonas Larsen. 2010. *Tourism, Performance and the Everyday: Consuming the Orient*. London and New York: Routledge.

Hall, Colin Michael. 2004. "Reflexivity and Tourism Research: Situating Myself and/with Others." In *Qualitative Research in Tourism: Ontologies, Epistemologies and Methodologies*, edited by Jenny Phillimore and Lisa Goodson, 137–155. London and New York: Routledge.

———. 2011. "Researching the Political in Tourism: Where Knowledge Meets Power." In *Fieldwork in Tourism: Methods, Issues and Reflections*, edited by Colin Michael Hall, 39–54. London and New York: Routledge.

Hall, Colin Michael, and Allan M. Williams, eds. 2002. *Tourism and Migration: New Relationships Between Production and Consumption*. Dordrecht: Kluwer Academic Publishers.

Hall, Colin Michael, and Hazel Tucker, eds. 2004. *Tourism and Postcolonialism: Contested Discourses, Identities and Representations*. London and New York: Routledge.

Hall, Ronald E. 1995. "The Color Complex. Bleaching Syndrome." *Race, Gender & Class* 2(2): 99–109.

Hamilton, Jill. 2005. *Thomas Cook: The Holiday Maker*. Cheltenham: History Press Limited.

Hampton, Mark P. 1998. "Backpacker Tourism and Economic Development." *Annals of Tourism Research* 25(3): 639–660.

———. 2010. "Not Such a Rough or Lonely Planet? Backpacker Tourism: An Academic Journey." In *Beyond Backpacker Tourism: Mobilities and Experiences*, edited by Kevin Hannam and Anya Diekmann, 8–20. Bristol: Channel View Publications.

Hannam, Kevin, and Diekmann A. Anya. 2010. "From Backpacking to Flashpacking: Developments in Backpacker Tourism Research." In *Beyond Backpacker Tourism: Mobilities and Experiences*, edited Kevin Hannam and Anya Diekmann, 1–7. Bristol: Channel View Publications.

Hannam, Kevin, and Irena Ateljevic, eds. 2008. *Backpacker Tourism: Concepts and Profiles*. Clevedon: Channel View Publications.

Hannerz, Ulf. 1996. *Transnational Connections: Culture, People, Places*. London and New York: Routledge.

Harris, Clare. 1999. *In the Image of Tibet: Tibetan Painting After 1959*. London: Reaktion Books.

Harrison, David, ed. 2001. *Tourism and the Less Developed World: Issues and Case Studies*. Wallingford: CABI Publishing.

———. 2008. "Pro-Poor Tourism: A Critique." *Third World Quarterly* 29(5): 851–868.

Harrison, Faye V., ed. 1991. *Decolonizing Anthropology: Moving Further Toward an Anthropology for Liberation*. Washington: American Anthropological Association.

Hartmann, J. 1991. "The Significance of Youth Tourism in a Post-Industrial Society: The Case of Sweden." Paper presented at the International Conference on Youth Tourism, New Delhi (India), November 18–21, 1991.

Harvey, David. 1989. *The Condition of Postmodernity: An Enquiry into the Origins of Cultural Change*. Oxford: Blackwell.

Hastrup, Kirsten. 1995. *A Passage to Anthropology: Between Experience and Theory*. London and New York: Routledge.

Hatton, Timothy J., and Jeffrey G. Williamson. 1992. *What Drove the Mass Migrations from Europe in the Late Nineteenth Century?* Cambridge: National Bureau of Economic Research.

Haviv, Ayana Shira. 2005. "Next Year in Kathmandu: Israeli Backpackers and the Formation of a New Israeli Identity." In *Israeli Backpackers: From Tourism to Rite of Passage*, edited by Chaim Noy and Erik Cohen, 45–88. Albany: State University of New York Press.

Hazan, Haim, and Esther Hertzog, eds. 2012. *Serendipity in Anthropological Research: The Nomadic Turn*. Farnham: Ashgate.

Headrick, Daniel R. 1981. *The Tools of Empire: Technology and European Imperialism in the Nineteenth Century*. Oxford: Oxford University Press.

Healy, Robert G. 1994. "Tourist Merchandise as a Means of Generating Local Benefits from Ecotourism." *Journal of Sustainable Tourism* 2(3): 137–151.

Heath, Sue. 2007. "Widening the Gap: Pre-University Gap Years and the 'Economy of Experience.'" *British Journal of Sociology of Education* 28(1): 89–103.

Herold, Edward, Rafael Garcia, and Tony DeMoya. 2011. "Female Tourists and Beach Boys: Romance or Sex Tourism?" *Annals of Tourism Research* 28(4): 978–997.

Herzfeld, Michael. 2005. *Cultural Intimacy: Social Poetics in the Nation-State*. New York and London: Routledge.

———. 2006. "Spatial Cleansing: Monumental Vacuity and the Idea of the West." *Journal of Material Culture* 11(1/2): 127–149.

Heuman, Daniel. 2005. "Hospitality and Reciprocity: Working Tourists in Dominica." *Annals of Tourism Research* 32(2): 407–418.

Higgins-Desbiolles, Freya. 2007. "Hostile Meeting Grounds: Encounters Between the Wretched of the Earth and the Tourist Through Tourism and Terrorism in the 21st Century." In *Tourism and Politics: Global Frameworks and Local Realities*, edited by Peter M. Burns and Marina Novelli, 309–332. Amsterdam: Elsevier.

———. 2010. "Justifying Tourism: Justice Through Tourism." In *Tourism and Inequality: Problems and Prospects*, edited by Stroma Cole and Nigel J. Morgan, 195–212. Oxford: CABI Publishing.

Hillman, Ben. 2003. "Paradise under Construction: Minorities, Myths, and Modernity in Northwest Yunnan." *Asian Ethnicity* 4(2): 175–188.

Hilton, James. 1933. *The Lost Horizon*. New York: W. Morrow and Co.

Hobbes, Thomas. 1968. *Leviathan*. Harmondsworth: Penguin Publishers.

Hobsbawm, Eric. 1983. "Introduction: Inventing Traditions." In *The Invention of Tradition*, edited by Eric Hobsbawm and Terence Ranger, 1–14. Cambridge: Cambridge University Press.

Høivik, Tord, and Turid Heiberg. 1980. "Center–Periphery. Tourism and Self-Reliance." *International Social Science Journal* 32(1): 69–98.

Hollinshead, Keith. 1992. "'White' Gaze, 'Red' People—Shadow Visions: The Disidentification of 'Indians' in Cultural Tourism." *Leisure Studies* 11(1): 43–64.

———. 1999. "Surveillance of the Worlds of Tourism: Foucault and the Eye of Power." *Tourism Management* 20(1): 7–23.

Horolets, Anna. 2012. "Codzienność jako atrakcja turystyczna w doświadczeniu polskich turystow niszowych do byłego ZSRR" [Everyday Life as a Tourist Attraction in the Experience of Polish Niche Tourists to the Former USRR]. *Kultura i Społeczeństwo* 3: 95–112.

———. 2013. *Konformizm, bunt, nostalgia: Turystyka niszowa z Polski do krajów byłego ZSRR* [Conformity, Rebellion, Nostalgia: Niche Tourism from Poland to the Former USRR Countries]. Kraków: Universitas.

Hottola, Petri. 1999. *The Intercultural Body: Western Woman, Culture Confusion and Control of Space in the South Asian Travel Scene*. Joensuu: University of Joensuu.

———. 2005. "The Metaspatialities of Control Management in Tourism: Backpacking in India." *Tourism Geographies* 7(1): 1–22.

Howard, Robert W. 2007. "Five Backpacker Tourist Enclaves." *International Journal of Tourism Research* 9(2): 73–86.

Hume, David L. 2009. "The Development of Tourist Art and Souvenirs—The Arc of the Boomerang: From Hunting, Fighting and Ceremony to Tourist Souvenir." *International Journal of Tourism Research* 11(1): 55–70.

———. 2014. *Tourism Art and Souvenirs: The Material Culture of Tourism*. London and New York: Routledge.

Hunter, Margaret L. 2002. "If You're Light, You're Alright: Light Skin Color as Social Capital for Women of Color." *Gender & Society* 16(2): 175–193.

Hustinx, Lesley, and Frans Lammertyn. 2003. "Collective and Reflexive Styles of Volunteering: A Sociological Modernization Perspective." *Voluntas: International Journal of Voluntary and Nonprofit Organizations* 14(2): 167–187.

Hutnyk, John. 1996. *The Rumour of Calcutta: Tourism, Charity, and the Poverty of Representation*. London: Zed Books.

Hutt, Michael. 1996. "Looking for Shangri-la: From Hilton to Lamichhane." In *The Tourist Image. Myths and Myth Making in Tourism*, edited by Tom Selwyn, 49–60. Chichester: Wiley.

Iyer, Pico. 2008. *The Open Road: The Global Journey of the Fourteenth Dalai Lama*. New York: Knopf Publishing Group.

Jafari, Jafar. 1982. "Comment." *Current Anthropology* 23(2): 137.

———, ed. 2000. *Encyclopedia of Tourism*. London and New York: Routledge.

Jaquette, Jane S., and Gale Summerfield, eds. 2006. *Women and Gender Equity in Development Theory and Practice: Institutions, Resources, and Mobilization*. Durham and London: Duke University Press.

Jarvis, Jeff, and Victoria Peel. 2010. "Flashpacking in Fiji: Reframing the 'Global Nomad' in a Developing Destination." In *Beyond Backpacker Tourism: Mobilities and Experiences*, edited by Kevin Hannam and Anya Diekmann, 21–39. Bristol: Channel View Publications.

Jha, Makhan, ed. 1985. *Dimensions of Pilgrimage: An Anthropological Appraisal*. New Delhi: Inter-India Publications.

Joseph, Christina A., and Anandam P. Kavoori. 2001. "Mediated Resistance: Tourism and the Host Community." *Annals of Tourism Research* 28(4): 998–1009.

Joy, Charlotte. 2010. "Heritage and Tourism: Contested Discourses in Djenne, a World Heritage Site in Mali." In *Tourism, Power and Culture: Anthropological Insights*, edited by Donald V. L. Macleod and James G. Carrier, 47–63. Bristol: Channel View Publications.

Kaczmarek, Łukasz. 2012. "Poza luksusowym więzieniem. Społeczne oddziaływania turystyki na Jamajce" [Outside of a Luxury Prison: Social Impact of Tourism in Jamaica]. *Lud* 96: 249–272.

Kajiwara, Kageaki. 1997. "Inward-Bound, Outward-Bound: Japanese Tourism Reconsidered." In *Tourism and Cultural Development in Asia and Oceania*, edited by Shinji Yamashita, Abdul Kadir H. J. Din, and Jeremy Seymour Eades, 164–177. Penerbit: Selangor.

Kamerkar, Mani, and Soonu Dhunjisha. 2002. *From the Iranian Plateau to the Shores of Gujarat: The Story of Parsi Settlements and Absorption in India*. Mumbai: Allied Publishers.

Kanemasu, Yoko. 2008. "Weapons of the Workers: Employees in the Fiji Hotel Scene." In *Tourism at the Grassroots: Villagers and Visitors in the Asia-Pacific*, edited by John Connell and Barbara Rugendyke, 114–130. London and New York: Routledge.

Kapadia, Karin. 1996. *Siva and Her Sisters: Gender, Caste, and Class in Rural South India*. Boulder: Westview Press.

Karko, Kate. 2000. *Namma: A Tibetan Love Story*. London: Hodder & Stoughton.

Karlberg, Michael. 2005. "The Power of Discourse and the Discourse of Power: Pursuing Peace Through Discourse Intervention." *International Journal of Peace Studies* 10(1): 1–23.

Kaur, Jagdish. 1985. *Himalayan Pilgrimages and the New Tourism*. New Delhi: Himalayan Books.

Kim, Young Yun. 2001. *Becoming Intercultural: Integrative Theory of Communication and Crosscultural Adaptation*. Thousand Oaks: Sage Publications.

Kishwar, Madhu. 2005. "A Horror of 'Isms': Why I Do Not Call Myself a Feminist." In *Feminism in India*, edited by Maitrayee Chaudhuri, 26–51. London: Zed Books.

Klieger, Paul Christiaan. 1992. *Tibetan Nationalism (The Role of Patronage in the Accomplishment of a National Identity)*. Berkeley: Folklore Institute.

Kotowa, Barbara. 2008. "Scjentyzm jako światopogląd nauki" [Scientism as a Worldview of Science]. *Nowa Krytyka*, December 13, 2008. http://www.nowakrytyka.pl/spip.php?article450.

Kotraiah, C. T. M. 2008. *Archaeology of Hampi-Vijayanagara*. Delhi: Bharatiya Kala Prakashan Publishers.

Kraft, Siv Elen. 2007. "Religion and Spirituality in Lonely Planet's India." *Religion* 37(3): 230–242.

Krippendorf, Jost. 1986. "The New Tourist: Turning Point for Leisure and Travel." *Tourism Management* 7(2): 131–135.

Krog, Stian E. 2007. "Living Homes and Dead Monuments: Cultural Heritage and the Construction of Space and Place in Hampi, India." MA thesis, University of Bergen.

Kruhse-MountBurton, Suzy. 1995. "Sex Tourism and Traditional Australian Male Identity." In *International Tourism: Identity and Change*, edited by Marie-Françoise Lanfant, John B. Allcock, and Edward M. Bruner, 192–204. London: Sage Publications.

Kurcz, Maciej. 2012. *Jak przeżyć w afrykańskim mieście? Człowiek wobec pograniczności i procesow urbanizacyjnych w południowosudańskim mieście* [*How to Survive in an African City? Man in the Face of Liminality and Urbanization Processes in a South Sudanese City*]. Katowice: Wydawnictwo Uniwersytetu Śląskiego.

Laidlaw, James. 2014. *The Subject of Virtue: An Anthropology of Ethics and Freedom*. Cambridge: Cambridge University Press.

Lanfant, Marie-Françoise. 1995. "International Tourism, Internationalization and the Challenge to Identity." In *International Tourism: Identity and Change*, edited by Marie-Françoise Lanfant, John B. Allcock, and Edward M. Bruner, 24–43. London: Sage Publications.

Lanfant, Marie-Françoise, and Nelson H. H. Graburn. 1992. "International Tourism Reconsidered: The Principle of the Alternative." In *Tourism Alternatives*, edited by Valene L. Smith and William R. Eadington, 88–112. Chichester: Wiley.

Larsen, Kjersti. 2000. "The Other Side of 'Nature': Expanding Tourism, Changing Landscapes, and Problems of Privacy in Urban Zanzibar." In *Producing Poverty and Nature in Africa*, edited by Vigdis Broch-Due and Richard A. Schroeder, 198–219. Stockholm: Nordiska Afrikainstitutet & Transaction Publishers.

Laxson, Joan D. 1991. "How 'We' See 'Them': Tourism and Native Americans." *Annals of Tourism Research* 18(3): 365–391.

Laythorpe, K. 2010. "Backpackers as Volunteer Tourists: Evidence from Tanzania." In *Beyond Backpacker Tourism: Mobilities and Experiences*, edited by Kevin Hannam and Anya Diekmann, 140–152. Bristol: Channel View Publications.

Lazarus, Neil. 2006. "Postcolonial Studies After the Invasion of Iraq." *New Formations* 59: 10–23.

Lea, John. 1988. *Tourism and Development in the Third World*. London and New York: Routledge.

Leacock, Eleanor B. 1975. "Class, Commodity, and the Status of Women." In *Women Cross-Culturally: Change and Challenge*, edited by Ruby Rohrlich-Leavitt, 601–616. The Hague and Paris: Mouton Publishers.

———. 1981. *Myths of Male Dominance: Collected Articles on Women Cross-Culturally*. New York: Monthly Review Press.

Lenz, Ramona. 2010. "'Hotel Royal' and Other Spaces of Hospitality: Tourists and Migrants in the Mediterranean." In *Thinking Through Tourism*, edited by Julie Scott and Tom Selwyn, 209–229. Oxford and New York: Berg.

Leon, Tamir. 2000. "Two Separate Social Groups Among the Israeli Backpackers in India." Paper presented at the Annual Conference of the Israeli Anthropological Association, Jerusalem, May 24–25, 2000.

Lévi-Strauss, Claude. 1969. *The Elementary Structures of Kinship*. Translated by James Harle Bell, John Richard von Sturmer, and Rodney Needham. Boston: Beacon Press.

Lie, John. 1996. "Sociology of Contemporary Japan." *Current Sociology* 44(1): 1–101.

Little, Walter E. 2004. *Mayas in the Marketplace: Tourism, Globalization, and Cultural Identity*. Austin: University of Texas Press.

Lobo, J. 1991. *The Repercussions of Luxury Tourism in Goa, India*. Bangalore: Mirlithu Publishing.

Loker, Laurie. 1993. *Backpacker Phenomenon II: More Answers to Further Questions*. Townsville: James Cook University of North Queensland.

Loker-Murphy, Laurie, and Philip L. Pearce. 1995. "Young Budget Travelers: Backpackers in Australia." *Annals of Tourism Research* 22(4): 819–843.

Loomba, Ania. 2015. *Colonialism/Postcolonialism*. London and New York: Routledge.

Lopez, Donald S. Jr. 1998. *Prisoners of Shangri-La: Tibetan Buddhism and the West*. Chicago and London: The University of Chicago Press.

Lozanski, Kristin. 2010. "Defining 'Real India': Representations of Authenticity in Independent Travel." *Social Identities* 16(6): 741–762.

Lyotard, Jean-François. 1984. *The Postmodern Condition: A Report on Knowledge*. Translated by Geoff Bennington and Brian Massumi. Minneapolis: University of Minnesota Press.

MacCannell, Dean. 1976. *The Tourist: A New Theory of the Leisure Class*. Berkeley: University of California Press.

———. 1992. *Empty Meeting Grounds: The Tourist Papers*. London and New York: Routledge.

MacDonald, Pauline. 2013. *Dharamsala Days, Dharamsala Nights: The Unexpected World of the Refugees from Tibet*. n.p.: Pauline MacDonald.

MacLean, Rory. 2006. *Magic Bus: On the Hippie Trail from Istanbul to India*. London: Viking.

Macleod, Donald V. L. 2004. *Tourism, Globalisation and Cultural Change: An Island Community Perspective*. Clevedon: Channel View Publications.

Macleod, Donald V. L., and James G. Carrier, eds. 2010. *Tourism, Power and Culture: Anthropological Insights*. Bristol: Channel View Publications.

Mahony, Sorcha. 2018. *Searching for a Better Life: Growing Up in the Slums of Bangkok*. New York and Oxford: Berghahn Books.

Mahrouse, Gada. 2008. "Race-Conscious Transnational Activists with Cameras: Mediators of Compassion." *International Journal of Cultural Studies* 11(1): 87–105.

Mallon, Florencia E. 1994. "The Promise and Dilemma of Subaltern Studies: Perspectives from Latin American History." *American Historical Review* 99(5): 1491–1515.

Maoz, Darya. 2004. "The Conquerors and the Settlers: Two Groups of Young Israeli Backpackers in India." In *The Global Nomad. Backpacker Travel in Theory and Practice*, edited by Greg Richards and Julie Wilson, 109–122. Clevedon: Channel View Publications.

———. 2005. "Young Adult Israeli Backpackers in India." In *Israeli Backpackers: From Tourism to Rite of Passage*, edited by Chaim Noy and Erik Cohen, 159–188. Albany: State University of New York Press.

———. 2006. "Mutual Gaze." *Annals of Tourism Research* 33(1): 221–239.

———. 2008. "The Backpacking Journey of Israeli Women in Mid-Life." In *Backpacker Tourism: Concepts and Profiles*, edited by Kevin Hannam and Irena Ateljevic, 188–198. Clevedon: Channel View Publications.

Marcus, George E. 1995. "Ethnography in/of the World System: The Emergence of Multi-Sited Ethnography." *Annual Review of Anthropology* 24: 95–117.

———. 1998. *Ethnography Through Thick and Thin*. Princeton: Princeton University Press.

Marschall, Sabine. 2004. "Commodifying Heritage: Post-Apartheid Monuments and Cultural Tourism in South Africa." In *Tourism and Postcolonialism: Contested Discourses, Identities and Representations*, edited by Colin Michael Hall and Hazel Tucker, 95–112. London and New York: Routledge.

Martin, Luther H., Huck Gutman, and Patrick H. Hutton, eds. 1988. *Technologies of the Self: A Seminar with Michel Foucault*. Amherst: University of Massachusetts Press.

Mason, Peter. 2002. "The 'Big OE': New Zealanders Overseas Experiences in Britain." In *Tourism and Migration: New Relationships Between Production and Consumption*, edited by Colin Michael Hall and Allan M. Williams, 87–101. Dordrecht: Kluwer Academic Publishers.

Matthiasson, Carolyn J., ed. 1974. *Many Sisters: Women in Cross-Cultural Perspective*. New York: The Free Press.

Mauss, Marcel. 1966. *The Gift: The Form and Reason for Exchange in Archaic Societies*. Translated by Ian Cunnison. London: Cohen & West.

Mbaiwa, Joseph E. 2005. "Enclave Tourism and Its Socio-Economic Impacts in the Okavango Delta, Botswana." *Tourism Management* 26(2): 157–172.

McGehee, Nancy Gard, and Carla Almeida Santos. 2005. "Social Change, Discourse and Volunteer Tourism." *Annals of Tourism Research* 32(3): 760–779.

McGuckin, Eric A. 1997. "Postcards from Shangri-la: Tourism, Tibetan Refugees, and the Politics of Cultural Production." PhD dissertation, New York University.

McKean, Philip F. 1989. "Towards a Theoretical Analysis of Tourism: Economic Dualism and Cultural Involution in Bali." In *Hosts and Guests: The Anthropology of Tourism* (2nd edition), edited by Valene L. Smith, 119–138. Philadelphia: University of Pennsylvania Press.

Mercille, Julien. 2005. "Media Effects on Image: The Case of Tibet." *Annals of Tourism Research* 32(4): 1039–1055.

Mevorach, Oded. 1997. "The Long Trip After the Military Service: Characteristics of the Travelers, the Effects of the Trip and Its Meaning." PhD dissertation, The Hebrew University.

Meyer, Dorothea. 2006. *Caribbean Tourism, Local Sourcing and Enterprise Development: Review of the Literature*. Sheffield: Sheffield Hallam University.

Minca, Claudio, and Rachele Borghi. 2009. "Morocco: Restaging Colonialism for Masses." In *Cultures of Mass Tourism: Doing the Mediterranean in the Age of Banal Mobilities*, edited by Pau Obrador Pons, Mike Crang, and Penny Travlou, 21–52. Farnham: Ashgate.

Ministry of Tourism. 2017. *Annual Report 2016–17*. New Delhi: Ministry of Tourism, Government of India.

Mishra, Neha, and Ronald Hall. 2017. "Bleached Girls: India and Its Love for Light Skin." *The Conversation*, July 21, 2017. http://theconversation.com/bleached-girls-india-and-its-love-for-light-skin-80655.

Mishra, Vijay, and Bob Hodge. 1991. "What Is Post(-)Colonialism?" *Textual Practice* 5(3): 399–414.

Moore, Henrietta L. 1988. *Feminism and Anthropology*. Cambridge: Polity Press.

Moore-Gilbert, Bart. 1997. *Postcolonial Theory: Contacts, Practices, Politics*. London: Verso.

Morgan, Nigel J., and Annette Pritchard. 1998. *Tourism Promotion and Power: Creating Images, Creating Identities*. Chichester: Wiley.

Mostafanezhad, Mary. 2013. "The Politics of Aesthetics in Volunteer Tourism." *Annals of Tourism Research* 43: 150–169.

Mountcastle, Amy. 1997. "Tibetans in Exile: The Construction of Global Identities." PhD dissertation, The State University of New Jersey.

Mowforth, Martin, and Ian Munt. 2003. *Tourism and Sustainability: Development and New Tourism in the Third World*. London and New York: Routledge.

Mróz, Lech. 1989. "Wędrowni kowale i handlarze bydła z Radżastanu – hipotetyczni przodkowie Cyganów" [Traveling Blacksmiths and Cattle Traders from Rajasthan—The Hypothetical Ancestors of the Gypsies]. *Etnografia Polska* 33(1): 77–108.

Munt, Ian. 1994. "The 'Other Postmodern' Tourism: Culture, Travel and the New Middle Classes." *Theory, Culture and Society* 11: 101–123.

Murphy, Laurie. 2001. "Exploring Social Interactions of Backpackers." *Annals of Tourism Research* 28(1): 50–67.

Muzaini, Hamzah. 2006. "Backpacking Southeast Asia: Strategies of 'Looking Local.'" *Annals of Tourism Research* 33(1): 144–161.

Naik, Dhanasing B. 2010. *The Art and Literature of Banjara Lambanis*. New Delhi: Abhinav Publications.

Nash, Dennison. 1977. "Tourism as a Form of Imperialism." In *Hosts and Guests: The Anthropology of Tourism* (1st edition), edited by Valene L. Smith, 33–47. Oxford: Blackwell.

Nash, Dennison, and Valene L. Smith. 1991. "Anthropology and Tourism." *Annals of Tourism Research* 18(1): 12–25.

Newlands, Ken. 2004. "Setting Out on the Road Less Travelled: A Study of Backpacker Travel in New Zealand." In *The Global Nomad: Backpacker Travel in Theory and Practice*, edited by Greg Richards and Julie Wilson, 217–236. Clevedon: Channel View Publications.

Newman, Robert S. 2001. *Of Umbrellas, Goddesses and Dreams: Essays on Goan Culture and Society*. Mapusa: Other India Press.

Nogués Pedregal, Antonio-Miguel. 2008. "A Contextual Approach to the Power Relations Between Tourism and Development." In *Tourism Development Growth, Myths and Inequalities*, edited by Peter M. Burns and Marina Novelli, 141–158. Wallingford: CABI International.

Norbu, Dawa. 2001. *China's Tibet Policy*. Richmond: Curzon Press.

Norman, Alex. 2011. *Spiritual Tourism: Travel and Religious Practice in Western Society*. London and New York: Continuum.

Nowicka, Ewa. 2009. "Co robimy tym ludziom? Wpływ trekkingu w Himalajach na społeczeństwo Nepalu" [What Are We Doing to These People? The Impact of Trekking in the Himalayas on the Nepalese Society]. In *Podroż i miejsce w perspektywie antropologicznej* [*Journey and Place in an Anthropological Perspective*], edited by Dorota Rancew-Sikora, 29–45. Gdańsk: Wydawnictwo Uniwersytetu Gdańskiego.

Noy, Chaim. 2005. "Israeli Backpackers: Narrative, Interpersonal Communication, and Social Construction." In *Israeli Backpackers: From Tourism to Rite of Passage*, edited by Chaim Noy and Erik Cohen, 111–157. Albany: State University of New York Press.

Noy, Chaim, and Erik Cohen, eds. 2005. *Israeli Backpackers: From Tourism to Rite of Passage*. Albany: State University of New York Press.

Nuñez, Theron A. Jr. 1963. "Tourism, Tradition, and Acculturation: Weekendismo in a Mexican Village." *Ethnology* 2(3): 347–352.

———. 1977. "Touristic Studies in Anthropological Perspective." In *Hosts and Guests: The Anthropology of Tourism* (1st edition), edited by Valene L. Smith, 207–216. Oxford: Blackwell.

O'Connell Davidson, Julia. 1996. "Sex Tourism in Cuba." *Race & Class* 38(1): 39–48.

———. 2001. "The Sex Tourist, the Expatriate, His Ex-Wife and Her 'Other': The Politics of Loss, Difference and Desire." *Sexualities* 41(1): 5–24.

Odzer, Cleo. 1995. *Goa Freaks: My Hippie Years in India*. New York: Blue Moon Books.

Ogden, Rebecca. 2015. "Understanding Cuban Tourism: Affect and Capital in Post-Special Period Cuba." PhD dissertation, The University of Manchester.

Oppermann, Martin. 1993. "Tourism Space in Developing Countries." *Annals of Tourism Research* 20(3): 535–556.

———. 1999. "Sex Tourism." *Annals of Tourism Research* 26(2): 251–266.

Oppermann, Martin, and Kye-Sung Chon. 1997. *Tourism in Developing Countries*. London: International Thomson Business Press.

O'Regan, Michael. 2010. "Backpacker Hostels: Place and Performance." In *Beyond Backpacker Tourism: Mobilities and Experiences*, edited by Kevin Hannam and Anya Diekmann, 85–101. Bristol: Channel View Publications.

O'Reilly, Camille C. 2006. "From Drifter to Gap Year Tourist: Mainstreaming Backpacker Travel." *Annals of Tourism Research* 33(4): 998–1017.

O'Rourke, Dennis. 1999. "On the Making of 'Cannibal Tours.'" 1–23. www.cameraworklimited.com/get/58.pdf.

Ortner, Sherry B. 1974. "Is Female to Male as Nature Is to Culture?" In *Woman, Culture, and Society*, edited by Michelle Zimbalist Rosaldo and Louise Lamphere, 68–87. Stanford: Stanford University Press.

———. 2016. "Dark Anthropology and Its Others: Theory Since the Eighties." *HAU: Journal of Ethnographic Theory* 6(1): 47–73.

Palmer, Catherine A. 1994. "Tourism and Colonialism: The Experience of the Bahamas." *Annals of Tourism Research* 21(4): 792–812.

Parameswaran, Radhika, and Kavitha Cardoza. 2009. "Melanin on the Margins: Advertising and the Cultural Politics of Fair/Light/White Beauty in India." *Journalism and Communication Monographs* 11(3): 213–274.

Paris, Cody. 2010. "The Virtualization of Backpacker Culture: Virtual Mooring, Sustained Interactions and Enhanced Mobilities." In *Beyond Backpacker Tourism:*

Mobilities and Experiences, edited by Kevin Hannam and Anya Diekmann, 40–63. Bristol: Channel View Publications.

Patullo, Polly. 1996. *Last Resorts: The Cost of Tourism in the Caribbean*. Kingston: Ian Randle Publishers.

Pawlak, Marek. 2018. *Zawstydzona tożsamość. Emocje, ideologie i władza w życiu polskich migrantów w Norwegii* [*Ashamed Identity: Emotions, Ideologies, and Power in the Life of Polish Migrants in Norway*]. Kraków: Wydawnictwo Uniwersytetu Jagiellońskiego.

Pawlik, Jacek J. 2011. "Uwarunkowania społeczno-kulturowe systemu politycznego w Indiach." [Socio-Cultural Conditions of the Political System in India]. In *Azja Wschodnia i Azja Południowa w stosunkach międzynarodowych* [*East and South Asia in International Relations*], edited by Justyna Nakonieczna and Jakub Zajączkowski, 19–29. Warszawa: Wydawnictwa Uniwersytetu Warszawskiego.

———. 2016. *Miasto kreatywnych ludzi. Mieszkańcy Lomé w obliczu wzrastającego ubóstwa* [*A City of Creative People: The Inhabitants of Lomé in the Face of Increasing Poverty*]. Olsztyn: Uniwersytet Warmińsko-Mazurski.

Pearce, Philip L. 1990. *The Backpacker Phenomenon: Preliminary Answers to Basic Questions*. Townsville: James Cook University of North Queensland.

Phuntso, Tsewang. 2004. "Government in Exile." In *Exile as Challenge: The Tibetan Diaspora*, edited by Dagmar Bernstorff and Hubertus von Welck, 125–149. New Delhi: Orient Longman.

Picard, Michel. 1990. "'Cultural Tourism' in Bali: Cultural Performances as Tourist Attraction." *Indonesia* 49: 37–74.

———. 1996. *Bali: Cultural Tourism and Touristic Culture*. Translated by Diana Darling. Singapore: Archipelago Press.

———. 2008. "Balinese Identity as Tourist Attraction: From 'Cultural Tourism' (pariwisata budaya) to Bali erect (ajeg Bali)." *Tourist Studies* 8(2): 155–173.

Piscitelli, Adriana. 2004. "On 'Gringos' and 'Natives': Gender and Sexuality in the Context of International Sex Tourism." *Vibrant: Virtual Brazilian Anthropology* 1(1–2): 87–114.

Pi-Sunyer, Oriol. 1977. "Through Native Eyes: Tourists and Tourism in a Catalan Maritime Community." In *Hosts and Guests: The Anthropology of Tourism* (1st edition), edited by Valene L. Smith, 149–156. Oxford: Blackwell.

Plange, Nii. 1989. "Politics and Tourism: Another Form of Dependency and Insecurity." *Review: Security and Development* 10(17): 22–31.

Podemski, Krzysztof. 2005. *Socjologia podróży* [*Sociology of Travel*]. Poznań: Wydawnictwo Naukowe UAM.

Poirier R. A. 1995. "Tourism and Development in Tunisia." *Annals of Tourism Research* 22(1): 157–171.

Poon, Auliana. 2003. "Competitive Strategies for a 'New' Tourism." In *Classic Review in Tourism*, edited by Chris Cooper, 130–142. Clevedon: Channel View Publications.

Prakash, Gyan. 1994. "Subaltern Studies as Postcolonial Criticism." *The American Historical Review* 99(5): 1475–1490.

Pratt, Mary Louise. 1992. *Imperial Eyes: Travel Writing and Transculturation*. London and New York: Routledge.

Prideaux, Bruce, and Hideki Shiga. 2007. "Japanese Backpacking: The Emergence of a New Market Sector—A Queensland Case Study." *Tourism Review International* 11(1): 45–56.

Pritchard, Annette, and Nigel J. Morgan. 2000. "Privileging the Male Gaze: Gendered Tourism Landscapes." *Annals of Tourism Research* 27(4): 884–905.

Pritchard, Annette, Nigel J. Morgan, and Irena Ateljevic. 2011. "Hopeful Tourism: A New Transformative Perspective." *Annals of Tourism Research* 38(3): 941–963.

Prost, Audrey. 2006. "The Problem with 'Rich Refugees,' Sponsorship, Capital, and the Informal Economy of Tibetan Refugees." *Modern Asian Studies* 40(1): 233–253.

Rabinow, Paul. 1977. *Reflections on Fieldwork in Morocco*. Berkeley: University of California Press.

Radcliffe-Brown, Alfred Reginald. 1922. *The Andaman Islanders: A Study in Social Anthropology*. Cambridge: Cambridge University Press.

Rakowski, Tomasz. 2016. *Hunters, Gatherers, and Practitioners of Powerlessness: An Ethnography of the Degraded in Postsocialist Poland*. New York and Oxford: Berghahn Books.

Rana, Pravin S., and Rana P. B. Singh. 2011. "Perceptions and Images of Tourists and Pilgrims in Banaras." In *Holy Places and Pilgrimages: Essays on India*, edited by Rana P. B. Singh, 165–206. Gurgaon: Shubhi Publications.

Rapport, Nigel J. 2000. "The Narrative as Fieldwork Technique: Processual Ethnography for a World in Motion." In *Constructing the Field: Ethnographic Fieldwork in the Contemporary World*, edited by Vered Amit, 71–95. London and New York: Routledge.

Reid, Alastair. 1992. "Reflections: Waiting for Colombus." *The New Yorker*, February 24, 1992: 57–75.

Reid, Donald G. 2003. *Tourism, Globalization, and Development: Responsible Tourism Planning*. London: Pluto Press.

Richards, Greg, and Julie Wilson, eds. 2004. *The Global Nomad: Backpacker Travel in Theory and Practice*. Clevedon: Channel View Publications.

Richter, Linda K. 1989. *The Politics of Tourism in Asia*. Honolulu: University of Hawaii Press.

Ricoeur, Paul. 1994. "Imagination in Discourse and in Action." In *Rethinking Imagination: Culture and Creativity*, edited by Gillian Robinson and John F. Rundell, 87–117. London and New York: Routledge.

Rivoal, Isabelle, and Noel B. Salazar. 2013. "Contemporary Ethnographic Practice and the Value of Serendipity." *Social Anthropology* 21(2): 178–185.

Robbins, Joel. 2013. "Beyond the Suffering Subject: Toward an Anthropology of the Good." *Journal of the Royal Anthropological Institute* 19(3): 447–462.

Robertson, Roland. 1995. "Glocalization: Time-Space and Homogeneity-Heterogeneity." In *Global Modernities*, edited by Mike Featherstone, Scott Lash, and Roland Robertson, 25–44. London: Sage Publications.

Robinson, Michael. 1999. "Cultural Conflicts in Tourism: Inevitability and Inequality." In *Tourism and Cultural Conflicts*, edited by Michael Robinson and Priscilla Boniface, 1–32. Wallingford: CABI Publishing.

Rodgers, Graeme. 2004. "'Hanging Out' with Forced Migrants: Methodological and Ethical Challenges." *Forced Migration Review* 21: 48–49.

Rojek, Chris. 2010. *The Labour of Leisure: The Culture of Free Time*. London: Sage Publications.

Romanowicz, Anna. 2017. *Unintended Revolution: Middle Class, Development and Non-Governmental Organizations*. Kraków: Jagiellonion University Press.

Rosaldo, Renato. 1993. *Culture and Truth: The Remaking of Social Analysis*. Boston: Beacon Press.

Rose, Gillian. 1997. "Situating Knowledges: Positionality, Reflexivities and Other Tactics." *Progress in Human Geography* 21(3): 305–320.

Ross, Robert J. S., and Kent C. Trachte. 1990. *Global Capitalism: The New Leviathan*. Albany: State University of New York Press.

Rossel, Pierre, ed. 1988. *Tourism: Manufacturing the Exotic*. Copenhagen: International Working Group for Indigenous Affairs.

Routledge, Paul. 2001. "'Selling the Rain,' Resisting the Sale: Resistant Identities and the Conflict over Tourism in Goa." *Social & Cultural Geography* 2(2): 221–240.

Rubin, Gayle S. 1975. "The Traffic of Women: Notes on the 'Political Economy' of Sex." In *Toward an Anthropology of Women*, edited by Rayna R. Reiter, 157–210. New York: Monthly Review Press.

Rukhsana, Syed, and Pirzada Mohd. Amin. 2012. *Purdah in Kashmir: A Sociological Study*. Saarbrücken: Lambert Academic Publishing.

Russell, Dale W., and Cristel Antonia Russell. 2010. "Experiential Reciprocity: The Role of Direct Experience in Value Perceptions." *Journal of Travel & Tourism Marketing* 27(6): 624–634.

Ryang, Sonia. 1997. "Native Anthropology and Other Problems." *Dialectical Anthropology* 22(1): 23–49.

Rybus, Agata, and Anna Wieczorkiewicz. 2014. "Po obu stronach straganu" [On Both Sides of the Stall]. *Konteksty* LXVIII(1): 149–162.

Rydzewski, Robert. 2020. "The 'Refugee Crisis': An Anthropological Case Study of the Serbian Part of the Western Balkan Route." PhD dissertation, Adam Mickiewicz University.

Sacramento, Octávio. 2015. "From Europe with Passion: Frameworks of the Touristic Male Desire of Ponta Negra in the North-East of Brazil." *Current Issues in Tourism* 21(2): 210–224.

Sahlins, Marshall. 1972. *Stone Age Economics*. Chicago and New York: Aldine Atherton.

Said, Edward W. 1979. *Orientalism*. New York: Vintage Books.

Salazar, Noel B. 2009. "Imaged or Imagined? Cultural Representations and the 'Tourismification' of Peoples and Places." *Cahiers d'E´tudes Africaines* 49(1–2): 49–71.

———. 2010. *Envisioning Eden: Mobilizing Imaginaries in Tourism and Beyond*. Oxford: Berghahn Books.

———. 2012. "Tourism Imaginaries: A Conceptual Approach." *Annals of Tourism Research* 39(2): 863–882.

Salazar, Noel B., and Nelson H. H. Graburn, eds. 2014. *Tourism Imaginaries: Anthropological Approaches*. Oxford: Berghahn Books.

Saldanha, Arun. 2007. *Psychedelic White: Goa Trance and the Viscosity of Race*. Minneapolis: University of Minnesota Press.

Samphel, Thubten. 2006. "Dharamshala! Dharamshala!" *Tibetan Review*, March–April: 28–29.

Sarkar, Sumit, and Tanika Sarkar, eds. 2007. *Women and Social Reform in Modern India: A Reader*. Ranikhet: Permanent Black.

Schedneck, Brooke. 2017. "Beyond the Glittering Golden Buddha Statues: Difference and Self-Transformation Through Buddhist Volunteer Tourism in Thailand." *Journeys: The International Journal of Travel and Travel Writing* 18(1): 57–78.

Schein, Louisa. 2002. "Mapping Hmong Media in Diasporic Space." In *Media Worlds: Anthropology on New Terrains*, edited by Faye D. Ginsburg, Lila Abu-Lughod, and Brian Larkin, 229–244. Berkeley: University of California Press.

Scheyvens, Regina. 2002a. "Backpacker Tourism and Third World Development." *Annals of Tourism Research* 29(1): 144–164.

———. 2002b. *Tourism for Development: Empowering Communities*. Harlow: Prentice-Hall.

———. 2007. "Exploring the Tourism-Poverty Nexus." *Current Issues in Tourism* 10(2–3): 231–254.

Schilcher, Daniela. 2007. "Growth Versus Equity: The Continuum of Pro-Poor Tourism and Neoliberal Governance." *Current Issues in Tourism* 10(2–3): 166–193.

Schneider, Arnd, and Christopher Wright, eds. 2006. *Contemporary Art and Anthropology*. Oxford and New York: Berg.

Selwyn, Tom. 1996a. "Introduction." In *The Tourist Image: Myths and Myth Making in Tourism*, edited by Tom Selwyn, 1–32. Chichester: Wiley.

———, ed. 1996b. *The Tourist Image: Myth and Mythmaking in Tourism*. Chichester: Wiley.

Shakabpa, Tsepon W. D. 1967. *Tibet: A Political History*. New Haven and London: Yale University Press.

Shakya, Tsering. 1993. "Whither the Tsampa Eaters?" *Himal* 6: 8–11.

Sharma, Aradhana. 2008. *Logics of Empowerment Development, Gender, and Governance in Neoliberal India*. Minneapolis: University of Minnesota Press.

Sharpley, Richard, and Priya Sundaram. 2005. "Tourism: A Sacred Journey? The Case of Ashram Tourism, India." *International Journal of Tourism Research* 7(3): 161–171.

Shaw, Gareth, and Allan M. Williams. 1998. "Entrepreneurship and Tourism Development." In *The Economic Geography of the Tourist Industry*, edited by Dimitri Ioannides and Keith G. Debbage, 235–255. London and New York: Routledge.

Shekhawat, Prahlad Singh. 1995. "McLeod Ganj: A Cross-Cultural Town in India." *Tibetan Review*, September: 14–17.

Sheller, Mimi, and John Urry. 2006. "The New Mobilities Paradigm." *Environment and Planning A* 38(2): 207–226.

Shepherd, Robert. 2002. "Commodification, Culture and Tourism." *Tourist Studies* 2(2): 183–201.

———. 2012. *Faith in Heritage: Displacement, Development, and Religious Tourism in Contemporary China*. Walnut Creek: Left Coast Press.

Silverman, Eric K. 1999. "Tourist Art as the Crafting of Identity in the Sepik River (Papua New Guinea)." In *Unpacking Culture: Art and Commodity in Colonial and Postcolonial Worlds*, edited by Ruth B. Phillips and Christopher B. Steiner, 51–66. Berkeley: University of California Press.

Simoni, Valerio. 2017. *Tourism and Informal Encounters in Cuba*. New York and Oxford: Berghahn Books.

Simpson, Kate. 2004. "'Doing Development': The Gap Year, Volunteer-Tourists and a Popular Practice of Development." *Journal of International Development* 16(5): 681–692.

Sin, Harng Luh. 2010. "Who Are We Responsible to? Locals' Tales of Volunteer Tourism." *Geoforum* 41(6): 983–992.

Singh, Rana P. B., ed. 2011. *Holy Places and Pilgrimages: Essays on India*. Gurgaon: Shubhi Publications.

Singh, Rana P. B., and Pravin S. Rana. 2016. "Indian Sacred Natural Sites: Ancient Traditions of Reverence and Conservation Explained from a Hindu Perspective." In *Asian Sacred Natural Sites: Philosophy and Practice in Protected Areas and Conservation*, edited by Bas Verschuuren and Naoya Furuta. London and New York: Routledge.

Singh, Sagar. 2019. *Rethinking the Anthropology of Love and Tourism*. Lanham: Lexington Books.

Singh, Sarina, Joe Bindloss, Paul Clammer, et al. 2005. *India* (11th edition). Melbourne: Lonely Planet Publications.

Singh, Shalini. 2001. "Indian Tourism: Policy, Performance, and Pitfalls." In *Tourism and the Less Developed World: Issues and Case Studies*, edited by David Harrison, 137–150. Wallingford: CABI Publishing.

Singh, Surabhi, Ulrike Repnik, Aditi Chanchani, Swathi Seshadri, and Rosemary Viswanath. 2012. *Women Street Vendors & Tourism: Negotiating Lives and Spaces*. Bangalore: Equations.

Skinner, Jonathan, and Dimitrios Theodossopoulos, eds. 2011. *Great Expectations: Imagination and Anticipation in Tourism*. Oxford: Berghahn Books.

Slocum, Sally. 1975. "Woman the Gatherer: Male Bias in Anthropology." In *Toward an Anthropology of Women*, edited by Rayna R. Reiter, 36–50. New York and London: Monthly Review Press.

Smith, Laurajane. 2006. *Uses of Heritage*. London and New York: Routledge.

Smith, Valene L., ed. 1977. *Hosts and Guests: The Anthropology of Tourism* (1st edition). Oxford: Blackwell.

———, ed. 1989. *Hosts and Guests: The Anthropology of Tourism* (2nd edition). Philadelphia: University of Pennsylvania Press.

———. 2005. "Anthropologists in the Tourism Workplace." *NAPA Bulletin* 23: 252–269.

Smith, Valene L., and Maryann Brent, eds. 2001. *Hosts and Guests Revisited: Tourism Issues of the 21st Century*. Elmsford: Cognizant.

Snepenger, David J., Jerry D. Johnson, and Raymond Rasker. 1995. "Travel-Stimulated Entrepreneurial Migration." *Journal of Travel Research* 34(1): 40–44.

Sørensen, Anders. 2003. "Backpacker Ethnography." *Annals of Tourism Research* 30(4): 847–867.

Speed, Clare, and Tony Harrison. 2004. "Backpacking in Scotland: Formal Public Sector Responses to an Informal Phenomenon." In *The Global Nomad: Backpacker Travel in Theory and Practice*, edited by Greg Richards and Julie Wilson, 149–167. Clevedon: Channel View Publications.

Spivak, Gayatri Chakravorty. 1993. "Can the Subaltern Speak?" In *Colonial Discourse and Post-Colonial Theory: A Reader*, edited by Patrick Williams and Laura Chrisman, 66–111. New York and Sydney: Harvester Wheatsheaf.

———. 1997. "In a Word: Interview with Ellen Rooney." In *The Second Wave: A Reader in Feminist Theory*, edited by Linda Nicholson, 356–378. London and New York: Routledge.

Srisang, K. 1992. "Third World Tourism: The New Colonialism." *Focus: Bulletin of Campaigning NGO Tourism Concern* 4: 2–3.

Steiner, Christopher B. 1999. "Authenticity, Repetition, and the Aesthetics of Seriality: The Work of Tourist Art in the Age of Mechanical Reproduction." In *Unpacking Culture: Art and Commodity in Colonial and Postcolonial Worlds*, edited by Ruth B. Phillips and Christopher B. Steiner, 87–103. Berkeley: University of California Press.

Stoller, Paul. 1997. "Globalizing Method: The Problems of Doing Ethnography in Transnational Spaces." *Anthropology and Humanism* 22(1): 81–94.

Strathern, Marilyn. 1988. *The Gender of the Gift: Problems with Women and Problems with Society in Melanesia*. Berkley: University of California Press.

Sturma, Michael. 1999. "Packaging Polynesia's Image." *Annals of Tourism Research* 26(3): 712–715.

Subba, Tanka B. 1990. *Flight and Adaptation: Tibetan Refugees in the Darjeeling-Sikkim Himalaya*. Dharamsala: Library of Tibetan Works and Archives.

Suryanarayan, Neelakshi. 2017. "From Yashwant Place to Yashka: A Case Study of Commodification of Russian in India." *International Journal of Bilingual Education & Bilingualism* 20(4): 428–442.

Tapontsang, Adhe. 1997. *Ama Adhe, the Voice that Remembers: The Heroic Story of a Woman's Flight to Free Tibet*. Boston: Wisdom Publications.

Teo, Peggy, and Sandra Leong. 2006. "A Postcolonial Analysis of Backpacking." *Annals of Tourism Research* 33(1): 109–131.

Thomas, Nicholas. 1996. *Out of Time: History and Evolution in Anthropological Discourse*. Ann Arbor: University of Michigan Press.

Trask, Haunani-Kay. 2000. "Tourism and the Prostitution of Hawaiian Culture." *Cultural Survival Quarterly* 24. https://www.culturalsurvival.org.023b8ezz0c2b.han.amu.edu.pl/publications/cultural-survival-quarterly/tourism-and-prostitution-hawaiian-culture.

Tribe, John. 2005. *The Economics of Recreation, Leisure and Tourism*. Oxford: Elsevier.

Trivedi, Harish. 1993. *Colonial Transactions: English Literature and India*. Calcutta: Papyrus.

Tsering, Tenzin. 1998. ". . . and If You Want to Meet a Yeti." *Tibetan Bulletin*, July–September: 16–17.

Tseten, Kesang. 1993. "Exile Politics and the 'Third Spread.'" *Himal*, March/April: 29–30.

Tsing, Anna L. 2005. *Friction: An Ethnography of Global Connection*. Princeton: Princeton University Press.

Tucker, Hazel. 2003a. *Living with Tourism: Negotiating Identities in a Turkish Village*. London and New York: Routledge.

———. 2003b. "The Host-Guest Relationship and Its Implications in Rural Tourism." In *New Directions in Rural Tourism*, edited by Lesley Roberts, Derek Hall, and Mitchell Morag, 80–89. Farnham: Ashgate.

———. 2009. "Recognizing Emotion and Its Postcolonial Potentialities: Discomfort and Shame in a Tourism Encounter in Turkey." *Tourism Geographies* 11(4): 444–461.

———. 2011. "Success and Access to Knowledge in the Tourist-Local Encounter: Confrontations with the Unexpected in a Turkish Community." In *Great Expectations: Imagination and Anticipation in Tourism*, edited by Jonathan Skinner and Dimitrios Theodossopoulos, 27–39. Oxford: Berghahn Books.

Turner, Louis, and John Ash. 1975. *The Golden Hordes: International Tourism and Pleasure Periphery*. London: Constable.

Turton, David, ed. 1997. *War and Ethnicity: Global Connections and Local Violence*. Woodbridge: University of Rochester Press.

Tuting, Ludmilla, and Kunda Dixit, eds. 1986. *Bikas-Binas? Development-Destruction? The Change in Life and Environment of the Himalaya: A Reader for Concerned People and Tourists About Ecology, So-Called Development and Tourism*. Munich: Geobuch.

UNWTO. 2009. *The Indian Outbound Travel Market with Special Insight into the Image of Europe as a Destination*. Madrid: United Nations World Tourism Organization.

———. 2015. *India: Country-specific: Basic Indicators (Compendium) 2009–2013*. Madrid: United Nations World Tourism Organization.

———. 2016. *India: Country-Specific: Arrivals of Non-Resident Tourists at National Borders, by Nationality 2011–2015*. Madrid: United Nations World Tourism Organization.

———. 2017. *India: Country-Specific: Outbound Tourism 1995–2015*. Madrid: United Nations World Tourism Organization.

Uriely, Natan, and Arie Reichel. 2000. "Working Tourists and Their Attitudes to Hosts." *Annals of Tourism Research* 27(2): 267–283.

Uriely, Natan, Yuval Yonay, and Dalit Simchai. 2002. "Backpacking Experiences: A Type and Form Analysis." *Annals of Tourism Research* 29(2): 520–538.

Urry, John. 1995. *Consuming Places*. London and New York: Routledge.

———. 2002. *The Tourist Gaze: Leisure and Travel in Contemporary Societies*. London: Sage Publications.

van Beek, Walter E. A. 2003. "African Tourist Encounters: Effects of Tourism on Two West African Societies." *Africa* 73(2): 251–289.

van Gennep, Arnold. 1960. *The Rites of Passage*. Translated by Monika Vizedom and Gabrielle L. Caffee. Chicago: University of Chicago Press.

Varma, Pavan K. 1999. *The Great Indian Middle Class*. New Delhi: Penguin Books.

Veissière, Samuel. 2011. *Ghosts of Empire: Sex, Mobility, and Violence in the Transatlantic Cultural Economy of Desire*. Berlin: LIT Verlag.

Verghese, Anila. 2002. *Hampi: Monumental Legacy*. New Delhi: Oxford University Press.

Vertovec, Steven, and Robin Cohen, eds. 2002. *Conceiving Cosmopolitanism: Theory, Context, and Practice*. Oxford: Oxford University Press.

Vogt, Jay. 1976. "Wandering: Youth and Travel Behavior." *Annals of Tourism Research* 4(1): 25–41.

Wagner, Anja. 2013. *The Gaddi Beyond Pastoralism: Making Place in the Indian Himalayas*. New York and Oxford: Berghahn Books.

Walker, Vanessa. 2006. *Mantras & Misdemeanours: An Accidental Love Story*. Sydney: Allen & Unwin.

Wallerstein, Immanuel. 2004. *World-Systems Analysis: An Introduction*. Durham and London: Duke University Press.

Wearing, Stephen. 2001. *Volunteer Tourism: Experiences that Make a Difference*. Wallingford: CABI Publishing.

Welk, Peter. 2004. "The Beaten Track: Anti-Tourism as an Element of Backpacker Identity Construction." In *The Global Nomad: Backpacker Travel in Theory and Practice*, edited by Greg Richards and Julie Wilson, 77–91. Clevedon: Channel View Publications.

———. 2010. "Town of 1770, Australia: The Creation of a New Backpacker Brand." In *Beyond Backpacker Tourism: Mobilities and Experiences*, edited by Kevin Hannam and Anya Diekmann, 169–186. Bristol: Channel View Publications.

Westwood, Sheena, Nigel J. Morgan, and Annette Pritchard. 2006. "Situation, Participation and Reflexivity in Tourism Research: Furthering Interpretive Approaches to Tourism Enquiry." *Tourism Recreation Research* 31(2): 33–42.

White, Hayden. 1987. *Tropics of Discourse: Essays in Cultural Criticism*. Baltimore and London: Johns Hopkins University Press.

Wieczorkiewicz, Anna. 2003. "Prawdziwe pamiątki z podroży. Kwestia autentyczności w antropologii turystyki" [Genuine Travel Souvenirs: Authenticity in the Anthropology of Tourism]. *Kultura i Społeczeństwo* 2: 145–177.

———. 2004. "Jeść jak tubylcy. Fragmenty turystycznej mitologii" [Eat Like the Natives: Aspects of Tourist Mythology]. In *Opowiadanie w perspektywie badań porównawczych* [*Storytelling from the Perspective of Comparative Research*], edited by Zofia Mitosek, 267–284. Kraków: Universitas.

———. 2008. *Apetyt turysty: O doświadczaniu świata w podróży* [*The Tourist Appetite: About Experiencing the World While Traveling*]. Kraków: Universitas.

Williams, Allan M., and Colin Michael Hall. 2002. "Tourism, Migration, Circulation and Mobility. The Contingencies of Time and Place." In *Tourism and Migration: New Relationships Between Production and Consumption*, edited by Colin Michael Hall and Allan M. Williams, 1–52. Dordrecht: Kluwer Academic Publishers.

Williams, Anthony V., and Wilbur Zelinsky. 1970. "On Some Patterns in International Tourist Flows." *Economic Geography* 46(4): 549–567.

Willis, Paul. 2000. *The Ethnographic Imagination*. Cambridge: Polity Press.

Willis, Paul, and Mats Trondman. 2000. "Manifesto for Ethnography." *Ethnography* 1(1): 5–16.

Wilson, David. 1994. "Probably as Close as You Can Get to Paradise: Tourism and the Changing Image of Seychelles." In *Tourism: The State of the Art*, edited by A. V. Seaton, Carson L. Jenkins, Roy C. Wood, Peter U. C. Dieke, Minerva M. Bennett, L. Rory MacLellan, and R. K. M. Smith, 765–774. Chichester: Wiley.

———. 1997a. "Paradoxes of Tourism in Goa." *Annals of Tourism Research* 24(1): 52–75.

———. 1997b. "Strategies for Sustainability: Lessons from Goa and the Seychelles." In *Tourism and Sustainability: Principles to Practice*, edited by Mike J. Stabler. New York and Wallingford: CABI Publishing.

Wilson, Julie, and Greg Richards. 2008. "Suspending Reality: An Exploration of Enclaves and the Backpacker Experience." In *Backpacker Tourism: Concepts and Profiles*, edited by Kevin Hannam and Irena Ateljevic, 9–25. Clevedon: Channel View Publications.

Withey, Lynne. 1998. *Grand Tours and Cook's Tours: A History of Leisure Travels, 1750 to 1915*. London: Aurum Press.

Woodward, Kathleen. 2004. "Calculating Compassion." In *Compassion: The Culture and Politics of an Emotion*, edited by Lauren Berlant, 59–86. London and New York: Routledge.

WTTC. 2013. *Benchmarking Travel & Tourism in India*. London: World Tourism and Travel Council.

———. 2017. *Travel & Tourism Economic Impact 2017*. India. London: World Tourism and Travel Council.

Wylie, Turrell V. 1959. "A Standard System of Tibetan Transcription." *Harvard Journal of Asian Studies* 22: 261–267.

Young, George. 1973. *Tourism: Blessing or Blight?* Harmondsworth: Penguin.

Yu, Hong, and Mary Ann Littrell. 2003. "Product and Process Orientations to Tourism Shopping." *Journal of Travel Research* 42(2): 140–150.

Zimbalist Rosaldo, Michelle. 1974. "Women, Culture, and Society: A Theoretical Overview." In *Women, Culture, and Society*, edited by Michelle Zimbalist Rosaldo and Louise Lamphere, 17–42. Stanford: Stanford University Press.

Index

Page references for figures are italicized

About the Author

Natalia Bloch is an anthropologist and associate professor in the Institute of Anthropology and Ethnology, Adam Mickiewicz University, Poznań, Poland. She specializes in anthropology of mobility and works particularly in the postcolonial context, with a special focus on colonial legacies in contemporary migration and tourism, mobility regimes, and the intersections of various forms of spatial mobility. She conducted research in Tibetan refugee settlements and among mobile workers and entrepreneurs in the informal tourism sector in India. She has also been involved in research and applied projects on migrant integration in Poland. She teaches courses in postcolonial theory, anthropology of migration, anthropology of tourism, and refugee studies. She has published, among others, in *Critique of Anthropology*, *Annals of Tourism Research*, *Journal of Refugee Studies*, *Critical Asian Studies*, and *Transfers: Interdisciplinary Journal of Mobility Studies*.

www.ingramcontent.com/pod-product-compliance
Lightning Source LLC
LaVergne TN
LVHW011807200926
848484LV00005B/1093
9781793624710